Image Segmentation

Image Segmentation

Principles, Techniques, and Applications

Tao Lei
Shaanxi University of Science and Technology
Xi'an, China

Asoke K. Nandi
Brunel University London
Uxbridge, UK

This edition first published 2023
© 2023 John Wiley & Sons Ltd

The right of Tao Lei and Asoke K. Nandi to be identified as the authors of this work has been asserted in accordance with law.

Registered Offices
John Wiley & Sons, Inc., 111 River Street, Hoboken, NJ 07030, USA
John Wiley & Sons Ltd, The Atrium, Southern Gate, Chichester, West Sussex, PO19 8SQ, UK

Editorial Office
The Atrium, Southern Gate, Chichester, West Sussex, PO19 8SQ, UK

For details of our global editorial offices, customer services, and more information about Wiley products visit us at www.wiley.com.

Wiley also publishes its books in a variety of electronic formats and by print-on-demand. Some content that appears in standard print versions of this book may not be available in other formats.

Library of Congress Cataloging-in-Publication data applied for
ISBN: 9781119859000

Cover Design: Wiley
Cover Image: © d3sign/Getty Images

Set in 9.5/12.5pt STIXTwoText by Straive, Pondicherry, India
Printed and bound by CPI Group (UK) Ltd, Croydon, CR0 4YY

C9781119859000_130922

To
My parents, my wife Yan Lu, and our daughter—Lu Lei.
Tao Lei

My wife, Marion, and our children—Robin, David, and Anita Nandi.
Asoke K. Nandi

Brief Contents

Contents

About the Authors

Tao Lei received a PhD in information and communication engineering from Northwestern Polytechnical University, Xi'an, China, in 2011. From 2012 to 2014, he was a postdoctoral research fellow with the School of Electronics and Information, Northwestern Polytechnical University, Xi'an, China. From 2015 to 2016, he was a visiting scholar with the Quantum Computation and Intelligent Systems group at University of Technology Sydney, Sydney, Australia. From 2016 to 2019, he was a postdoctoral research fellow with the School of Computer Science, Northwestern Polytechnical University, Xi'an, China. He has authored and coauthored 80+ research papers published in IEEE TIP, TFS, TGRS, TGRSL, ICASSP, ICIP, and FG.

He is currently a professor with the School of Electronic Information and Artificial Intelligence, Shaanxi University of Science and Technology. His current research interests include image processing, pattern recognition, and machine learning. Professor Lei is an associate editor of *Frontiers in Signal Processing*; he is also a guest editor of *Remote Sensing* and *IEEE JESTAR*. He is a senior member of IEEE and CCF.

Asoke K. Nandi received the degree of PhD in physics from the University of Cambridge (Trinity College), Cambridge. He held academic positions in several universities, including Oxford, Imperial College London, Strathclyde, and Liverpool, as well as Finland Distinguished Professorship in Jyvaskyla (Finland). In 2013 he moved to Brunel University London, to become the Chair and Head of Electronic and Computer Engineering.

In 1983 Professor Nandi jointly discovered the three fundamental particles known as W^+, W^-, and Z^0, providing the evidence for the unification of the electromagnetic and weak forces, for which the Nobel Committee for Physics in 1984 awarded the prize to his two team leaders for their decisive contributions.

His current research interests lie in the areas of signal processing and machine learning, with applications to communications, gene expression data, functional magnetic resonance data, machine condition monitoring, and biomedical data. He has made many fundamental theoretical and algorithmic contributions to many aspects of signal processing and machine learning. He has much expertise in big data, dealing with heterogeneous data and extracting information from multiple data sets obtained in different laboratories and different times. Professor Nandi has authored over 600 technical publications, including 260 journal papers as well as five books, entitled *Condition Monitoring with Vibration Signals* (Wiley, 2020), *Automatic Modulation Classification: Principles, Algorithms and Applications* (Wiley, 2015), *Integrative Cluster Analysis in Bioinformatics* (Wiley, 2015), *Blind Estimation Using Higher-Order Statistics* (Springer, 1999), and *Automatic Modulation Recognition of Communications Signals* (Springer, 1996). The h-index of his publications is 80 (Google Scholar) and his Erdös number is 2.

Professor Nandi is a Fellow of the Royal Academy of Engineering (UK) and of seven other institutions. Among the many awards he received are the Institute of Electrical and Electronics Engineers (USA) Heinrich Hertz Award in 2012, the Glory of Bengal Award for his outstanding achievements in scientific research in 2010, from the Society for Machinery Failure Prevention Technology, a division of the Vibration Institute (USA) in 2000, the Water Arbitration Prize of the Institution of Mechanical Engineers (UK) in 1999, and the Mountbatten Premium of the Institution of Electrical Engineers (UK) in 1998. Professor Nandi is an IEEE Distinguished Lecturer (2018–2019). Professor Nandi is the Field Chief Editor of *Frontiers in Signal Processing* journal.

Preface

Image segmentation is one of the most challenging frontier topics in computer vision. It provides an important foundation for image analysis and image description, as well as image understanding. The basic task of image segmentation is to segment an image into several regions that are nonoverlapping, with these regions having accurate boundaries. The current task of image segmentation not only requires accurate region division but also requires a label output on different regions, that is, semantic segmentation. With the development of computer vision and artificial intelligence techniques, the roles and importance of image segmentation have grown significantly. Image segmentation has been applied widely in various fields, such as industry of detection, intelligent transportation, biological medicine, agriculture, defense, and remote sensing. At present, a large number of image segmentation techniques have been reported, and many of them have been successfully applied to actual product development. However, the fast development of imaging and artificial intelligence techniques requires image segmentation to deal with increasingly more complex tasks. These complex tasks require more effective and efficient image segmentation techniques. For that reason, there is a growing body of literature resulting from efforts in research and development by many research groups around the world. Although there are many publications on image segmentation, there is only a few collections of recent techniques and methods devoted to the field of computer vision.

This book attempts to summarize and improve principles, techniques, and applications of current image segmentation. It can help researchers, postgraduate students, and practicing engineers from colleges, research institutes, and enterprises to understand the field quickly. Firstly, based on this book, researchers can quickly understand the basic principles of image segmentation and related mathematical methods such as clustering, mathematical morphology, and convolutional neural networks. Secondly, based on classic image processing and machine learning theory, the book introduces a serious of recent methods to achieve fast and accurate image segmentation. Finally, the book introduces the effect of image segmentation in various application scenarios such as traffic, medicine, remote sensing, and materials. In brief, the book aims to inform, enthuse, and attract more researchers to enter the field and thus develop further image segmentation theory and applications.

Chapter 1 is a brief introduction to image segmentation and its applications in various fields including industry, medicine, defense, and environment. Besides, an example is presented to help readers understand image segmentation quickly.

Chapter 2 is concerned with principles of clustering. Three clustering approaches that are concerned closely with image segmentation, are presented, i.e. k-means clustering, fuzzy c-means clustering, spectral clustering, and gaussian mixed model.

Chapter 3 is concerned with principles of mathematical morphology since it is important in image processing, especially watershed transform, which is popular for image segmentation. In this chapter, morphological filtering, morphological reconstruction, and the watershed transform are presented. Besides, multivariate mathematical morphology is presented since it is important for multichannel image processing, which can help image segmentation for multichannel images.

Chapter 4 is concerned with principles of neural networks since they are important in image processing, especially convolutional neural networks, which are popular for image segmentation. In this chapter, artificial neural networks, convolutional neural networks, and graph convolutional networks are presented.

Chapter 5 introduces a fast image segmentation approach based on fuzzy clustering. This chapter illustrates related works with improved FCM and presents two strategies: local spatial information integration and membership filtering, which achieves better segmentation results.

Chapter 6 introduces a fast and robust image segmentation approach based on the watershed transform. This chapter illustrates related works with seeded image segmentation and presents an adaptive morphological reconstruction method that can help the watershed transform to achieve better segmentation results.

Chapter 7 introduces a fast image segmentation approach based on superpixel and the Gaussian mixed model (GMM). This chapter illustrates related works with superpixel algorithms and presents the idea of combing superpixel and GMM for image segmentation.

Chapter 8 introduces the application of image segmentation for traffic scene segmentation. This chapter illustrates related works with traffic scene semantic segmentation and presents the idea of multi-scale information fusion combing nonlocal network for traffic scene semantic segmentation.

Chapter 9 introduces the application of image segmentation for medical images. This chapter illustrates related works with liver and liver-tumor segmentation and presents two approaches for liver and liver-tumor segmentation including lightweight V-net and deformable context encoding network.

Chapter 10 introduces the application of image segmentation for remote sensing. This chapter illustrates related works with change detection and presents two approaches for change detection including unsupervised change detection and end-to-end change detection for very high resolution (VHR) remote sensing images.

Chapter 11 introduces the application of image segmentation for material analysis. This chapter presents three applications for different material analysis including metallic materials, foam materials, and ceramics materials.

This book is up-to-date and covers a lot of the advanced techniques used for image segmentation, including recently developed methods. In addition, this book provides new methods, including unsupervised clustering, watershed transform, and deep learning for image segmentation, which covers various topics of current research interest. Additionally, the book will provide several popular image segmentation applications including traffic scene, medical images, remote sensing images, and scanning electron microscope images. A work of this magnitude will, unfortunately, contain errors and omissions. We would like to take this opportunity to apologize unreservedly for all such indiscretions in advance. We welcome comments and corrections; please send them by email to a.k.nandi@ieee.org or by any other means.

Acknowledgment

It should be remarked that some of the research results reported in this book have been sourced from refereed publications, arising from projects originally funded by the Royal Society (UK) grant (IEC\NSFC\170 396) and NSFC (China) grant (61811530325).

SEPTEMBER 2022

Tao Lei and Asoke K. Nandi
Xi'an, China, and London, UK

List of Symbols and Abbreviations

Chapter 1

Symbols and Abbreviations

p_{ii}	The number of pixels of i-th class predicted as belonging to i-th class
p_{ij}	The number of pixels of i-th class predicted as belonging to j-th class
MPA	Mean pixel accuracy
TP	The true positive fraction
FP	The false positive fraction
FN	The false negative fraction
F1-Score	The harmonic mean of precision and recall
$S(A)$	The set of surface voxels of A

Chapter 2

Symbols and Abbreviations

c	The number of clusters
N	The number of samples
ϖ	The number of iterations
u_{ik}	The strength of the i-th sample x_i relative to the k-th cluster center v_k
$\|x_i - v_k\|$	The Euclidean distance between sample x_i and cluster center v_k
η	The convergence threshold
B	The maximum number of iterations
$V^{(0)}$	The initialize cluster center
v_k	The cluster center
λ_j	The Lagrange multiplier
$U^{(0)}$	The initialized membership matrix

U	The anti fuzzy membership matrix
S_A	A similarity matrix
σ	The scale parameters of Gaussian kernel function
D	A degree matrix
F_e	The feature vector
$N(\boldsymbol{X} \mid v_k, \Sigma_k)$	The k-th Gaussian density function
Σ_k	The covariance matrix
π_k	The Prior Probability

Chapter 3

Symbols and Abbreviations

A	A set
E	A structuring element
f	An image
δ	The dilation operation
ε	The erosion operation
C_l	A complete lattice
Γ	A complete lattice
γ	Opening operator
ϕ	Closing operator
ψ	Idempotent
s^f	The filter of size
$\delta_g^{(1)}$	Geodesic dilation
g	A masker image
$\varepsilon_i^{(1)}(f)$	The geodesic erosion
$\mathbf{x}$	A pixel in an image
$t_{\max}$	The maximum value
f^c	The complement of a grayscale image f
ψ^*	The dual operator of ψ
P_1	The function of the variant r
P_2	The function of the variant g
P_3	The function of the variant b

Chapter 4

Symbols and Abbreviations	
w_i	The weight
b	The bias term
z	A hyperplane
f_a	An activation function
a^r	An activated result
x_i	An input
$\sigma(\cdot)$	A sigmoid function
β	A learnable parameter or a fixed hyperparameter
$\boldsymbol{M}$	The maximum number of iterations
$*$	The convolution operation
F	The convolution kernel space size
K	The number of convolution kernels
S^c	The convolution kernel sliding step
P	The filling size of input tensors
N	The number of categories of classification tasks
l	The number of convolutional layers
$\otimes$	The cross-correlation operation
err	The error term
p	Feature maps
$f'_l(\cdot)$	The derivative of the activation function
$\widetilde{\otimes}$	The wide convolution
$\text{rot}180\,(\cdot)$	The 180° of rotation
$F\{\boldsymbol{f}\}$	The corresponding spectral domain signal
$\boldsymbol{L}$	A laplace matrix
$\boldsymbol{x}$	A graph signal
$\boldsymbol{g}$	A filter
$\odot$	Hadamard Product
$\boldsymbol{g}_\theta$	A diagonal matrix
$\boldsymbol{X}$	A feature Matrix
$\boldsymbol{x}_i^l$	The i-th column of matrix
T_k	The chebyshev polynomial
$\boldsymbol{\Theta}$	A parameter matrix
$\boldsymbol{Z}$	The output after graph convolution
$\boldsymbol{H}^l$	The node vector of the l-th layer
$\boldsymbol{W}^l$	The parameters of the corresponding layer
M_A	An adjacency matrix

Chapter 5

Symbols and Abbreviations

$g = \{x_1, x_2, \cdots, x_N\}$	A grayscale image
x_i	The gray value of the i-th pixel
pv_k	The prototype value of the k-th cluster
v_{ki}	The fuzzy membership value
$U = [u_{ki}]^{c \times N}$	The membership partition matrix
N	The total number of pixels
c	The number of clusters
m	The weighting exponent
G_{ki}	The fuzzy factor
x_r	The neighbor of x_i
N_i	The set of neighbors within a window around x_i
d_{ir}	The spatial Euclidean distance
$\hat{x}_i$	A mean value or median value
u_{kl}	The fuzzy membership
ξ_l	The gray level
τ	The number of the gray levels
R^C	The morphological closing reconstruction
f_o	The original image
λ_L	The lagrange multiplier
η	A minimal error threshold
g_v	The grayscale value
c	The cluster prototypes value
m	The fuzzification parameter
w	The size of filtering window
η	The minimal error threshold
f_n	A new image
$U^{(0)}$	The initialized membership partition matrix
$R_g^\delta(f)$	The morphological dilation reconstruction
δ	The dilation operation
$\wedge$	The pointwise minimum
$R_g^\varepsilon(f)$	The morphological erosion reconstruction
ε	The erosion operation
$\vee$	The pointwise maximum
$R^C(g)$	The morphological close reconstruction
E	A structuring element
CV	Overlap measure
PRI	Similarity measure
ϖ	The iteration number
w	The size of the filtering window

Chapter 6

Symbols and Abbreviations

f	A marker image
g	A mask image
R^δ	The morphological dilation reconstruction
$\wedge$	The pointwise minimum
R^ε	The morphological erosion reconstruction
$\vee$	The pointwise maximum
y	An output result
r_S	The radius of a structuring element
ω_{cj}	The weighted coefficient on the jth scale result
i	The scale parameter of a structuring element
H	A chain of nested partitions
s	The number of small regions in segmentation results

Chapter 7

Symbols and Abbreviations

u_{ki}	The strength of the i-th sample x_i relative to the k-th cluster center v_k
f^{ϖ}	The degree of fuzziness of u_{ki}
D_{ki}	The combination of the pixel dissimilarity and region dissimilarity
d_{ki}	The dissimilarity distance between the i-th pixel and the k-th clustering center
d_{kR_i}	The region dissimilarity between the region R_i obtained by mean-shift and the k-th clustering center
R_i	The region that contains the i-th pixel
ω_j	A weighting parameter of neighborhood pixels
N_i	The neighborhood of the i-th pixel
ζ	The region-level iterative strength
E_{R_i,R_j}	The Euclidean distance between the mean values of region R_i and R_j
Z	A normalized constant
u_{kl}	The fuzzy membership of gray value l with respect to the k-th clustering center v_k
m	The weighting exponent
ξ	A grayscale image
ξ_l	The grayscale level
γ_l	The number of pixels whose gray level equals to ξ_l
h_s	The spatial bandwidth
h_r	The range bandwidth
h_k	The minimum size of final output regions

R^O	The morphological opening reconstruction
R^C	The morphological closing reconstruction
r_{SE}	The radius of a structure element
R^{MC}	The MMGR operation
s	The size of the minimal region
t	The size of the maximal region
η'	A minimal error threshold
s_k	The number of desired superpixels
s_m	The weighting factor between color and spatial differences
s_s	The threshold used for region merging
l	The color level
τ	The number of regions of a superpixel image
λ	A Lagrange multiplier
$\tilde{J}_m$	The partial differential equation
η_c	The convergence condition used for SFFCM
$U^{(o)}$	The initialized membership partition matrix
$U^{(t)}$	The membership partition matrix
A_k	The set of pixels belonging to the k-th class
C_k	The set of pixels belonging to the class in the GT
S	The mean value
SA	The root mean square error

Chapter 8

Symbols and Abbreviations

y_i	The output feature map
$C(x)$	The response factor
x	The input feature map
s_{cp}	The similarity of the corresponding positions of i and j
g_f^j	A representation of the feature map at position j
z_i	The output feature map
$W_z y_i$	The $1 \times 1 \times 1$ convolution
A_j	A local feature map
E_i	The output feature map
$S \in R^{N \times N}$	The spatial attention map
s_{ji}	The i-th position's impact on j-th position
$R^{C \times H \times W}$	Three-dimensional space
α	A scale parameter
$\exp()$	Perform a matrix multiplication

$\sum_{i=1}^{N}$	Element-wise sum operation
$X \in R^{C \times C}$	The channel attention map
x_{ji}	The i-th channel's impact on the j-th channel
β	A learnable parameter
H	A local feature map
Q	One of the two feature maps is generated using two 1×1 convolutions
K	The other of the two feature maps is generated using two 1×1 convolutions
u	The Position
Q_u	A vector
C'	The number of channel
Ω_u	A feature vector set
$\Omega_{i,u}$	The i-th element of Ω_u
$d_{i,u}$	The degree of correlation between Q_u and $\Omega_{i,\,u}$
V	The output feature maps via 1×1 convolution
Φ_u	A collection of feature vectors in V
$\Phi_{i,u}$	The i-th element of Φ_u
$A_{i,u}$	A scalar value at channel i and position u
H'_u	A feature vector in H'_u at position u

Chapter 9

Symbols and Abbreviations

ε	The expansion rate of convolution operation
C	The number of channels
k	The size of convolution kernel
L_{main}	The loss function
X	The training samples
W	The parameters of backbone network
t_i	The label of x_i
$\hat{w}_i$	The parameters of point-wise convolution
η_i	The weight of the i-th auxiliary loss function
λ	The decay coefficient
x	The input feature maps
w	The weight of a network
e_0	The location of a pixel
e_n	The location of neighboring pixels
$\tilde{L}$	The deformation result of L
Δe_n	The offset locations
$x(e)$	The pixel value

q_j	The four surrounding pixels involved in the computation at the irregular sampling position
$B(\cdot,\cdot)$	The bilinear interpolation kernel
D_K	The space-resolution of convolution kernels
M	The dimension of input feature maps
N	The dimension of output feature maps
Y	The output feature map
y_1	The global pooling result
y_2	The output from the module of ladder autrous convolution
$GP_S(x)$	The global pooling operation
B	The normalization of feature weight
$G_{K,\varepsilon}$	The output of densely connected pyramid pooling
K^p	The level of pyramid
$\oplus$	The concatenation operation
p	The ground truth
$\hat{p}$	The predicted segmentation
L_{cross}	The cross entropy loss
L_{dice}	The dice loss
lr	Learning rate
i	The number of iterations of this training
t_i	The total number of iterations

Chapter 10

Symbols and Abbreviations

x_i	The gray value of the ith pixel
v_k^p	The prototype value of the k-th cluster
u_{ki}	The fuzzy membership value of the i-th pixel with respect to cluster k
N_f	The total number of pixels in an image f
c	The number of clusters
m	The weighting exponent for FCM
J_m	The objective function of FCM algorithms
G_{ki}	The fuzzy factor
L^{pre}	The candidate landslide regions of pre-event images
L^{post}	The candidate landslide regions of post-event images
L^d	The difference image
L^{db}	The binarized difference image
T_1	The grayscale value of an image
S	A set whose elements
X_0	An initialized array

E	A structuring element
δ	The morphological dilation operation
F	The operation of filling holes
E_c	A structuring element that is used for connecting leaking regions
r	The radius of a structuring element
W_{bi}	The width of bitemporal images
H_{bi}	The height of bitemporal images
f_m	A marker image
R^e	The morphological erosion
R^δ	The dilation reconstructions
R^O	The morphological opening
R^C	The closing reconstructions
σ	The standard deviation of the Gaussian filter used for preprocessing
ΔT	The time step
c_w^b	The weighting coefficient used to balance the pairwise potential
P_{lm}	The total pixel number of the identified landslides that are matched with the corresponding ground truth
P_r	The total pixel number of the ground truth
P_l	The total pixel number of the identified landslides
P_{rum}	The total pixel number of the corresponding ground truth that is not matched with the identified landslides
P_{over}	The total pixel number of detected false landslides
p_m^d	The multiple of down-sampling
$\leq_{PCA}$	The lexicographic order based on principle component analysis
$\mathbf{v}(R, G, B)$	A color pixel
$\mathbf{v'}(P_f, P_s, P_t)$	The transformed color pixel using the principal component analysis (PCA)
$\vec{\varepsilon}$	The multivariate morphological erosion
$\vec{\delta}$	The multivariate morphological dilation
$\vee_{PCA}$	The supremum based on lexicographical ordering $\leq_{PCA}$
$\wedge_{PCA}$	The infimum based on lexicographical ordering $\leq_{PCA}$
f_{co}	A color image

Chapter 11

Symbols and Abbreviations

$F_n(x, y)$	A gaussian wrap function
μ_N	The normalization factor
c_n	The gaussian surrounding space constant
r_{cs}	The radius of circular structure element

f^m	The mask image of a marked image
R^γ	The morphological open operation
R^ϕ	The morphological closed operation
m	The scale of the largest structural element
H	A gradient image
I	The regional minimum image of a gradient image
W	The segmentation result of watershed transformation
I_j	The j-th connected component in I
W_j	The j-th segmentation area in W
x_p	The p-th pixel in W
x_q	The q-th pixel in I
k	The structural element parameters
Acc	Accuracy
R	Recall rate
A_k	The type k-th hole obtained by the test algorithm
G_k	The standard segmentation result of the i-th hole
$\cap$	The intersection operation
$\cup$	The union operation
B_1	The initialized structural element
Θ	The erosion operation
$\oplus$	The dilate operation
D_g	The distance on the map
D_a	The actual distance
j	The region number in a superpixel image
∂_j	The j-th region in a superpixel image
$\boldsymbol{v}_j$	The average gray-scale value of pixels in ∂_j
A_m	The affinity matrix
σ^2	The scaling parameter of A
D_m	The degree matrix
λ	The eigenvalue set of A
L_m	The laplacian matrix
y_k	The prototype value of the k-th cluster
u_{kj}	The membership value of the j-th sample with respect to k-th cluster
U	Membership partition matrix
c	The number of clusters
m'	A weighting exponent
$C(y_1)$	The number of elements classified into y_1
$C(y_2)$	The number of elements classified into y_2

List of Acronyms

ADAM	Adaptive moment estimation
AFCF	Automatic fuzzy clustering framework for image segmentation
AlexNet	ImageNet classification with deep convolutional neural networks
AMR	Adaptive morphological reconstruction
AMR-RMR-WT	Fast and Automatic Image Segmentation Using Superpixel-Based Graph Clustering
AMR-SC	Spectral clustering based on pre-segmentation of AMR-WT
ASD	Average symmetric surface distance
ASPP	Atrous spatial pyramid pooling
ASSD	Average symmetric surface distance
AutoML	Automatic machine learning
BDE	Boundary displacement error
BP	Back-Propagation
CNN	Convolutional neural network
CPU	Central processing unit
CS	Comparison scores
CT	Computed Tomography
CV	Segmentation covering
DC	Deformable convolution
DSC	Depthwise separable convolution
DWT	Discrete wavelet transform
EGC	Eigen-value gradient clustering
ELU	Exponential Linear Unit
ELSE	Edge-based level-set
EM	Expectation–maximization
EnFCM	Enhanced Fuzzy c-means clustering
FAS-SGC	Fast and automatic image segmentation algorithm employing superpixel-based graph clustering
FCM	Fuzzy c-means clustering
FCN	Full convolution network
FCN-PP	Fully convolutional network within pyramid pooling
FN	False negative fraction
FP	False positive fraction
GB	Gigabyte
GCE	Global consistency error
GELU	Gauss error linear element

GFLOPs	Giga Floating-point Operations Per Second
GL-graph	Global/regional affinity graph
GMM	Gaussian mixture model
GPU	Graphics Processing Unit
GT	Ground truths
HC	Hierarchical clustering
HD	Hausdorff distance
H-DenseUNet	Hybrid densely connected UNet
HMRF	Hidden Markov random field
HSR	High spatial resolution
HSV	Hue-Saturation-Value
HU	Hounsfield
IDWT	Inverse discrete wavelet transform
IOU	Intersection over Union
IRB	Inverted residual bottleneck
ISBI	International symposium on biomedical imaging
Ladder-ASPP	Ladder-atrous-spatial-pyramid-pooling
LIM	Landslide inventory mapping
LiTS	Liver Tumor Segmentation Challenge
LM	Landslide mapping
LMSE	Linear Mean Squared Error
LSC	Linear Spectral Clustering
LV-Net	Lightweight V-Net
MCG	Multiscale combinatorial grouping
MGR	Morphological gradient reconstruction
MGR-WT	Morphological Gradient Reconstruction based Watershed Transform
MIoU	Mean intersection over union
MMF	A new approach to morphological color image processing
MMG	Multiscale morphological gradient algorithm
MMG-WT	Multiscale morphological gradient for watersheds
MMGR	Multiscale morphological gradient reconstruction
MMGR-WT	Novel WT based on MMGR
MMR	Morphological reconstruction
MP	McCulloch Pitts
MPA	Mean Pixel Accuracy
MR	Morphological reconstruction
MRI	Magnetic Resonance Imaging
MRF	Markov random fields
MRI	Magnetic Resonance Imaging
MSD	Maximum symmetric surface distance
MSE	Mean square error
MSR	Multi Scale Retinex
MSSD	Maximum symmetric surface distance
NAS	Neural Architecture Search
NLP	Natural language processing
NP	Non-deterministic Polynomial
OBEM	Object-based expectation maximization
OE	Overall error

OEF	Oriented edge forests
OG	Original gradient
PA	Pixel Accuracy
PC	Personal Computer
PCA	Principal components analysis
PDE	Partial differential equation
PP	Pyramid pooling
PReLU	Parametric ReLU
PRI	Probabilistic rand index
PW	Power watershed
ReLU	Rectified linear unit
RGB	Red-Green-Blue
RGD	Random gradient descent
RLSE	Region-based level-set
RMSE	Root mean square error
RMSD	Root mean square deviation
RReLU	Randomized ReLU
RVD	Relative volume difference
RW	Random Walker
RWT	Robust watershed transform
SC	Spectral clustering
SCG	Single-scale combinatorial grouping
SE	Structured edge
SEM	Scanning electron microscope
SEs	Structuring elements
SGD	Stochastic Gradient Descent
SH	Superpixel Hierarchy
SLIC	Simple linear iterative clustering
SOTA	State-of-the-art
SPP	Spatial Pyramid Pooling
SVMs	Support vector machines
TM	Trademark
TP	True positive fraction
UAV	Unmanned Aerial Vehicle
VHR	Very high-resolution
VI	variation of information
ViT	Vision Transformer
VO	Vector ordering
VOC	Visual Object Classes
VOE	Volume Overlap Error
WT	Watershed transformation
YIQ	National Television Standards Committee
3-D	Three-dimensional

Part I

Principles

1

Introduction

Since the birth of computers in 1950s, people have been using computers to process multimedia information such as text, graphics, and images. With the rapid development of computer technology, signal processing, and imaging technology, image processing technology has been boosted in recent years. Humans use their senses such as vision, hearing, and touch to obtain external information. Among all these senses, vision is the most momentous way for obtaining information since it can often capture more information than hearing and touch. Therefore, images play a vital role in people's perception of the world.

Image processing technology mainly includes image transformation [1], image restoration [2], image compression [3], image segmentation [4], target detection and recognition [5], and image classification [6]. In image processing, image segmentation plays a crucial role. The main reason is that image segmentation is the foundation of high-level tasks in computer vision. With the continuous improvement of image processing technology, image segmentation is developed from the early edge detection, region merging and division, and contour extraction to the current semantic segmentation. Currently, image segmentation technology has been widely used in industrial detection [7], medical image analysis [8], remote sensing earth observation [9], intelligent driving [10], and other fields. In view of the expeditious development and comprehensive application of image segmentation technology, this book will focus on the basic theoretical knowledge related to image segmentation, the introduction of mainstream algorithms of image segmentation, and specific applications of image segmentation, as shown in Figure 1.1.

1.1 Preliminary Concepts

Image segmentation is dividing an image into nonoverlapping regions. Early image segmentation is relatively simple, which mainly tackles some problems coming from industrial defect detection, target detection, and image enhancement tasks. Due to the limitation of early imaging technology, early digital images have some characteristics of low resolution, blur, noise corruption, and so on. However, since image segmentation can effectively extract the contour and other details of an image, it can be used to enhance visual effects of the image. For example, in industrial detection, the industrial camera is deployed on the pipeline, so the camera imaging results have a fixed background. Under this circumstance, it is relatively easy to use an image segmentation technique to obtain the target and defect area.

For the above example, although this binary image segmentation technology is easy to implement, it is not universal, which means it is unsuitable for images with complex background. The main reason is that this kind of image segmentation method only depends on low-level features of images to make decisions. As low-level features of images have a weak ability of representation for real semantic information, these binary image segmentation methods usually provide poor

Image Segmentation: Principles, Techniques, and Applications, First Edition. Tao Lei and Asoke K. Nandi.
© 2023 John Wiley & Sons Ltd. Published 2023 by John Wiley & Sons Ltd.

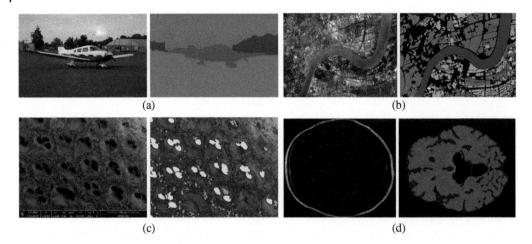

Figure 1.1 The application of image segmentation in different scenes. (a) Natural image. (b) Remote sensing image. (c) Scanning electron microscope image. (d) Medical image.

segmentation results in practical applications. With the rapid development of imaging technology, the image quality has improved, such as higher image resolution, more bands, and richer colors. However, with the improvement of artificial intelligence technology, the task of image analysis has become more complex requiring higher segmentation accuracy, faster inference speed, and less resource consumption. For example, in automatic driving, the semantic understanding of the scene is the basic requirement, along with the speed and video information, which requires that the segmentation time of an image should be less than 1/24 second. Therefore, the current task of image segmentation is more sophisticated and thus face many challenges.

At present, with the advent of the era of artificial intelligence, image segmentation ushered in new development opportunities. Furthermore, robots, automatic driving, intelligent medical diagnosis, and so on derive more types of image segmentation tasks. Especially in the field of medical image analysis, the early image segmentation methods need more human-computer interaction to obtain fairly satisfactory image segmentation results. The main reason is that medical images rely on x-ray scanning, magnetic resonance, ultrasound, and other imaging methods, which are easily interfered by noise, and the imaging effect is obviously inferior to that of optical imaging. Therefore, medical image segmentation has always been one of the most challenging frontier topics in the field of computer vision, as shown in Figure 1.2.

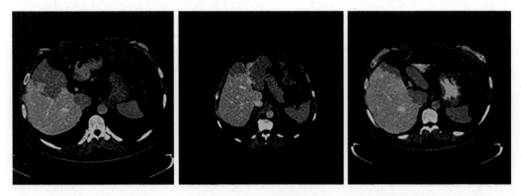

Figure 1.2 The results of liver segmentation in CT abdominal images using traditional model.

So far, a large number of image segmentation algorithms have emerged, from the initial contour-based [11], pixel-based [12], and region-based image segmentation [13] to the current mainstream semantic image segmentation. Though these methods have been successfully applied in different fields to some extent, image segmentation still encounters many challenges, such as the difficulty of sample labeling and model parameter adjustment, poor generalization, and high resource usage. Tackling these problems requires continuous research and innovation.

1.2 Foundations of Image Segmentation

Image segmentation involves multidisciplinary knowledge, including imaging principles, color theory, cognitive science, and machine learning techniques, among which machine learning is the most important. Popular machine learning algorithms are grouped into three categories: supervised learning, unsupervised learning, and weakly supervised learning. The difference between supervised learning and unsupervised learning is whether supervised information (labeled data) is necessary or not. Unsupervised learning does not need labeled data while supervised learning requires a large number of labeled data. In practical applications, supervised learning is more common since it can provide higher accuracy than unsupervised learning. For example, students' learning is carried out under the supervision and guidance of teachers, which is obviously more effective than study by oneself. However, supervised learning is at the expense of cost, requiring considerable human labor and computing resources. Compared to supervised learning and unsupervised learning, weakly supervised learning only requires a small number of labeled data or coarse labels since it can effectively utilize a large number of unlabeled data to learn models. Thus, unsupervised learning, supervised learning, and weakly supervised learning have respective advantages and thus are used for different tasks.

Early image segmentation methods, such as threshold method, edge detection, and watershed, do not need labeled samples and model training, so these methods are efficient, but they often produce poor results. Although the later methods based on partial differential equations, clustering, and spectrograms can achieve better image segmentation results, they often suffer from low computational efficiency. As there are many kinds of unsupervised image segmentation methods, it is difficult to classify them accurately. Based on the low-level features of images, unsupervised image segmentation is often roughly divided into pixel-based, contour-based, and region-based segmentation methods.

1.2.1 Pixel-Based Image Segmentation

Pixel-based image segmentation mainly includes thresholding methods and clustering methods. Since thresholding methods heavily depend on the number of thresholds and the parameter selection, the actual segmentation results are relatively rough. Compared with thresholding methods, clustering methods are widely used because of their high robustness. Clustering methods mainly involve hierarchical clustering [14], c-means clustering [15], fuzzy c-means clustering [16], density clustering [17], and spectral clustering [18]. The hierarchical clustering algorithm utilizes the similarity between pixels to achieve hierarchical decomposition. This kind of algorithm is simple and easy to implement, but the corresponding segmentation results depend on the tree construction and threshold selection; hence, image segmentation methods based on hierarchical clustering are not common.

Both c-means clustering and fuzzy c-means clustering belong to prototype clustering. These methods firstly initialize the clustering prototype and then iteratively optimize the prototype to obtain the optimal solution. In comparison with fuzzy c-means clustering, general c-means clustering is considered as hard clustering, which means that each sample completely belongs to or not belongs to a certain class, and its membership degree value is 1 or 0, respectively. Fuzzy c-means clustering defines its membership degree value ranging from 0 to 1, which can achieve better clustering results than c-means clustering. Usually, a fuzzy clustering algorithm uses the minimum error criterion to construct its objective function, then obtains the membership degree of each pixel to the clustering center by optimizing some objective function, and finally uses the membership degree to classify pixels in an image to achieve image segmentation. Common fuzzy c-means clustering algorithms rely on Euclidean distance calculations since a multichannel image is usually defined in a complex high-dimensional manifold space, and so it is easy to cause large errors using Euclidean distance measurement directly. Consequently, Mahalanobis distance is introduced into high-dimensional complex data measurement, which forms the basis of the Gaussian mixture model. Applying the Gaussian mixture model to multichannel image segmentation has clear advantages over fuzzy c-means clustering. Although prototype clustering can be widely used in image segmentation, these methods regard a pixel as a single independent sample, thus ignoring the spatial structure information of the image, so the segmentation results are easily affected by noise.

In recent years, a new clustering idea has emerged, which determines the data classification by considering the spatial distribution structure of all samples, especially their density, and is usually called density clustering. Density clustering uses the different density of samples to determine the initial clustering center, and then updates the cluster according to the association between samples. At the same time, it updates the clustering center and finally realizes the classification of the sample set. On the other hand, applying density clustering to image segmentation can achieve a better overall segmentation effect as the density clustering focuses on the relationship between all samples. Density clustering can predict the number of clusters in advance according to the density distribution of data and then obtains automatic image segmentation results. Therefore, in recent years, image segmentation based on density has received more attention by scholars, and more research results have been reported.

1.2.2 Contour-Based Image Segmentation

Contour-based image segmentation is to achieve image segmentation by searching the boundary of each region in an image. In the early stages, this type of method was mainly based on the image gradient computation such as the well-known Sobel, Robert, Canny operators. However, it is very hard to transform the image gradient into a closed contour. Hence in the traditional edge detection, we usually use a threshold method to obtain contour detection results according to different task requirements.

With the development of the applications of partial differential equations in image processing, the active contour model has become widely used in image segmentation. Image segmentation based on the active contour model usually adopts the energy functional method, whose basic idea is to use a continuous curve to express the target edge and define an energy function so that its independent variables include the edge curve. Thus, the question of image segmentation is transformed into formulating the minimum value solving process of the energy function. This kind of method mainly involves the parametric active contour model and geometric active contour model.

For the parametric active contour model, a deformable parametric curve and its corresponding energy function are firstly constructed, and then a closed curve with minimum energy is regarded as the objective contour by controlling the deformation of the parametric curve to minimize the energy objective function. The representative method of parametric contour model is the Snake model [19], whose core idea is to initialize a variable energy curve around a target to be segmented and then evolve and deform the curve to convergence under the constraint of the minimum value of the energy function. The evolution rule of the curve is based on the calculation of the internal energy and external energy of the curve. In fact, the internal energy of the curve is used for smoothing, while the external energy guides the active contour model moving to image features, so as to obtain a more accurate target contour.

The geometric active contour model is represented by a level set method. Level set is a digital method to track the contour and surface motion. It does not operate the contour directly but sets the contour as a zero-level set of a high-dimensional function, which is called the level set function. Then this function is differentiated, and the contour of the motion is obtained by extracting the zero-level set from the output. The main strength of using the level set is that any complex structure can be modeled and topologically transformed [20].

1.2.3 Region-Based Image Segmentation

Region-based image segmentation firstly initializes the seeds and then finds the approximate pixels that are similar to seeds, these pixels are usually regarded as a class, and finally updates the seeds until convergence is reached. Region-based image segmentation methods mainly include region growing [21], region splitting and merging [22], and superpixel [23].

Region growing aims to extract connected regions in an image according to some predefined growth criteria such as similar grayscale, edge, or texture information [24]. Its basic idea is, starting from one or several seeded pixels and following the growth rule, to merge the pixels iteratively into the adjacent region until all the pixels are merged. The region growing method is not often used solely, but as one step in image analysis, it is used to depict small and simple regions (such as tumors and lesions) in an image. The main drawback of region merging is that the seed points need to be determined manually.

The basic idea of the region splitting and merging algorithm is to determine a criterion of splitting and merging, or a measure of the consistency of region features. If the features of a region in an image are inconsistent, the region is divided into four equal sub regions. After that, only when the adjacent two sub regions meet the features consistency, they can be merged into a large region. The above steps are performed iteratively until all regions no longer meet the conditions of splitting and merging. The splitting ends when the splitting is no longer possible. Furthermore, a check of whether there are similar features or not in adjacent regions is performed; if there are similar features in adjacent regions, the similar regions will be merged. After this process is completed, image segmentation is achieved.

In fact, region growing and region splitting and merging algorithms have a similar principle, and they complement and promote each other. Region splitting in the extreme is to divide an image into a single pixel, and then merge these pixels according to some measurement criteria. To a certain extent, it can be considered as a region growing method of a single pixel [24]. Compared with region splitting and merging, region growing can save the process of splitting. Region splitting and merging can do similar merging in a larger region, while region growing can only start from a single pixel.

Region merging based on superpixel firstly uses superpixel algorithms to segment an image obtaining many small regions, and then uses the over-segmentation results to merge the regions to generate the final segmentation results. Due to the efficiency of superpixel that can reduce the number of different pixels in an image, this kind of method has obvious computational advantages for high-resolution image segmentation, which is of great significance for achieving fast image segmentation.

However, this kind of methods depends on the result of superpixel segmentation. The mainstream superpixel algorithms include SLIC [23], turbopixel [25], DBSCAN [26], LSC [26], DEL [27], GMM [28], and ISF [29]. Most of these superpixel algorithms adopt the contour iterative optimization strategy in a local grid to achieve superpixel segmentation. The advantages of these algorithms are that they can obtain the superpixel segmentation results based on the preset number of regions, which can obtain more accurate boundaries in the local grid. The drawback is that the optimization strategy only works in the local grid, so it is hard to capture the real object contour because of the loss of global information.

1.2.4 Neural Network–Based Image Segmentation

Although unsupervised image segmentation methods have been successfully applied in many fields, it is difficult to apply these methods to image segmentation tasks in complex scenes. In recent years, as data collection and storage have become easier than before, a large number of annotated large-scale data sets have emerged, such as ImageNet [30], COCO, and VOC. In addition, the emergence of high-performance parallel computing systems and the rapid development of GPUs promote the rapid development of deep learning. Applying deep learning to image segmentation can effectively overcome the defects of depending on handcrafted features for traditional methods and leads to better image segmentation results.

Deep convolutional neural network (CNN) [31] is one of the most representative models in deep learning. The structure and number of convolution kernels, pooling layers, and full connection layers are all designed to adapt to different visual tasks. According to different tasks, different activation functions are added as the last layer to obtain the specific conditional probability of each output neuron. Finally, the whole network is optimized on the objective function (such as mean square error or cross entropy loss) by the random gradient descent (RGD) algorithm.

In comparison to the traditional methods, CNN can achieve hierarchical feature representation, that is to say, learning the multilevel representation from pixels to high-level semantic features through hierarchical multilevel structure. Owing to CNN's powerful learning ability, some classic computer vision tasks can be redefined as a high-dimensional data conversion problem that can be easily solved from multiple perspectives.

Different from the image classification task, image segmentation is an encoding and decoding problem. Due to the lack of decoder in a CNN, its performance in image segmentation task is mediocre. Full convolution network (FCN) [32] realizes end-to-end image semantic segmentation using an encoder-decoder network structure for the first time. The basic idea of FCN is to use a 1×1 convolution kernel to generate low-resolution prediction, and then a pixel-dense prediction is obtained by upsampling or deconvolution.

This encoder-decoder structure has become the latest paradigm in the field of image segmentation based on deep learning. The encoder extracts image features by learning useful convolution kernels, and the decoder realizes the final output by learning the deconvolution kernels. Based on an FCN network, a large number of improvements have emerged to promote the accuracy of image segmentation, such as deformable encoder-decoder network [33], multiscale feature fusion network [34], dense connection network [35], and attention network [36].

Although image segmentation technology based on deep learning is developing rapidly, it still encounters many challenges. The first is the challenge from data. Being different from other image processing tasks, image semantic segmentation needs complex annotation information, especially for the pixel-level segmentation task. Since pixel-by-pixel labeling usually requires considerable labor and material resources, how to use less labeling information to obtain a better image semantic segmentation effect is a challenge. The second is the model challenge. The current mainstream segmentation models often have fixed backbone architecture, and engineers only adjust some function modules to make sure the models are available for different segmentation tasks. Therefore, these models often have poor robustness to noise, poor migration for different data sets, and do not have strong generalization ability. In addition, current neural network models have poor interpretability and flexibility. If a model is completely unexplainable, its application in many fields will be limited due to the lack of reliability. The third is the challenge of computational resources. Most of the current image segmentation models are complex in structure and consume considerable computational resources to process efficiently a large data set. Therefore, how to build a lightweight and compact segmentation model to better deploy in mobile devices is a challenge.

1.3 Examples: Image Segmentation

1.3.1 Automatic Drive

With recent innovations in science and technology, autonomous driving has become a research trend and a hot spot in the artificial intelligence and automotive fields. Autonomous driving based on the perception system can promote the development of transportation systems and effectively broaden the application field of science and technology. It is a great achievement and innovation in the history of transportation systems.

Autonomic driving firstly obtains a lot of data from the vehicle and external environment with the help of a perception system and then makes corresponding driving decisions through system analysis and judgment [37]. Compared with the traditional automobile driving technology, autonomous driving technology has a significant number of practical applications. The core of autonomous driving technology is image understanding. First, the vehicle camera can capture the actual scene information, and then computer vision technology is used to image semantic annotation, through combining with the data obtained by other sensors, and then intelligent decisions can be made by the onboard computer. Figure 1.3 shows the semantic segmentation of a natural image.

Figure 1.3　Natural image scene understanding.

1.3.2 Medical Image Analysis

Art health care is based on the existing medical information platform, integrating the health information resources as well as the data of health systems at all levels into a big data application platform by using advanced internet of things technology, artificial intelligence technology, and sensor technology. It achieves intelligent integration among the patients, medical staff, medical institutions, and medical devices by using techniques such as data analysis, knowledge mining, and artificial intelligence.

Artificial intelligence technology has been widely used in the field of medical image analysis. Automatic analysis of medical images can comprehensively level up the accuracy of medical diagnosis and treatment, especially in areas where medical resources are scarce. Also, deploying artificial intelligence technology in medical images analysis can effectively reduce the task difficulty for doctors, save medical resources, and quickly assist physicians to make a diagnosis.

Image segmentation is one of the most critical technologies in medical image analysis and is also the core required to realize intelligent medical treatments. With the development and popularization of medical imaging equipment, x-rays, computed tomography (CT), magnetic resonance imaging (MRI), and ultrasound imaging have become four important imaging aids to help clinicians diagnose diseases, evaluate prognosis, and plan operations in medical institutions. In order to help clinicians making accurate diagnoses, it is necessary to segment some key objects in medical images and extract features from the segmented regions. Common medical image segmentation tasks include liver and liver tumor segmentation [38], brain and brain tumor segmentation [39], optic disc segmentation [40], cell segmentation [41], and lung and lung nodule segmentation [42]. Figure 1.4 shows the segmentation results of liver, spleen, and liver tumors in an abdominal CT image using an AI technique.

1.3.3 Remote Sensing

Remote sensing is a noncontact, long-distance detection technology that often detects the distribution of above ground objects as well as underground natural resources in three dimensions in a comprehensive, quick, and effective way. Due to the above-mentioned advantages, it has gradually become an important method for understanding the world from a multidimensional and macro perspective. In recent years, remote sensing technology has been developed dramatically and relatively complete remote sensing monitoring systems have been developed for different data from low altitude, aviation, and aerospace satellites. It has been widely used in land, planning, water

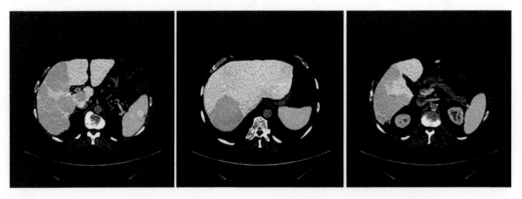

Figure 1.4 Segmentation results of organs and tumors in abdominal CT images.

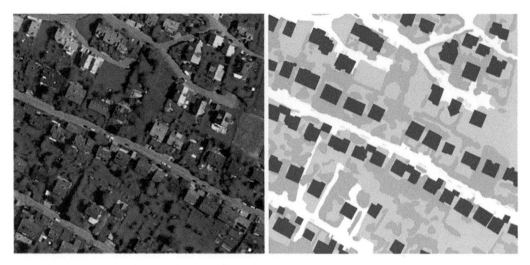

Figure 1.5 Remote sensing image feature classification using image semantic segmentation.

conservation, agriculture, forestry, marine, surveying and mapping, meteorology, ecological environment, aerospace, geology, military, mining fields, and so forth.

One of the main purposes of analyzing remote sensing images is to grasp the actual situation of the ground from a macro perspective, such as extracting roads from remote sensing images and obtaining actual geographical coordinates to better understand the overall situation of the roads. In remote sensing images, image segmentation technology is one of the crucial means to detect and analyze targets, especially the use of the image semantic segmentation technology that can accomplish the classification of the remote sensing images. In addition, image segmentation technology is also widely used in change detection of multi-temporal remote sensing images, which is very important for land change detection and disaster assessment. Figure 1.5 shows the classification result of a remote sensing image using image segmentation.

At present, the space-aeronautics integrated remote sensing based on low altitude UAV, aerial photography, and aerospace sensors can provide significant data for natural resources investigation and monitoring in a multidimensional and macro way. In the future, the development of an image semantic understanding technique and its applications will definitely strengthen the capacity of land, atmosphere, and marine remote sensing monitoring as well as the service and support capability on any country's resources, environment, ecological protection, emergency disaster reduction, mass consumption, and global observation.

1.3.4 Industrial Inspection

Defect detection is a crucial link in the industrial assembly line. Generally, due to the variety of defects, through traditional human detection it is difficult to completely identify the defect characteristics, which not only requires manual detection to distinguish the working conditions but also will waste a lot of labor. Compared with manual image detection, image segmentation easily realizes automatic defect detection and classification, along with the advantages of high reliability, high detection accuracy, fast detection, low cost, and wide applications.

In practical industrial detection applications, due to the influence of sensor material properties, factory electrical equipment interference and other factors, the images collected in industrial fields usually contain noise, which reduces the accuracy of target defect detection, so it is necessary to suppress noise for image enhancement in practical applications. According to the various distribution range of noise energy, image enhancement strategies such as Gaussian filter, mean filter, or median filter can be selected to suppress noise. Besides noise reduction, image enhancement can also purposefully enhance some image features, such as color, brightness, and contrast, by means of histogram equalization and gamma transform.

To obtain the defect region, it is usually natural to use image segmentation algorithms to decompose the image into several independent regions according to different characteristic attributes and task requirements. In defect detection, we are interested in whether there are abnormal phenomena in the image, such as scratches, dirt, and so on. After effective segmentation, the features in same region are the same or similar, while the features between different regions are obviously dissimilar. Common image segmentation methods include threshold-based segmentation method, region-based segmentation method, watershed transform, and level-set. Figure 1.6 shows the defect detection result for a magnetic tile surface.

In addition to the above applications, image segmentation is also widely used in security. Currently, intelligent security includes access control, alarm, and monitoring. These are the three most basic parts required by a complete intelligent security system. The application of image segmentation in the field of security is derived from the information redundancy caused by the massive image and video information generated by the security system every day. Without human intervention, the automatic image segmentation technology is used to analyze automatically the image sequence captured by the camera, including target segmentation and extraction, target recognition, and target tracking, as well as the understanding and description of the target behavior in the monitoring scene. According to the image analysis result, security guards can make decisions for some emergency circumstances.

As one of the most important research initiatives in the field of computer vision, image segmentation has a very broad range of applications, and thus it faces many challenges due to the requirement of various complex tasks. The current image semantic segmentation is a kind of pixel-level classification result, which weakens the utilization of underlying structural features of images to a certain extent. How to effectively integrate the spatial structure information and high-level semantic information of images is of great significance to the development of image segmentation techniques in the future.

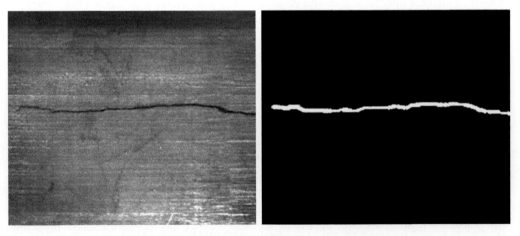

Figure 1.6   Defect detection for magnetic tile surface.

1.4   Assessment of Image Segmentation

In order to evaluate the advantages and disadvantages of image segmentation algorithms, researchers have developed many evaluation indexes. By imitating the human visual perception, these evaluation indexes mainly depend on error rate, segmentation contour quality, region quality, and other similar factors, and they are introduced in detail in the following sections.

In this section, A and B denote the ground truth and the predicted segmentation maps, respectively.

1) **Accuracy evaluation index**
   a) **Pixel Accuracy (PA)**
   Pixel accuracy simply finds the ratio of pixels properly classified divided by the total number of pixels. For $(K + 1)$ classes (K foreground classes and the background) pixel accuracy is defined as:

$$PA = \frac{\sum_{i=0}^{K} p_{ii}}{\sum_{i=0}^{K} \sum_{j=0}^{K} p_{ij}},$$ (1.1)

   where p_{ii} is the number of pixels of i-th class predicted as belonging to i-th class, and p_{ij} is the number of pixels of i-th class predicted as belonging to j-th class.
   b) **Mean Pixel Accuracy (MPA)**
   MPA is an extended version of PA that calculates the proportion of correctly classified pixels within each class and then averages the correct rate across all classes:

$$MPA = \frac{1}{K+1} \sum_{i=0}^{K} \frac{p_{ii}}{\sum_{j=0}^{K} p_{ij}}.$$ (1.2)

   c) **Precision/Recall/F1 score**
   Precision/Recall/F1 score are popular as they are often employed by the classical image segmentation models. Precision and recall can be defined for each class, as well as at the aggregate level, as follows:

$$Precision = \frac{TP}{TP + FP}, \quad Recall = \frac{TP}{TP + FN},$$ (1.3)

   where TP refers to the true positive fraction, FP refers to the false positive fraction, and FN refers to the false negative fraction. Usually, we are interested in a combined version of precision and recall rates. A popular such a metric is called the F1 and is defined as the harmonic mean of precision and recall:

$$F1 = \frac{2 \times Precision \times Recall}{Precision + Recall}.$$ (1.4)

2) **Regional quality evaluation index**
   a) **Dice coefficient**
   Dice coefficient is another popular metric for image segmentation (and is more commonly used in medical image analysis) that can be defined as the overlap area of predicted and ground-truth maps divided by the total number of pixels in both images:

$$Dice = \frac{2|A \cap B|}{|A| + |B|}.$$ (1.5)

b) **Intersection Over Union (IOU)**

IOU, or the Jaccard index, is one of the most popular metrics in semantic segmentation. It is defined as the area of intersection between the predicted segmentation map and the ground truth divided by the area of union between the predicted segmentation map and the ground truth:

$$IOU = J(A, B) = \frac{|A \cap B|}{|A \cup B|}, \tag{1.6}$$

where the value of IOU ranges from 0 to 1. In fact, the mean intersection over union (MIoU) of all classes is widely used to evaluate the performance of image segmentation algorithms for multiclass segmentation problems.

c) **Volume Overlap Error (VOE)**

VOE is the complement of the Jaccard index:

$$VOE(A, B) = 1 - \frac{|A \cap B|}{|A \cup B|}. \tag{1.7}$$

d) **Relative Volume Difference (RVD)**

The RVD is an asymmetric measure defined as:

$$RVD(A, B) = \frac{|B| - |A|}{|A|}. \tag{1.8}$$

3) **Contour quality evaluation index**

Let $S(A)$ denote the set of surface voxels of A. The shortest distance of an arbitrary voxel v to $S(A)$ is defined as:

$$d(v, S(A)) = \min_{s_A \in S(A)} v - s_A, \tag{1.9}$$

where $\|.\|$ denotes the Euclidean distance.

a) **Hausdorff distance (HD)**

Hausdorff distance is a measure that describes the degree of similarity between two sets of points, and it is defined as:

$$HD(A, B) = \max(d(A, B), d(B, A)). \tag{1.10}$$

b) **Average symmetric surface distance (ASD)**

The ASD is given by:

$$ASD(A, B) = \frac{1}{|S(A)| + |S(B)|} \left(\sum_{s_A \in S(A)} d(s_A, S(B)) + \sum_{s_B \in S(B)} d(s_B, S(A)) \right). \tag{1.11}$$

c) **Maximum symmetric surface distance (MSD)**

The MSD, also known as the symmetric Hausdorff distance, is similar to ASD except that the maximum distance is taken instead of the average:

$$MSD(A, B) = max \left\{ \max_{s_A \in S(A)} d(s_A, S(B)), \max_{s_B \in S(B)} d(s_B, S(A)) \right\}. \tag{1.12}$$

1.5   Discussion and Summary

Image segmentation is the key step of target detection and recognition in computer vision. However, early image segmentation technology only focuses on simple image statistical analysis, far from deep image understanding. Nowadays, with the rapid development of data sensing technology, computer technology, artificial intelligence, and other technologies, the tasks of target detection and recognition in images become more and more complex. Therefore, image segmentation technology also continues to evolve, from the early image segmentation based on pixel and regional features to the later image segmentation based on semantic features. At present, image semantic segmentation technology mainly depends on deep convolution neural networks. Although scholars have proposed a large number of convolution network models for different segmentation tasks, such as U-shaped networks for medical image segmentation, Siamese networks for remote sensing change detection, and multibranch networks for multi-modal data, there are still many challenges.

On the one hand, because the current depth network models heavily depend on the quality of training data, an important challenge in image segmentation is how to realize the small sample training, weak supervised learning, and automatic labeling of data. On the other hand, the unclarity of depth network model's interpretability and working mechanism also raises a challenge that is how to design an interpretable depth network model according to different tasks of image segmentation. In addition, the current network optimization algorithm is a general algorithm that ignores the difference of segmentation tasks. How to associate the image segmentation task with the optimization algorithm and design a special optimization algorithm that is suitable for the requirement of segmentation task is another challenge for the development of image segmentation theory. The specific analysis is given as follows.

1) **Data Challenges**

   Although deep learning has achieved great success in various tasks, many tasks in practical applications still face a serious shortage of labeled data. For example, full scene annotation in medical images and remote sensing images require very experienced experts. Therefore, medical data annotation is high cost and time consuming, and it is difficult to form massively labeled data sets. How to use only a small amount of labeled data to generate high-precision models is often called "few-shot learning," especially in the tasks of multi-objective segmentation of medical images and semantic segmentation of remote sensing images.

   In addition, the other extreme case is how to effectively improve the performance of models by using a large number of unlabeled data. For example, wrong image recognition may cause a disaster in autonomous driving. Therefore, researchers have built some very large data sets that contain hundreds of millions of images with rich annotations, and they hope to significantly improve the accuracy of models by effectively using these data. However, when the amount of data is large enough, the model's performance improvement will be limited through increasing the number of training samples.

2) **Model Challenges**

   In recent years, a large number of network structures have emerged for different image segmentation tasks. Nevertheless, most of them are designed based on human experience, which requires engineers to spend considerable time and energy. To solve this problem, scholars have proposed the neural architecture search theory, by automatically searching the optimal network structure according to the task requirement in a certain area to replace the artificial design network structure. Though the neural architecture search solves the problem to a certain extent, they still face many challenges. For example, the search area of the current neural architecture is very narrow because they only look for the locally optimal combination of existing network modules (such as deep separable convolution and identity connection), so some better modules are hard to find.

3) **Algorithmic Challenges**

In recent years, although many image segmentation algorithms based on deep learning have been reported, most of them are improved by designing new network architectures. The current network optimization algorithm is a general algorithm that cannot effectively cope with complex task requirements, such as the dramatic change of size, shape, and position of different segmentation targets. Therefore, how to associate image segmentation tasks with the selected optimization algorithm is still a challenging problem.

As a significant research topic, optimization problems are of great importance in theoretical research as well as algorithm research. The basis of optimization problems is mathematics, which is to find the maximum or minimum value of an objective function under certain constraints. According to the development of image segmentation technology and optimization theory, we give several development directions of designing optimization algorithms based on image segmentation tasks.

a) **Evaluation-Based Optimization Algorithms**

The purpose of segmentation evaluation is to guide, improve, and promote the performance of segmentation algorithms. The results of segmentation can be more accurate by combining evaluation with segmentation applications, using evaluation results to induce reasoning and design optimization algorithm. In addition, designing more accurate image segmentation evaluation indexes can more accurately measure the performance of the segmentation model for the feature extraction of the image to be segmented. Through such evaluation index, we can further obtain the knowledge related to the characteristics of the image to be segmented and the performance of the segmentation model. With the further guidance of this association, according to the feature analysis of images to be segmented, a new optimization algorithm can be designed for the specific image segmentation task.

b) **Prior Knowledge–Based Optimization Algorithms**

Usually, the objects to be segmented in the same scene contain similar shapes, textures, colors, and positions, especially in medical images, in which different human organs usually have similar anatomical prior knowledge. After the prior knowledge is obtained by calculating the relationship between pixels or encoding the image itself, it is added to the optimization algorithm as an incentive or penalty constraint. Then the optimal value is obtained by using gradient descent, which can make the segmentation algorithm converge quickly and promote the segmentation accuracy. In addition to the prior knowledge of the training samples, the relationship between samples can also be added to the optimization function as a constraint. For example, if the number of samples is unbalanced, it can be optimized by increasing the weighted value of the unbalanced sample segmentation results to improve the robustness of segmentation models.

With the continuous development and improvement of image segmentation theory, the application of image segmentation will be further expanded, especially in biology, chemistry, geography, and other fields. According to the challenges of image segmentation technology and the development direction of machine learning theory, we give the future development direction of image segmentation theory.

4) **Integrating Common Sense**

At present, the most popular technology used for image segmentation is deep convolution neural networks. As one of the core data-driven technologies, deep convolution neural networks purely rely on data to train different models. They use labeled samples in a training set to learn a nonlinear function, and then, during the test, the learned function is applied to pixels in the image to get predicted results, which completely ignores the common sense or the prior knowledge beyond the training samples.

In contrast, human beings can recognize objects not only depending on the samples they have seen, but also depending on their common sense about the real world. By inferring about what they have seen, human beings can avoid illogical recognition results. In addition, when encountering new or unexpected categories, humans can quickly adjust their knowledge to explain this new sense. How to obtain and represent common sense in deep convolution neural network and use common sense to perform inferring is one of the future research directions.

5) **Modeling Relationships**
In order to systematically understand a scene, it is vital to model the relationship and interaction between targets in the scene. The underlying scene structure extracted by relationship modeling can be used to solve the fuzzy uncertainty caused by limited data in current deep learning methods. Although researchers have been trying to go deep to the relational modeling, this research is still in the preliminary stage, which leaves a large space for further exploration. The effective use of relational modeling is an important research direction to promote the accuracy of image segmentation in the future.

6) **Learning to Learn**
Meta-learning is known as learning-to-learn [43]. Neural architecture search can be considered as a typical application of meta-learning. Because the mechanism, representation, and algorithm of learning process modeling are still relatively elementary, the research of meta-learning is still in the early stage. For example, the current neural architecture search [44] performs the optimal search in a narrow range, that is, the combination of existing network modules, and it cannot design new network modules automatically and then realize the fully automatic network structure design. However, with the progress of meta-learning, the potential of automatic network architecture design may be effectively released, and then a network structure far beyond a manual design can be obtained.

References

1 Jia, K., Wang, X., and Tang, X. (2012). Image transformation based on learning dictionaries across image spaces. *IEEE Trans. Pattern Anal. Mach. Intell.* **35** (2): 367–380.

2 Wen, B., Li, Y., and Bresler, Y. (2020). Image recovery via transform learning and low-rank modeling: the power of complementary regularizers. *IEEE Trans. Image Process.* **29**: 5310–5323.

3 Hu, Y., Yang, W., and Liu, J. (2020). Coarse-to-fine hyper-prior modeling for learned image compression. In: *Proceedings of the AAAI Conference on Artificial Intelligence*, vol. **34** No. 07, 11013–11020.

4 Minaee, S., Boykov, Y. Y., Porikli, F. et al. (2021). Image segmentation using deep learning: A survey. *IEEE Transactions on Pattern Analysis and Machine Intelligence*.

5 Tan, M., Pang, R., and Le, Q.V. (2020). Efficientdet: Scalable and efficient object detection. In: *Proceedings of the IEEE/CVF conference on computer vision and pattern recognition*, 10781–10790.

6 Wang, F., Jiang, M., Qian, C. et al. (2017). Residual attention network for image classification. In: *Proceedings of the IEEE conference on computer vision and pattern recognition*, 3156–3164.

7 Gao, Y., Lin, J., Xie, J., and Ning, Z. (2020). A real-time defect detection method for digital signal processing of industrial inspection applications. *IEEE Trans. Industrial Informatics* **17** (5): 3450–3459.

8 Isensee, F., Jaeger, P.F., Kohl, S.A. et al. (2021). nnU-Net: a self-configuring method for deep learning-based biomedical image segmentation. *Nat. Methods* **18** (2): 203–211.

9 Hao, S., Wang, W., and Salzmann, M. (2020). Geometry-aware deep recurrent neural networks for hyperspectral image classification. *IEEE Trans. Geosci. Remote Sens.* **59** (3): 2448–2460.

10 Xie, W., Xu, X., Liu, R., et al. (2020). Living in a simulation? An empirical investigation of the smart driving simulation test system. *An Empirical Investigation of the Smart Driving Simulation Test System (December 22, 2020). Forthcoming at Journal of the Association for Information Systems.*

11 Boykov, Y., Veksler, O., and Zabih, R. (2001). Fast approximate energy minimization via graph cuts. *IEEE Trans. Pattern Anal. Mach. Intell.* **23** (11): 1222–1239.

12 Vicente, S., Kolmogorov, V., and Rother, C. (2008). Graph cut based image segmentation with connectivity priors. In: *2008 IEEE Conference on Computer Vision and Pattern Recognition*, 1–8. IEEE.

13 Wong, D., Liu, J., Fengshou, Y. et al. (2008). A semi-automated method for liver tumor segmentation based on 2D region growing with knowledge-based constraints. In: *MICCAI Workshop*, vol. **41**, No. 43, 159.

14 Yang, W., Wang, X., Lu, J. et al. (2020). Interactive steering of hierarchical clustering. *IEEE Trans. Vis. Comput. Graph.*.

15 Lei, T., Jia, X., Zhang, Y. et al. (2018). Significantly fast and robust fuzzy c-means clustering algorithm based on morphological reconstruction and membership filtering. *IEEE Trans. Fuzzy Syst.* **26** (5): 3027–3041.

16 Lei, T., Liu, P., Jia, X. et al. (2019). Automatic fuzzy clustering framework for image segmentation. *IEEE Trans. Fuzzy Syst.* **28** (9): 2078–2092.

17 Lotfi, A., Moradi, P., and Beigy, H. (2020). Density peaks clustering based on density backbone and fuzzy neighborhood. *Pattern Recogn.* **107**: 107449.

18 Sun, G., Cong, Y., Wang, Q. et al. (2020). Lifelong spectral clustering. In: *Proceedings of the AAAI Conference on Artificial Intelligence*, vol. **34**, No. 04, 5867–5874.

19 Bresson, X., Esedoğlu, S., Vandergheynst, P. et al. (2007). Fast global minimization of the active contour/snake model. *J. Math. Imaging Vision* **28** (2): 151–167.

20 Osher, S. and Sethian, J.A. (1988). Fronts propagating with curvature-dependent speed: algorithms based on Hamilton-Jacobi formulations. *J. Comput. Phys.* **79** (1): 12–49.

21 Oghli, M.G., Fallahi, A., and Pooyan, M. (2010). Automatic region growing method using GSmap and spatial information on ultrasound images. In: *2010 18th Iranian Conference on Electrical Engineering*, 35–38. IEEE.

22 Deng, W., Xiao, W., Deng, H., and Liu, J. (2010). MRI brain tumor segmentation with region growing method based on the gradients and variances along and inside of the boundary curve. In: *2010 3rd International Conference on Biomedical Engineering and Informatics*, vol. **1**, 393–396. IEEE.

23 Achanta, R., Shaji, A., Smith, K. et al. (2012). SLIC superpixels compared to state-of-the-art superpixel methods. *IEEE Trans. Pattern Anal. Mach. Intell.* **34** (11): 2274–2282.

24 Shan, X., Gong, X., and Nandi, A.K. (2018). Active contour model based on local intensity fitting energy for image segmentation and bias estimation. *IEEE Access* **6**: 49817–49827.

25 Levinshtein, A., Stere, A., Kutulakos, K.N. et al. (2009). Turbopixels: fast superpixels using geometric flows. *IEEE Trans. Pattern Anal. Mach. Intell.* **31** (12): 2290–2297.

26 Ester, M., Kriegel, H.P., Sander, J., and Xu, X. (1996). A density-based algorithm for discovering clusters in large spatial databases with noise. *kdd* **96** (34): 226–231.

27 Liu, Y., Jiang, P.T., Petrosyan, V. et al. (2018). DEL: Deep Embedding Learning for Efficient Image Segmentation. In: *IJCAI*, vol. **864**, 870.

28 Zivkovic, Z. (2004). Improved adaptive Gaussian mixture model for background subtraction. In: *Proceedings of the 17th International Conference on Pattern Recognition*. ICPR 2004, vol. **2**, 28–31. IEEE.

29 Falcão, A.X., Stolfi, J., and de Alencar Lotufo, R. (2004). The image foresting transform: theory, algorithms, and applications. *IEEE Trans. Pattern Anal. Mach. Intell.* **26** (1): 19–29.

30 Deng, J., Dong, W., Socher, R. et al. (2009). Imagenet: A large-scale hierarchical image database. In: *2009 IEEE Conference on Computer Vision and Pattern Recognition*, 248–255. IEEE.

31 Sainath, T.N., Kingsbury, B., Mohamed, A.R. et al. (2013). Improvements to deep convolutional neural networks for LVCSR. In: *2013 IEEE Workshop on Automatic Speech Recognition and Understanding*, 315–320. IEEE.

32 Long, J., Shelhamer, E., and Darrell, T. (2015). Fully convolutional networks for semantic segmentation. In: *Proceedings of the IEEE Conference on Computer Vision and Pattern Recognition*, 3431–3440.

33 Dai, J., Qi, H., Xiong, Y. et al. (2017). Deformable convolutional networks. In: *Proceedings of the IEEE International Conference on Computer Vision*, 764–773.

34 Zhao, H., Shi, J., Qi, X. et al. (2017). Pyramid scene parsing network. In: *Proceedings of the IEEE Conference on Computer Vision and Pattern Recognition*, 2881–2890.

35 Huang, G., Liu, Z., Van Der Maaten, L., and Weinberger, K.Q. (2017). Densely connected convolutional networks. In: *Proceedings of the IEEE Conference on Computer Vision and Pattern Recognition*, 4700–4708.

36 Hu, J., Shen, L., and Sun, G. (2018). Squeeze-and-excitation networks. In: *Proceedings of the IEEE Conference on Computer Vision and Pattern Recognition*, 7132–7141.

37 Birrell, S.A. and Young, M.S. (2011). The impact of smart driving aids on driving performance and driver distraction. *Transport. Res. Part F: Traffic Psychol. Behav.* **14** (6): 484–493.

38 Lei, T., Wang, R., Zhang, Y. et al. (2021). Defed-net: deformable encoder-decoder network for liver and liver tumor segmentation. *IEEE Transactions on Radiation and Plasma Medical Sciences*.

39 Menze, B.H., Jakab, A., Bauer, S. et al. (2014). The multimodal brain tumor image segmentation benchmark (BRATS). *IEEE Trans. Med. Imaging* **34** (10): 1993–2024.

40 Cheng, J., Liu, J., Xu, Y. et al. (2013). Superpixel classification based optic disc and optic cup segmentation for glaucoma screening. *IEEE Trans. Med. Imaging* **32** (6): 1019–1032.

41 Ronneberger, O., Fischer, P., and Brox, T. (2015). U-net: convolutional networks for biomedical image segmentation. In: *International Conference on Medical Image Computing and Computer-Assisted Intervention*, 234–241. Cham: Springer.

42 Onishi, Y., Teramoto, A., Tsujimoto, M. et al. (2020). Multiplanar analysis for pulmonary nodule classification in CT images using deep convolutional neural network and generative adversarial networks. *Int. J. Comput. Assist. Radiol. Surg.* **15** (1): 173–178.

43 Pal, M., Kumar, M., Peri, R. et al. (2021). Meta-learning with latent space clustering in generative adversarial network for speaker diarization. *IEEE/ACM Trans. Audio Speech Language Process.* **29**: 1204–1219.

44 Mehrotra, A., Ramos, A. G. C., Bhattacharya, S. et al. (2020). NAS-Bench-ASR: Reproducible Neural Architecture Search for Speech Recognition. *International Conference on Learning Representations*.

2

Clustering

Today's society has entered the digital age. Much of the psychology and behavior of individuals, groups, and societies can be presented through data. With the help of big data, we can explore the internal connections between seemingly unrelated factors, not only to understand the overall operation of a society but also to analyze hidden potential problems. Data have now become a basic element of human production. To play the role of the engine of massive data, it is inseparable from the support of artificial intelligence technology.

There is no doubt that the rapid development of artificial intelligence can make mining of massive data possible. Artificial intelligence technology has emerged in many industries. In order to promote the deep integration of the digital economy and the real economy, many core technologies still need to be improved further. Clustering analysis should not be underestimated. Using clustering can divide massive data into meaningful category arrays. By mining the characteristics of the arrays, the distribution structure of data being tested can be revealed, thereby assisting the final decision. Clustering is not only a basic algorithm in the fields of pattern recognition, machine learning, and deep learning, but is also widely used in business site selection, product recommendation, and information retrieval.

This chapter focuses on the basic principles of mainstream clustering algorithms and their applications in image data. By analyzing their applications in different data, we can show the advantages and disadvantages of different algorithms and lay a theoretical foundation for the follow-up in-depth research.

2.1 Introduction

Clustering is the process of dividing the observed samples into several subclasses through the clustering model according to feature similarity. Intuitively, similar samples are clustered into one category, while dissimilar samples are scattered into various categories. The purpose of clustering is to mine the sample attributes by using the obtained category information, so as to reveal the essential laws hidden in the samples and to provide powerful help for subsequent judgment and decision-making [1].

Suppose $X = \{x_1, x_2, \cdots, x_n\} \in \mathbb{R}^{n \times D}$ is the sample set to be processed. Now, the sample set X is divided into c subclasses, namely, $X_1, X_2, \cdots, X_c$, which must satisfy the following conditions:

$$X_1 \cup X_2 \cup \cdots \cup X_c = X, \tag{2.1}$$

$$X_i \cap X_j = \varnothing, \quad 1 \le i \ne j \le c, \tag{2.2}$$

$$X_i \ne \varnothing, \quad X_i \ne X \quad 1 \le i \le c, \tag{2.3}$$

Image Segmentation: Principles, Techniques, and Applications, First Edition. Tao Lei and Asoke K. Nandi.
© 2023 John Wiley & Sons Ltd. Published 2023 by John Wiley & Sons Ltd.

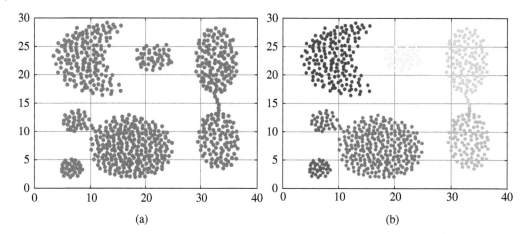

Figure 2.1 Schematic diagram of clustering. (a) Original data. (b) Clustering result.

According to the above conditions, each sample in the sample set can only belong to a certain class at most, and the class cannot be empty. Figure 2.1 shows the schematic diagram of clustering algorithm. Of course, the robustness of different clustering models to sample distribution varies greatly.

Generally, clustering algorithms include three steps: feature extraction, similarity measurement, and clustering. Among them, feature extraction is the abstract essence of mining samples, similarity measurement reflects the different degrees of samples, and clustering divides samples into different categories according to similarity measurement, mines the internal correlation between samples according to the division results, and evaluates the potential information of test samples [2].

This section will mainly introduce five mainstream clustering algorithms: k-means, fuzzy c-means clustering, spectral clustering (SC), hierarchical clustering (HC), and Gaussian mixture model, to show their respective clustering mechanisms by analyzing typical representative algorithms, analyze the advantages and disadvantages of corresponding algorithms, and finally focus on the application of fuzzy clustering in image segmentation.

2.2 K-Means

Clustering is to divide automatically the given samples into several subsets according to a certain clustering criterion, so that the corresponding samples can be divided into the same class as much as possible, while the dissimilar samples can be put into different classes. Since the clustering process does not require any prior information of the samples, clustering belongs to the category of unsupervised pattern recognition [3].

The simplest and a fairly effective clustering algorithm is the k-means algorithm, which uses the sum of squares of the distances from each sample point to the center as the optimization objective function and obtains the adjustment rules of iterative operation by calculating the extreme value of the objective function, so as to divide the input data X into c subsets. For the k-th cluster set, the criterion function is defined as:

$$J = \sum_{i=1}^{N}\sum_{k=1}^{c} u_{ik}\|x_i - v_k\|^2,$$

(2.4)

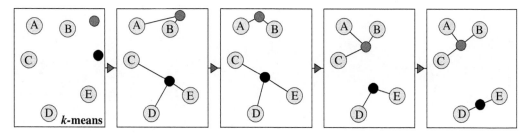

Figure 2.2 Clustering diagram of *k*-means algorithm.

where, *c* represents the number of clusters, *N* represents the number of samples, u_{ik} represents the strength of the *i*-th sample x_i relative to the *k*-th cluster center v_k, $u_{ik} = 0$ or 1, with a strict and clear demarcation boundary. As shown in Figure 2.2, $\|x_i - v_k\|$ represents the Euclidean distance between sample x_i and the cluster center v_k. The clustering result of "minimum within class and maximum between classes" can be achieved through this metric.

It can be seen from Figure 2.2 that the *k*-means algorithm assigns the test data to different categories through repeated iteration, but the process adopts the hard division criterion, that is, the test samples only belong to one category, which destroys the uncertainty of the samples to the category, so it limits the classification effect of the *k*-means algorithm to a certain extent.

The *k*-means optimizes its objective function by alternately optimizing u_{ik} and v_k. The specific process of the algorithm is shown as follows:

Step 1: Set the convergence threshold η, the number of clusters *c*, the number of iterations $\varpi = 0$, and the maximum number of iterations *B*, and initialize the membership matrix $V^{(0)}(v_k \in V^{(0)})$;

Step 2: Calculate $u_{ik} = \begin{cases} 1 & \|x_i - v_k\| = \min_{1 \leq j \leq c} \|x_i - v_j\| \\ 0 & otherwise \end{cases}$;

Step 3: Calculate the cluster center $v_k = \dfrac{\sum_{i=1}^{N} u_{ik} x_i}{\sum_{i=1}^{N} u_{ik}}$;

Step 4: If $|J^{(\varpi)} - J^{(\varpi+1)}| < \eta$ or $\varpi > B$, stop, otherwise, update $\varpi = \varpi + 1$, and go to step 2.

Figure 2.3 shows the effect of *k*-means algorithm on image segmentation. In Figure 2.3, with the increase of the number of clustering, more details and contour information of the image are presented in the segmentation results. However, because *k*-means adopts the hard partition criterion (i.e. binary values of membership either 0 or 1), it is unable to obtain high-precision segmentation results.

The significant advantages of *k*-means are its simple implementation, low computational complexity, and low memory consumption [4]. For an image with *N* pixels to be processed, the time complexity of *k*-means is $O(N \times c \times \varpi)$, where *c* represents the preset number of clusters, and ϖ represents the number of iterations of the algorithm.

It is known that *k*-means uses the mean value to calculate the center in each category, which makes the algorithm more sensitive to outliers. To solve this problem, researchers have proposed *k*-medians, which determines the center of the category by sorting the median value, thus reducing the sensitivity to outliers to a certain extent [5]. However, for a large data set, the running time of

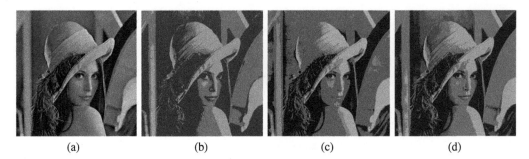

(a)	(b)	(c)	(d)

Figure 2.3 Comparison of image segmentation results using *k*-means under different numbers of clusters. (a) Test image. (b) *c* = 3. (c) *c* = 4. (d) *c* = 5.
Source line: Conor Lawless / Flickr

this algorithm is too long. In view of the problem that *k*-means is sensitive to the initialization center, researchers have put forward *k*-means++ to further improve the stability of *k*-means [6]. Of course, the most obvious deficiency of *k*-means is that it adopts the hard partition criteria, which leads to lack of global perception of data.

2.3 Fuzzy C-means Clustering

Fuzzy c-means clustering (FCM) was first proposed by Dunn and extended by Bezdek. It is one of the most widely used clustering algorithms [7]. FCM algorithm is evolved from *k*-means algorithm. Compared with the hardened sub-attribute of *k*-means algorithm, FCM algorithm can better perceive the global sample distribution through fuzzy soft characteristics, so as to obtain the optimal decision classification. The process of FCM algorithm is shown in Figure 2.4.

By comparing Figures 2.3 and 2.4, it can be seen that the *k*-means algorithm strictly divides the categories to which each sample belongs, while the FCM algorithm establishes the uncertain relationship between the samples for all categories through the fuzzy membership function, which is in line with humans' cognitive structure of the objective world. FCM and *k*-means both adopt the sum of error squares criterion as the objective function, but FCM uses the membership function to expand the universality of classification, and its objective function is:

$$J = \sum_{i=1}^{N} \sum_{k=1}^{c} u_{ik}^{m} \|x_i - v_k\|^2, \tag{2.5}$$

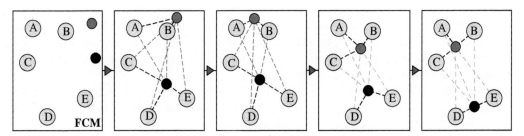

Figure 2.4 Clustering diagram of FCM algorithm.

where c represents the number of clusters, N represents the number of samples, u_{ik} represents the strength of membership degree of the i-th sample x_i relative to the k-th clustering center v_k, and its constraint condition is $\sum_{k=1}^{c} u_{ik} = 1$ and $u_{ik} \in [0, 1]$, $m > 1$, m represents the fuzzy weight index (usually set $m = 2$), $\|x_i - v_k\|$ represents the Euclidean distance between sample x_i and cluster center v_k. FCM generally adopts Lagrange multiplier method to minimize the objective function. Now, a new objective function can be constructed by associating the membership constraint with the objective function:

$$L = \sum_{i=1}^{N} \sum_{k=1}^{c} u_{ik}^{m} \|x_i - v_k\|^2 + \sum_{j=1}^{n} \lambda_j \left(1 - \sum_{k=1}^{c} u_{ik}\right), \tag{2.6}$$

where λ_j represents the j-th Lagrange multiplier. Taking the partial derivatives of the function L with respect to u_{ik}, v_k, λ_j, we obtain the following equations:

$$\frac{\partial L}{\partial v_k} = 2 \sum_{i=1}^{n} u_{ik}^{m} (x_i - v_k) = 0, \tag{2.7}$$

$$\frac{\partial L}{\partial u_{ik}} = m u_{ik}^{m-1} \|x_i - v_k\|^2 - \lambda_j = 0, \tag{2.8}$$

$$\frac{\partial L}{\partial \lambda_j} = 1 - \sum_{k=1}^{c} u_{ik} = 0, \tag{2.9}$$

according to (2.7), the cluster center v_k can be obtained as:

$$v_k = \frac{\sum_{i=1}^{N} u_{ik}^{m} x_i}{\sum_{i=1}^{N} u_{ik}^{m}}. \tag{2.10}$$

From (2.8), we can get:

$$u_{ik} = \left(\frac{\lambda_j}{m}\right)^{\frac{1}{m}} \|x_i - v_k\|^{-\frac{2}{m-1}}, \tag{2.11}$$

and substituting (2.11) into $\sum_{k=1}^{c} u_{ik} = 1$ we can obtain:

$$\sum_{i=1}^{c} u_{ik} = \sum_{j=1}^{c} \left(\frac{\lambda_j}{m}\right)^{\frac{1}{m}} \|x_i - v_k\|^{-\frac{2}{m-1}} = 1. \tag{2.12}$$

By rearranging, one can get:

$$\left(\frac{\lambda_j}{m}\right)^{\frac{1}{m}} = \frac{1}{\sum_{j=1}^{c} \|x_i - v_k\|^{-\frac{2}{m-1}}}, \tag{2.13}$$

Then by substituting (2.13) into (2.11), the degree of membership u_{ik} can be obtained as:

$$u_{ik} = \frac{\|x_i - v_k\|^{-2/(m-1)}}{\sum_{j=1}^{c} \|x_i - v_j\|^{-2/(m-1)}}. \tag{2.14}$$

The specific calculation process of FCM is as follows:

Step 1: Set the convergence threshold η, the number of clusters c, the number of iterations $\varpi = 1$ and the maximum number of iterations B, and initialize the membership matrix $U^{(0)}(u_{ik} \in U^{(0)})$;

Step 2: Calculate the cluster center $v_k = \dfrac{\sum_{i=1}^{N} u_{ik}^m x_i}{\sum_{i=1}^{N} u_{ik}^m}$;

Step 3: Calculate the membership degree matrix $u_{ik} = \dfrac{\|x_i - v_k\|^{-2/(m-1)}}{\sum_{j=1}^{c} \|x_i - v_j\|^{-2/(m-1)}}$;

Step 4: If $|J^{(\varpi)} - J^{(\varpi+1)}| < \eta$ or $\varpi > B$, the algorithm ends; otherwise, $\varpi = \varpi + 1$, and continue to step 2;

Step 5: De-fuzzify the membership matrix U, namely $L_j = \arg_i\{\max\{u_{ij}\}\}$, and assign each sample to the category corresponding to the maximum membership value to obtain the classification result (label).

Through the analysis of step 2.3, it can be seen that the cluster center can feed back the position offset of all samples under the traction of subclass membership, and the membership matrix can record the global correlation of each sample to all cluster centers. The two complement each other and cooperate together, which can not only deeply excavate the potential characteristics of samples but also improve the optimization efficiency of the algorithm. This is the essential difference between the FCM algorithm and the k-means algorithm.

When FCM is applied to an image, the segmentation results are shown in Figure 2.5. It can be seen from Figure 2.5, as the number of clusters increases, the segmentation results show more details and contour edges. FCM has become a mainstream image segmentation algorithm because of its simplicity and efficiency. For an image with N pixels to be processed, the time complexity of the FCM algorithm is $O(N \times c \times \varpi)$, where, c represents the preset number of clusters, and ϖ represents the number of iterations of the algorithm.

Although the fuzzy clustering algorithm can realize image segmentation, because image data has some distinct characteristics compared to traditional structured data, there are the following shortcomings in applying FCM algorithm directly to image segmentation: firstly, the FCM algorithm ignores the spatial information of an image and is sensitive to the uneven distribution of pixels and noise, resulting in poor image segmentation effects; secondly, the FCM algorithm for image segmentation needs to traverse all pixels, so the time complexity of the

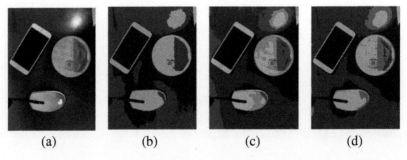

(a)   (b)   (c)   (d)

Figure 2.5 Comparison of segmentation results using the FCM under different numbers of clusters. (a) Test image. (b) $c = 3$. (c) $c = 4$. (d) $c = 5$.

algorithm will increase rapidly with the increase of image resolution [8]; and, finally, the FCM algorithm does not consider the correlation between channnels when processing color images or multi-channel images (high-dimensional images), resulting in poor robustness to high-dimensional images.

In view of the fact that FCM ignores the spatial information of images, researchers have proposed an FCM based on the neighborhood information [9–11]. This algorithm mainly uses the relationship between neighborhood pixels and cluster center to modify the membership intensity of the center pixel and complete image segmentation at the same time. However, with the increase of noise intensity, it is difficult to use the neighborhood information of images to correct the category of the central pixel. To solve this problem, researchers have proposed FCM based on adaptive neighborhood information [12, 13]. These algorithms balance detail information and interference noise mainly by mining the similarity between pixel blocks, so as to avoid losing too much detail information while removing noise [14]. However, the time complexity of the algorithm is high, which is not conducive to the extension of the algorithm.

For the problem that FCM needs to go through all pixels in an image, researchers have proposed FCM based on histogram information [15, 16]. Firstly, the original image is filtered, and then the number of pixels is replaced by the number of gray levels of the image to effectively reduce data redundancy, thus greatly reducing the time complexity of the algorithm and realizing a fast FCM for image segmentation [17]. However, FCM based on histogram information is only suitable for gray level images, and it is difficult to extend this directly to color image processing.

By analyzing the correlation between the dimensions of high-dimensional images, researchers have proposed the FCM based on dimension weight [18, 19]. This kind of algorithm mainly uses a new measurement method to replace the Euclidean measurement method in the traditional algorithms, so as to improve the classification effect of the algorithm on high-dimensional images. The mainstream improvement method is to use Markov distance to replace the Euclidean distance in FCM [20], reflect the correlation between image dimensions through covariance, and improve the segmentation effect of multichannel images. However, the introduction of covariance information increases the time complexity of algorithms. In order to reduce the number of iterations of algorithms, the clustering results of k-means are typically used as the input data in this kind of algorithms to reduce the running time of the algorithm.

2.4 Hierarchical Clustering

HC is a kind of classification using a series of nested tree structures generated by samples [21]. According to the partitioning strategy, it can be divided into top-down split HC and bottom-up condensed HC. Split-stratified clustering refers to taking all the data as a sample and splitting and refining the sample layer by layer until the convergence condition is satisfied. In contrast, convergent stratified clustering refers to placing the samples at the lowest level of the tree, merging and condensing layer by layer, and finally clustering into a class [22]. The algorithm process of convergent stratified clustering is mainly introduced below:

Step 1: Regard each data as an independent sample, and then calculate the minimum distance between samples;

Step 2: Combine the two samples with the smallest distance into a new sample;

Step 3: Calculate the distance between the new sample and other samples again;

Step 4: If all samples are merged into one category, the iteration is terminated; otherwise, step 3 is executed;

Step 5: Use the formed clustering tree to interactively select the clustering results at the appropriate level.

The above algorithm uses the most commonly used average join method. When this algorithm is applied to synthetic data, the segmentation result is shown in Figure 2.6.

It can be seen from Figure 2.6 that by traversing the tree structure, the hierarchical relationship of the original data can be effectively mined, combined with human experience, so as to achieve the ideal clustering effect. However, the HC algorithm needs to calculate repeatedly the relative distance between the data, resulting in high time complexity. The HC algorithm is more sensitive to singular values; at the same time, it is easy to produce a domino effect and it lacks certain practicability. To solve this problem, researchers have proposed a variety of improvement strategies, which can effectively expand the practicability of HC.

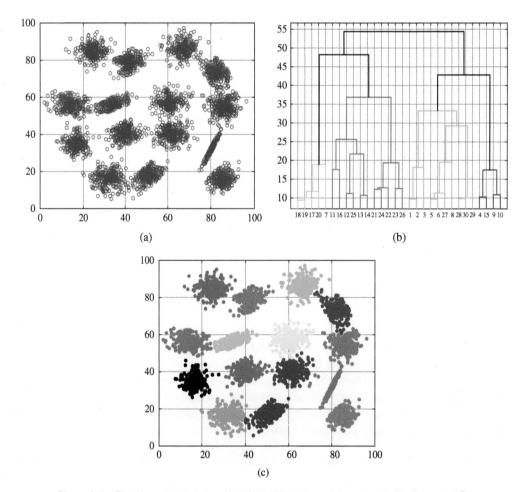

Figure 2.6 The clustering result using hierarchical clustering on synthetic data (c = 15). (a) Synthetic data. (b) Tree structure. (c) The clustering result.

2.5   Spectral Clustering

SC is a clustering algorithm based on atlas theory, whose essence is to transform the clustering task into a graph partition task [23]. The algorithm first constructs an undirected graph by using sample data points, then attains the eigenvalues and eigenvectors of the Laplacian matrix corresponding to the graph, and finally selects the appropriate eigenvectors for clustering, thus dividing the graph into c disjoint subsets to achieve the maximum similarity within the subgraph and the minimum similarity between the subgraphs.

The earliest partition criterion belongs to the min cut criterion [24], which cuts the minimum weight of two subgraphs and thus achieve data partition. However, the min cut criterion ignores the similarity within the subgraph, so it is easy to produce the skew partition problem. In order to solve this problem, researchers have successively proposed the ratio cut criterion [25], norm cut criterion [26], and minimum maximum cut criterion [27], which alleviate the emergence of oblique division to a certain extent. Due to the low computational efficiency and unstable performance of these methods, many improved methods such as multichannel rate cut criterion [28], multichannel specification cut criterion [29], multichannel minimum and maximum cut criterion [27] are reported. However, the optimal solution of the multichannel spectral division criterion is a NP-hard problem, which needs to find its approximate solution in the relaxed real number field. Subsequently, the eigenvectors corresponding to the first few maximum features of the Laplace matrix are typically used to approximate the original spectral space.

Among many SC algorithms, the NJW algorithm, proposed by Ng et al., is the most widely used [30], which can be understood as mapping high-dimensional spatial data to low-dimensional feature data and then using the heuristic clustering algorithm to achieve the final graph partition in low-dimensional space. The NJW divides the data $X = \{x_1, x_2, \cdots, x_N\} \in \mathbb{R}^{N \times d}$ containing n samples into c subsets as follows:

Step 1: Construct a similarity matrix $S_A \in \mathbb{R}^{N \times N}$; the elements in matrix S_A use a Gaussian kernel function to calculate $S_{Aij} = exp\left(-\|x_i - x_j\|^2/2\sigma^2\right)$, and make $S_{Aii} = 0$, where $\|x_i - x_j\|$ represents the Euclidean distance between the i-th sample x_i and the j-th sample x_j, and σ represents the scale parameter of the Gaussian kernel function;

Step 2: Calculate the normalized Laplacian matrix $L = D^{-1/2}S_A D^{1/2}$, where D is the degree matrix, and the diagonal element D_{ii} is the sum of the i-th row elements of matrix S_A;

Step 3: Find the eigenvector $F_e = \{e_1, e_2, \cdots, e_c\} \in \mathbb{R}^{N \times c}$ corresponding to the first c largest eigenvalues of the matrix L;

Step 4: Normalize the row vector of matrix F_e to obtain matrix Y, namely $Y_{ik} = F_{eik}/\left(\sum_{l=1}^{c} F_{eil}^2\right)^{1/2}$, where Y_{ik} represents the k-th eigenvector corresponding to i samples $(1 \leq k \leq c)$;

Step 5: Regard each row in the matrix Y as a point in the $\mathbb{R}^c$ space, and cluster it into c categories through k-means or other clustering algorithms;

Step 6: When the i-th row in the matrix Y is assigned as the l-th label, then the corresponding original data x_i will also be assigned as the l-th category $(1 \leq l \leq c)$.

NJW can transform complex data into low-dimensional feature vectors, making the linearly inseparable original data show obvious feature differences, thereby easily realizing data classification. The classification result of NJW applied to synthetic data is shown in Figure 2.7.

As can be seen from Figure 2.7, good classification results can be obtained only when NJW selects appropriate scale parameters. In order to overcome the sensitivity of the SC algorithm to the parameter, Zelnik-Manor and Perona [31] proposed a self-tuning SC algorithm, which adaptively selects

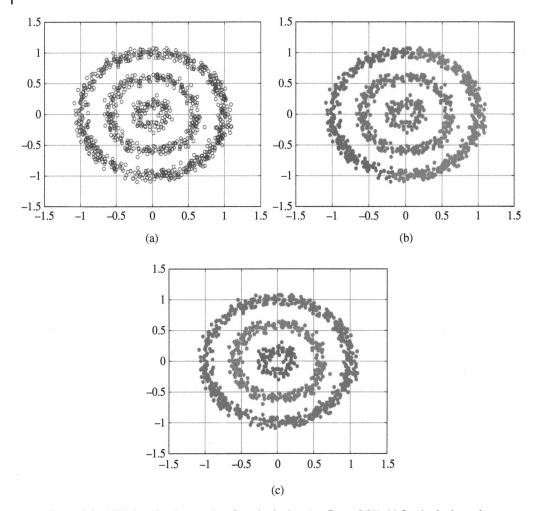

Figure 2.7 NJW classification results of synthetic data ($c = 3$, $\sigma = 0.01$). (a) Synthetic three-ring data. (b) Segmentation result of NJW ($\sigma = 0.1$). (c) Segmentation result of NJW ($\sigma = 0.01$).

the size of scale parameters according to the local structure of data, avoiding selecting a single scale parameter on the clustering result. However, this algorithm still needs to build the similarity matrix, which means it cannot be applied to massive data (including image data). In order to make up for the shortage, Fowlkers [32] proposed a strategy of using the Nystrom method to greatly reduce the complexity of eigenvector solution and expanding the applicability of the algorithm.

2.6 Gaussian Mixed Model

Gaussian mixed model (GMM) refers to the linear combination of multiple Gaussian distribution functions [33], as shown in Figure 2.8. It can be seen from Figure 2.8 that the probability distribution of observation data can be presented through multiple Gaussian functions, and the mean, variance, and linear combination coefficient can be learned in the process of model generation, so as to ensure the approximation of the continuous probability distribution curve with arbitrary accuracy.

Theoretically, GMM can fit any type of distribution and is typically used to solve the problem that data in the same set contains multiple different distributions [33]. The Gaussian mixture model can be written in the form of linear superposition of c Gaussian distributions

$$p(\boldsymbol{X}) = \sum\nolimits_{k=1}^{c} \pi_k N(\boldsymbol{X} \,|\, v_k, \Sigma_k), \qquad (2.15)$$

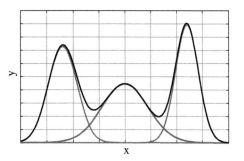

where $N(X \,|\, v_k, \Sigma_k)$ represents the k-th Gaussian density function, v_k is the mean value of the density function, Σ_k is the covariance matrix of the density function, and π_k represents the prior probability of

Figure 2.8 Linear combination of Gaussian functions.

the k-th density function, and satisfies $\sum_{k=1}^{c} \pi_k = 1$ and $\pi_k \in [0, 1]$. For data with n independent samples $X = \{x_1, x_2, \cdots, x_n\} \in \mathbb{R}^{N \times D}$, the above formula can be written as the joint probability distribution of multiple Gaussian models

$$p(\boldsymbol{X}) = \prod\nolimits_{i=1}^{N} \sum\nolimits_{k=1}^{c} \pi_k N(x_i \,|\, v_k, \Sigma_k), \qquad (2.16)$$

and the density function of the i-th sample is:

$$N(x_i \,|\, v_k, \Sigma_k) = \frac{1}{(2\pi)^{D/2} |\Sigma_k|^{1/2}} e^{-\frac{1}{2}(x_i - v_k)^T \Sigma_k^{-1} (x_i - v_k)}. \qquad (2.17)$$

As the probability value is small but the value of sample number N is large, the joint probability distribution is prone to floating point underflow. Therefore, the maximum likelihood function is typically used to change the numerical distribution, and the constraint condition $\sum_{k=1}^{c} \pi_k = 1$ can be used to construct a new expression

$$J = \sum\nolimits_{i=1}^{N} \log \left(\sum\nolimits_{k=1}^{c} \pi_k N(x_i \,|\, v_k, \Sigma_k) \right) + \lambda \left(\sum\nolimits_{k=1}^{c} \pi_k - 1 \right). \qquad (2.18)$$

The Gaussian distribution of the new object function is mainly controlled by π_k, v_k, and Σ_k. By solving $\frac{\partial J}{\partial v_k} = 0$, we can get:

$$\begin{aligned}
\frac{\partial J}{\partial v_k} &= \sum\nolimits_{i=1}^{N} \log \frac{\partial \sum_{k=1}^{c} \pi_k N(x_i | v_k, \Sigma_k)}{\partial v_k} \\
&= \sum\nolimits_{i=1}^{N} \frac{1}{\sum_{k=1}^{c} \pi_k N(x_i | v_k, \Sigma_k)} \frac{\partial \pi_k N(x_i | v_k, \Sigma_k)}{\partial v_k}. \\
&= \sum\nolimits_{i=1}^{N} \frac{\pi_k N(x_i | v_k, \Sigma_k)(x_i - v_k)}{\sum_{k=1}^{c} \pi_k N(x_i | v_k, \Sigma_k)} = 0
\end{aligned} \qquad (2.19)$$

Supposing $u_{ik} = \dfrac{\pi_k N(x_i \,|\, v_k, \Sigma_k)}{\sum_{k=1}^{c} \pi_k N(x_i \,|\, v_k, \Sigma_k)}$, the above equation can be solved

$$\frac{\partial J}{\partial v_k} = \sum\nolimits_{i=1}^{N} u_{ik}(x_i - v_k) = 0, \qquad (2.20)$$

and by shifting the term, the clustering center can be obtained as:

$$v_k = \frac{\sum_{i=1}^{N} u_{ik} x_i}{\sum_{i=1}^{N} u_{ik}}. \qquad (2.21)$$

The covariance matrix in the iteration is obtained by solving $\dfrac{\partial J}{\partial \Sigma_k} = 0$

$$\frac{\partial J}{\partial \Sigma_k} = \sum_{i=1}^{N} \frac{\pi_k \left(-N(x_i \mid v_k, \Sigma_k) \Sigma_k^{-1} + N(x_i \mid v_k, \Sigma_k)(x_i - v_k) \Sigma_k^{-2} (x_i - v_k)^T \right)}{2 \sum_{k=1}^{c} \pi_k N(x_i \mid v_k, \Sigma_k)} = 0, \tag{2.22}$$

the covariance matrix of the k-th category can be obtained by shifting and merging:

$$\Sigma_k = \frac{\sum_{i=1}^{N} u_{ik}(x_i - v_k)(x_i - v_k)^T}{\sum_{i=1}^{N} u_{ik}}, \tag{2.23}$$

and by solving $\dfrac{\partial J}{\partial \pi_k} = 0$, the covariance matrix is obtained, namely:

$$\frac{\partial J}{\partial \pi_k} = \sum_{i=1}^{N} \frac{N(x_i \mid v_k, \Sigma_k)}{\sum_{k=1}^{c} \pi_k N(x_i \mid v_k, \Sigma_k)} + \lambda = 0. \tag{2.24}$$

By shifting items, we can get:

$$\sum_{i=1}^{N} \frac{\pi_k N(x_i \mid v, \Sigma_k)}{\sum_{k=1}^{c} \pi_k N(x_i \mid v_k, \Sigma_k)} = -\lambda \pi_k, \tag{2.25}$$

that is: $\sum_{i=1}^{N} u_{ik} = -\lambda \pi_k$; it is known that $\sum_{i=1}^{N} \sum_{k=1}^{c} u_{ik} = \sum_{i=1}^{N} \dfrac{\sum_{k=1}^{c} \pi_k N(x_i \mid v_k, \Sigma_k)}{\sum_{k=1}^{c} \pi_k N(x_i \mid v_k, \Sigma_k)} = N = -\lambda \sum_{i=1}^{c} \pi_k = -\lambda$, substituting it into (2.25), we can get:

$$\pi_k = \frac{\sum_{i=1}^{N} u_{ik}}{N}. \tag{2.26}$$

As we all know, the covariance matrix is a symmetric matrix, where the elements on the diagonal represent the variance of each type of sample. The smaller the value, the more compact the distribution of data in that category of samples and vice versa. The nondiagonal elements represent the covariance between different categories. If the change trend of samples in the two categories is the same, its value is positive. But if the change trend of samples in the two categories is opposite, its value is negative. Therefore, the measurement based on the covariance matrix can better mine the numerical distribution of test samples [34], so as to improve the segmentation effect from the GMM algorithm.

For Gaussian mixture model, its goal is to maximize the likelihood function of mean value, covariance, and prior probability. The algorithm process of the Gaussian mixture model is:

Step 1: According to the current model parameters, calculate the posterior probability of the observation data x_i belonging to the sub-model k:

$$u_{ik} = \frac{\pi_k N(x_i \mid v_k, \Sigma_k)}{\sum_{k=1}^{c} \pi_k N(x_i \mid v_k, \Sigma_k)}. \tag{2.27}$$

Step 2: Calculate the parameters of a new round of iterative model according to u_{ik}:

$$v_k^{new} = \frac{\sum_{i=1}^{N} u_{ik} x_n}{\sum_{i=1}^{N} u_{ik}}, \tag{2.28}$$

$$\Sigma_k^{new} = \frac{\sum_{i=1}^{N} u_{ik}(x_i - v_k)(x_i - v_k)^T}{\sum_{i=1}^{N} u_{ik}}, \tag{2.29}$$

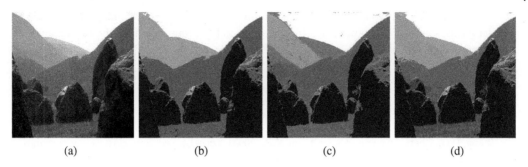

(a)	(b)	(c)	(d)

Figure 2.9   Comparison of segmentation results using the GMM under different numbers of clusters. (a) Test image. (b) $c = 4$. (c) $c = 5$. (d) $c = 6$.

$$\pi_k^{new} = \frac{\sum_{i=1}^{N} u_{ik}}{N}. \tag{2.30}$$

Step 3: Repeat step 1 and step 2 until convergence.

Since the Gaussian measurement criterion with covariance is adopted in the GMM algorithm, GMM can capture complex data, as shown in Figure 2.9. Of course, similarly to the FCM algorithm, the GMM algorithm also captures more details with the change of the number of clusters. However, the GMM algorithm ignores the spatial information of images, so the algorithm is still sensitive to the distribution of pixels. In order to improve this deficiency, similarly to the FCM algorithm, researchers have also proposed a variety of improved algorithms by integrating image spatial information into object functions.

2.7   Discussion and Summary

People often say that "birds of a feather flock together and people flock together," and its core idea is clustering. Through clustering, people can mine the sample distribution and understand its attribute relationship. At present, clustering algorithms have been widely used in many fields, including pattern recognition, data analysis, image processing, and market research [35].

In this chapter, we have introduced the basic principle of clustering algorithms, have focused on five kinds of clustering algorithms including k-means, fuzzy clustering, SC, HC, and the Gaussian mixture model, deeply analyzed the working mechanism of each algorithm, and explained the advantages and disadvantages of each algorithm. Because SC and HC need to build a large adjacency matrix, the image segmentation effects from these two algorithms are not given, and only the classification result on synthetic data is shown. K-means, fuzzy clustering, and the Gaussian mixture model can still converge quickly when processing image data, but they all ignore the spatial structure information of images, resulting in coarse segmentation effect.

Through the analysis of this chapter, it is not difficult to find that the mainstream clustering algorithms have advantages on different data classification tasks. For example, SC has good robustness to nonlinear data, while fuzzy clustering can better perceive spherical data. However, they still face many challenges, such as: (i) how to effectively fuse the spatial information of images; (ii) how to improve the execution efficiency of algorithms; and (iii) how to determine adaptively the number of clusters of the algorithm. In order to expand further the practical applications of clustering algorithms, researchers should analyze the advantages and disadvantages of relevant algorithms from the bottom, give full play to their advantages, and make up for their shortcomings by introducing regular terms, sparse theory, and data structure, so as to lay a foundation for the implementation of clustering algorithms.

References

1 Zhang, D.Q., Chen, S.C., Pan, Z.S., and Tan, K.R. (2003). Kernel-based fuzzy clustering incorporating spatial constraints for image segmentation. In: *Proceedings of the 2003 International Conference on Machine Learning and Cybernetics (IEEE Cat. No. 03EX693)*, vol. **4**, 2189–2192. IEEE.

2 Huang, D., Wang, C.D., Peng, H. et al. (2018). Enhanced ensemble clustering via fast propagation of cluster-wise similarities. *IEEE Trans. Syst. Man Cybern. Syst.*.

3 Bai, L., Cheng, X., Liang, J. et al. (2017). Fast density clustering strategies based on the k-means algorithm. *Pattern Recogn.* **71**: 375–386.

4 Mignotte, M. (2011). A de-texturing and spatially constrained K-means approach for image segmentation. *Pattern Recogn. Lett.* **32** (2): 359–367.

5 Grønlund, A., Larsen, K.G., Mathiasen, A., et al. (2017). Fast exact k-means, k-medians and Bregman divergence clustering in 1D. *arXiv preprint arXiv*:1701.07204.

6 Arthur, D. and Vassilvitskii, S. (2006). *K-Means++: The Advantages of Careful Seeding*. Stanford.

7 Bezdek, J.C., Ehrlich, R., and Full, W. (1984). FCM: the fuzzy c-means clustering algorithm. *Comput. Geosci.* **10** (2–3): 191–203.

8 Zhang, H., Wang, Q., Shi, W., and Hao, M. (2017). A novel adaptive fuzzy local information c-means clustering algorithm for remotely sensed imagery classification. *IEEE Trans. Geosci. Remote Sens.* **55** (9): 5057–5068.

9 Ahmed, M.N., Yamany, S.M., Mohamed, N. et al. (2002). A modified fuzzy c-means algorithm for bias field estimation and segmentation of MRI data. *IEEE Trans. Med. Imaging* **21** (3): 193–199.

10 Chen, S. and Zhang, D. (2004). Robust image segmentation using FCM with spatial constraints based on new kernel-induced distance measure. *IEEE Trans. Syst. Man Cybern. B Cybern.* **34** (4): 1907–1916.

11 Guo, Y. and Sengur, A. (2015). NCM: neutrosophic c-means clustering algorithm. *Pattern Recogn.* **48** (8): 2710–2724.

12 Zhao, F., Fan, J., and Liu, H. (2014). Optimal-selection-based suppressed fuzzy c-means clustering algorithm with self-tuning non local spatial information for image segmentation. *Expert Syst. Appl.* **41** (9): 4083–4093.

13 Gong, M., Liang, Y., Shi, J. et al. (2012). Fuzzy c-means clustering with local information and kernel metric for image segmentation. *IEEE Trans. Image Process.* **22** (2): 573–584.

14 Gong, M., Zhou, Z., and Ma, J. (2011). Change detection in synthetic aperture radar images based on image fusion and fuzzy clustering. *IEEE Trans. Image Process.* **21** (4): 2141–2151.

15 Szilagyi, L., Benyo, Z., Szilágyi, S.M., and Adam, H.S. (2003). MR brain image segmentation using an enhanced fuzzy c-means algorithm. In: *Proceedings of the 25th Annual International Conference of the IEEE Engineering in Medicine and Biology Society (IEEE Cat. No. 03CH37439)*, vol. **1**, 724–726. IEEE.

16 Cai, W., Chen, S., and Zhang, D. (2007). Fast and robust fuzzy c-means clustering algorithms incorporating local information for image segmentation. *Pattern Recogn.* **40** (3): 825–838.

17 Guo, F.F., Wang, X.X., and Shen, J. (2016). Adaptive fuzzy c-means algorithm based on local noise detecting for image segmentation. *IET Image Process.* **10** (4): 272–279.

18 Chatzis, S.P. (2013). A Markov random field-regulated pitman–yor process prior for spatially constrained data clustering. *Pattern Recogn.* **46** (6): 1595–1603.

19 Chatzis, S.P. and Varvarigou, T.A. (2008). A fuzzy clustering approach toward hidden Markov random field models for enhanced spatially constrained image segmentation. *IEEE Trans. Fuzzy Syst.* **16** (5): 1351–1361.

20 Liu, G., Zhang, Y., and Wang, A. (2015). Incorporating adaptive local information into fuzzy clustering for image segmentation. *IEEE Trans. Image Process.* **24** (11): 3990–4000.

21 Liu, A.A., Su, Y.T., Nie, W.Z., and Kankanhalli, M. (2016). Hierarchical clustering multi-task learning for joint human action grouping and recognition. *IEEE Trans. Pattern Anal. Mach. Intell.* **39** (1): 102–114.

22 Johnson, S.C. (1967). Hierarchical clustering schemes. *Psychometrika* **32** (3): 241–254.

23 Chen, J., Li, Z., and Huang, B. (2017). Linear spectral clustering superpixel. *IEEE Trans. Image Process.* **26** (7): 3317–3330.

24 Fiedler, M. (1973). Algebraic connectivity of graphs. *Czech. Math. J.* **23** (2): 298–305.

25 Hagen, L. and Kahng, A.B. (1992). New spectral methods for ratio cut partitioning and clustering. *IEEE Trans. Computer-Aided Design Integr. Circuits Syst.* **11** (9): 1074–1085.

26 Shi, J. and Malik, J. (2000). Normalized cuts and image segmentation. *IEEE Trans. Pattern Anal. Mach. Intell.* **22** (8): 888–905.

27 Ding, C.H., He, X., Zha, H. et al. (2001). A min-max cut algorithm for graph partitioning and data clustering. In: *Proceedings 2001 IEEE International Conference on Data Mining*, 107–114. IEEE.

28 Chan, P.K., Schlag, M.D., and Zien, J.Y. (1994). Spectral k-way ratio-cut partitioning and clustering. *IEEE Trans. Computer-Aided Design Integr. Circuits Syst.* **13** (9): 1088–1096.

29 Meila, M. and Xu, L. (2003). Multiway cuts and spectral clustering.

30 Ng, A.Y., Jordan, M.I., and Weiss, Y. (2002). On spectral clustering: Analysis and an algorithm. In: *Advances in Neural Information Processing Systems*, 849–856.

31 Zelnik-Manor, L. and Perona, P. (2004). Self-tuning spectral clustering. *Adv. Neural Inform. Process. Syst.* 1601–1608.

32 Fowlkes, C., Belongie, S., Chung, F., and Malik, J. (2004). Spectral grouping using the Nystrom method. *IEEE Trans. Pattern Anal. Mach. Intell.* **26** (2): 214–225.

33 Rasmussen, C.E. (1999). The infinite Gaussian mixture model. In: *NIPS*, vol. **12**, 554–560.

34 Zivkovic, Z. (2004). Improved adaptive Gaussian mixture model for background subtraction. In: *Proceedings of the 17th International Conference on Pattern Recognition, 2004*, ICPR 2004, vol. **2**, 28–31. IEEE.

35 Sinaga, K.P. and Yang, M.S. (2020). Unsupervised K-means clustering algorithm. In: *IEEE Access*, vol. **8**, 80716–80727.

3

Mathematical Morphology

3.1 Introduction

Mathematical morphology is a mathematical theory and technique conceived in 1964 by J. Serra and his supervisor G. Matheron from Paris, France. They introduced the conceptual framework, expressions, and operations of mathematical morphology such as opening and closing operations, the Boolean model, and binary operators and techniques during their quantitative petrological analysis and the exploitation value level prediction of iron cores; this laid the theoretical basis of the mathematical morphology. The basic idea of mathematical morphology is to measure and extract the corresponding shape in an image with some morphological "structuring" elements. A structuring element is a simple, predefined shape used to probe an image, drawing conclusions on whether this shape fits or misses the shapes in the image, and so achieving the purpose of image analysis and recognition such as noise suppression, feature extraction, edge detection, image segmentation, shape recognition, texture analysis, image restoration and reconstruction, and image compression. Given all of that, mathematical morphology is an image analysis subject based on topology and it is the basic theory employed for image processing [1].

Mathematical morphology theory contains a set of morphological algebraic operators including erosion, dilation, opening, and closing operations. Based on these basic operators, researchers can probe into superior and more effective algorithms that can be applied to image analysis and processing.

Mathematical morphology is a subject based on mathematical theory. Its basic ideas and methods have great importance to the theory and technology of digital image processing. Over the years, it has been widely used in many fields such as optical character recognition [2], scanning electron microscope image analysis [3], natural image segmentation [4], and medical image processing [5]. At present, mathematical morphology is an active area of research and development due to its important practical applications. In the following section, we will discuss the basic operations of mathematical morphology in detail.

3.2 Morphological Filtering

3.2.1 Erosion and Dilation

Erosion and dilation are the most basic morphological operations in morphological image processing. [6] We will define and illustrate these basic operations in this section.

Image Segmentation: Principles, Techniques, and Applications, First Edition. Tao Lei and Asoke K. Nandi.
© 2023 John Wiley & Sons Ltd. Published 2023 by John Wiley & Sons Ltd.

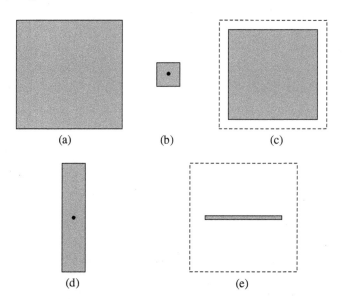

Figure 3.1 Erosion diagram: (a) The set. (b) A square structuring element. (c) The erosion result using the square structuring element. (d) A strip structuring element and (e) The erosion result using the strip structuring element.

ε_E refers to the erosion of a structural element E of a set A, which is defined as:

$$\varepsilon_E = A \ominus E = \{z \,|\, (E)_z \subseteq A\} \tag{3.1}$$

This formula indicates that the erosion of a structural element E to a set A is the set of all the points z that are z-shifted by E in set A (E is called a structuring element)[7]. Figure 3.1 is an example of an erosion operation where elements A and E are shades of gray with a white background. The solid line in Figure 3.1c is the boundary where the origin of E moves further. If this boundary is exceeded, the structuring element E will no longer be contained completely in A. The origin track in the boundary constitutes the erosion of E to A. When the structuring elements are different, the erosion results on the set A are also different. When we consider long strip structuring elements, the original set A changes to a straight line.

In Figure 3.2, we need to remove the thin line connecting the central area and the boundary in Figure 3.2a. If we use different sizes of structuring elements to perform an erosion operation on the image, we will obtain different results. When the structuring element is a disk of size 3×3, the thick line of the central cross line is tapered off, but the surrounding lines are not removed. When the size of the structuring element is 5×5, the two cross lines are completely removed, yet the surrounding lines still exist. When the structuring element is 10×10, all lines are removed. Therefore, different-sized structural elements will lead to different erosion results.

δ denotes the dilation of a structuring element E to a set A, which is defined as:

$$\delta_E = A \oplus E = \left\{z \,\middle|\, \left(\hat{E}\right)_z \cap A \neq \varnothing\right\} \tag{3.2}$$

The reflection of a structuring element E about its origin is denoted by $\hat{E}$. $\hat{E}$ and A have at least one nonzero common element. Unlike the erosion operation, the dilation operation will "increase" and "thicken" the objects in the binary image. The set in Figure 3.3a and that in Figure 3.1a are the same, and Figure 3.3b shows a square structuring element. The dotted line in Figure 3.3c shows the

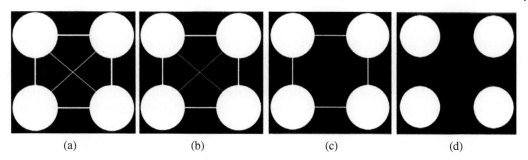

(a)   (b)   (c)   (d)

Figure 3.2 Removing some areas of the image using erosion operation: (a) A binary image of a wire-bond mask in which foreground pixels are shown in white. (b) Erosion result using a disk structuring element of size 3 × 3. (c) Erosion result using a square structuring element of size 5 × 5. (d) Erosion result using a square structuring element of size 10 × 10.

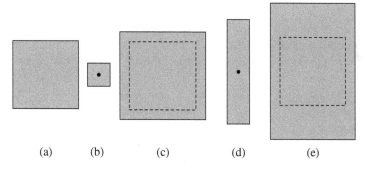

(a)   (b)   (c)   (d)   (e)

Figure 3.3 Dilation diagram: (a) Set. (b) A square structuring element. (c) The dilation result using the square structuring element. (d) A strip structuring element. (e) The dilation result using the strip structuring element.

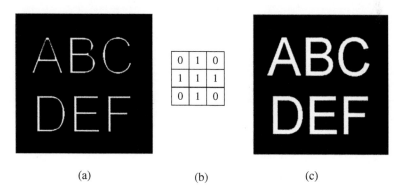

(a)   (b)   (c)

Figure 3.4 Use of the dilation operation to expand the broken letters. (a) The input image. (b) A disk structuring element of size 3 × 3. (c) The dilation result.

original set, and the solid line is the boundary. All points on or within the boundary constitute the dilation result of E to A.

Figure 3.4a shows a binary image with incomplete letters. Using the dilation operation, all background points that contact the object can be combined into the object to expand the boundary and fill the holes in the object. In Figure 3.4b, it can be seen that the broken pixels between letters are connected.

3.2.2 Opening and Closing

Besides the two basic operations of erosion and dilation, two combined operations in morphological image processing play a very important role: opening and its dual operation, closing [8]. This section will focus on the characteristics of the opening operation and the closing operation, which can be obtained through duality (i.e. through the analysis of the complement). Although the two operations are defined according to the combination of erosion and dilation, they have a more intuitive geometric form, from the perspective of structuring element filling.

The opening operation of the structuring element E on set A is expressed as $A \circ E$ and is defined as:

$$A \circ E = (A \ominus E) \oplus E \tag{3.3}$$

In order to better understand the role of the opening operation in image processing, we discuss the following equivalent equation:

$$A \circ E = \cup \{E + x : E + x \subset A\} \tag{3.4}$$

Where $\cup\{\cdot\}$ refers to the union of all sets in braces. This equation shows that the opening operation can be obtained by calculating the union of the translation track of all structuring elements that can be filled in an image [9]. The detailed process is to mark every filling position and to calculate the union when the structuring element is moved to each marked position, after which the result can be obtained. In other words, the opening operation of E to A firstly performs the erosion operation, then performs the dilation operation on the previous erosion result, and then expands E with the erosion result [10].

Figure 3.5a shows the set A and the structuring element E. Figure 3.5b shows some translations where E exactly matches area with A. The union of translation track is the shaded area in Figure 3.5c, which is the result of the opening operation. The white in the figure denotes the area where structuring elements cannot be completely matched with areas in A, i.e. this white area

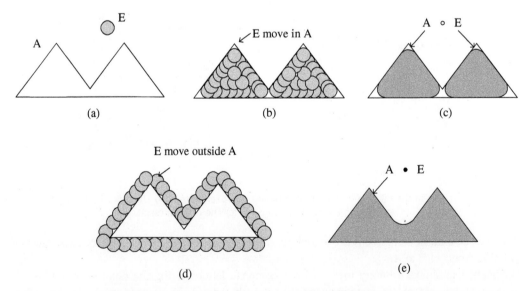

Figure 3.5 Opening operation and closing operation are union sets of moved structuring elements: (a) The set *A* and structuring element *E*. (b) The translation of *E* is exactly matched with set *A*. (c) Opening operation result (shaded area). (d) *E* moves outside set *A*. (e) Closing operation result (shaded area).

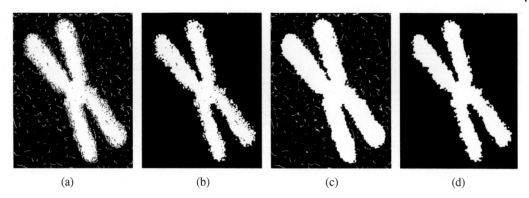

(a)   (b)   (c)   (d)

Figure 3.6 Experimental results of opening and closing operation: (a) Original image. (b) Opening operation result. (c) Closing operation result. (d) Opening operation first and then closing operation result.

cannot be obtained with the opening operation. The morphological opening operation completely removes the object areas that do not contain structuring elements, smooths the object contour, disconnects the narrow connections, and removes the small protruding areas.

Closing is the dual operation of opening, which performs a dilation operation first and then performs an erosion operation. The closing operation of E to A is expressed as $A \cdot E$:

$$A \cdot E = (A \oplus E) \ominus B \tag{3.5}$$

Geometrically, $A \cdot E$ is the union of all translations of E that do not overlap with A. Figure 3.5d shows some translations of E that do not overlap with A. By completing the above translation union operation, we get the shadow area shown in Figure 3.5e. Like the opening operation, the morphological closing operation can smooth the outline of the object. However, the closing operation is different from the opening operation, as it connects the narrow gaps together to form a slender bend and fills the holes that are smaller than the structuring elements [11].

Figure 3.6 shows that the opening operation eliminates the noise in an original image, as demonstrated in Figure 3.6(b). For the closing operation, as shown in Figure 3.6c, it can fill the small holes in objects, connect adjacent objects, and smooth their boundaries without significantly changing their area. But unfortunately, it introduces more noise. Contrary to the closing operation, the opening operation can eliminate small objects, isolate objects in thin places, and smooth the larger objects' boundaries. In the binary image, the opening and closing operations are combined to obtain the final result, exhibited in Figure 3.6d, which shows a better vision effect.

3.2.3 Basic Morphological Operation for Grayscale Images

In mathematical morphology, all the morphological operators are expressed in terms of the combinations of the two fundamental morphological operators: erosion and dilation. In grayscale images, the erosion operation of an image f by a structuring element E is denoted by ε_E and is defined as: [12, 13]

$$\varepsilon_E(x) = \min_{e \in E} \{f(x + e)\}. \tag{3.6}$$

The dilation operation is denoted by δ_E and is defined as:

$$\delta_E(x) = \max_{e \in E} \{f(x + e)\}. \tag{3.7}$$

It is well known that $(\delta_E, \varepsilon_E)$ is a pair of dual operators; we define the duality principle next.

Definition 3.1 Assume that Γ and C_l are two complete lattices. A pair of operators $(\varepsilon_E, \delta_E)$ is called an adjunction or dual operators between Γ and C_l if $(\delta_E(Y) \le x \Leftrightarrow \varepsilon_E(x))$, where $\varepsilon : \Gamma \to C_l$ and $d : C_l \to \Gamma$, and $X \in \Gamma$, $Y \in C_l$. In addition, $(\delta_E, \varepsilon_E)$ is also an adjunction or dual operator between Γ' and C_l', where Γ' and C_l' are the dual complete lattices of Γ and C_l, respectively [14, 15].

$$d(Y) = MIN\{X' \in \Gamma \,|\, Y \le \varepsilon(X')\} \tag{3.8}$$

$$\varepsilon_E(X) = MAX\{Y' \in C_l \,|\, \delta(Y') \le X\} \tag{3.9}$$

In practice, opening and closing based on the compositions of erosion and dilation operations are the two important morphological operators, represented by γ and ϕ, respectively, and defined as: $\gamma = \delta\varepsilon_E$ and $\phi = \varepsilon_E\delta$.

In order to research the properties of morphological operators, firstly, we give some properties of image transformation [16, 17]. An operator ψ acting on an input f is:

a) Idempotent if $\psi^2 = \psi$, where ψ^2 denotes the composition $\psi\psi$;
b) Extensive if the transformed images are greater than or equal to the original image: $\forall f, \psi(f) \ge f$;
c) Anti-extensive if the transformed image is always less than or equal to the input image: $\forall f, \psi(f) \le f$;
d) Increasing if ψ preserves the ordering relation between images, that is, for all images f and g, $f \le g \Rightarrow \psi(f) \le \psi(g)$.

Based on the concept of morphological transformation, we give the definitions of morphological filters and filters-derivates [8].

Definition 3.2 An increasing operator ψ acting on f is called:

a) A filter if it is idempotent: $\psi^2 = \psi$;
b) An over-filter if it is extensive: $\psi^2 \ge \psi$;
c) An under-filter if it is anti-extensive: $\psi^2 \le \psi$.

Now, we can see that an opening operator is an anti-extensive morphological filter while a closing operator is an extensive morphological filter.

3.2.4 Composed Morphological Filters

In practice, an opening filter filters out bright image structures, while a closing filter has the same filtering effect but on the dark image structures. Therefore, if an image is corrupted by a symmetrical noise function, it is interesting to use a sequential combination such as an opening filter followed by the dual closing filter or vice versa. Hence, we can construct four basic types of sequence filters: opening-closing, closing-opening, opening-closing-opening, and closing-opening-closing filters, which can be denoted by $\phi\gamma$, $\gamma\phi$, $\gamma\phi\gamma$, and $\phi\gamma\phi$, respectively [13]. Indeed, based on the sequence morphological filters, an alternative sequential filter of size s^f is defined as the sequential combination of one of these filters, starting the sequence with the filter of size 1 and terminating with the filter of size s^f:

$$ASF_{\phi\gamma}^{s^f} = \phi_{E_i}\gamma_{E_i}\cdots\phi_{E_2}\gamma_{E_2}\phi_{E_1}\gamma_{E_1}, \quad ASF_{\gamma\phi}^{s^f} = \gamma_{E_i}\phi_{E_i}\cdots\gamma_{E_2}\phi_{E_2}\gamma_{E_1}\phi_{E_1},$$

$$ASF_{\gamma\phi\gamma}^{s^f} = \gamma_{E_i}\phi_{E_i}\gamma_{E_i}\cdots\gamma_{E_2}\phi_{E_2}\gamma_{E_2}\gamma_{E_1}\phi_{E_1}\gamma_{E_1}, \quad ASF_{\phi\gamma\phi}^{s^f} = \phi_{E_i}\gamma_{E_i}\phi_{E_i}\cdots\phi_{E_2}\gamma_{E_2}\phi_{E_2}\phi_{E_1}\gamma_{E_1}\phi_{E_1}.$$

Proposition 3.1   A pair of morphological dual operators (ψ, ψ') is called a pair of morphological dual filters if (ψ, ψ') are increasing and both are idempotent: $\psi^2 = \psi$, $(\psi')^2 = \psi$.

Using Proposition 3.1, we can now get three pairs of basic morphological dual filters $(\gamma, \phi), (\phi\gamma, \gamma\phi)$ and $(\gamma\phi\gamma, \phi\gamma\phi)$, and two pairs of alternative sequential filters $\left(ASF_{\phi\gamma}^{sf}, ASF_{\gamma\phi}^{sf}\right), \left(ASF_{\gamma\phi\gamma}^{sf}, ASF_{\phi\gamma\phi}^{sf}\right)$. Here, we need to note that (ε_i, d) is a pair of dual operators but not a pair of dual filters. The filtering performance of $(\phi\gamma, \gamma\phi)$ is similar to that of $(\gamma\phi\gamma, \phi\gamma\phi)$, and $\left(ASF_{\phi\gamma}^{sf}, ASF_{\gamma\phi}^{sf}\right)$ is similar to $\left(ASF_{\gamma\phi\gamma}^{sf}, ASF_{\phi\gamma\phi}^{sf}\right)$. However, $(\gamma\phi\gamma, \phi\gamma\phi)$ and $\left(ASF_{\gamma\phi\gamma}^{sf}, ASF_{\phi\gamma\phi}^{sf}\right)$ are time-consuming; thus $(\gamma, \phi), (\phi\gamma, \gamma\phi)$ and $\left(ASF_{\phi\gamma}^{sf}, ASF_{\gamma\phi}^{sf}\right)$ are commonly used dual filters.

3.3   Morphological Reconstruction

Morphological reconstruction is based on the morphological gradient images, using morphological opening and closing operations to reconstruct the gradient images while preserving the important areas of an image and removing the noise in the image [18]. Reconstruction involves two images and a structuring element. One image is a marker containing the starting point of the transformation. The other image is a mask to constrain the transformation. The structuring element is used to define continuity. The core of morphology is geodesic erosion and geodesic dilation [19], which will be briefly introduced below.

3.3.1   Geodesic Dilation and Erosion

Usually, a test image is used as the mask image f, and the morphological transformation of the test image is used as the marker image g. f and g are binary images, and $f \subseteq g$. $\delta_g^{(1)}$ represents the geodesic dilation of the marker image of size 1 corresponding to the mask image defined as: [20]

$$\delta_g^{(1)}(f) = (f \oplus b) \cap g, \tag{3.10}$$

where $\cap$ represents the intersection of sets, and the geodesic dilation of f is:

$$\delta_g^{(n)}(f) = \delta_g^{(1)}\left[\delta_g^{(n-1)}(f)\right], \tag{3.11}$$

where $\delta_g^{(0)}(f) = f$. Figure 3.7 shows the geodesic dilation process of size 1.

The expanded morphological reconstruction of the marker image f to the mask image g is expressed as $R_g^\delta(f)$ and is defined as the geodesic dilation of f with respect to g. It can be calculated by iterating k times:

$$R_g^\delta(f) = \delta_g^{(k)}(f). \tag{3.12}$$

Similar to the geodesic dilation, the geodesic erosion marker f with size 1 and corresponding mask g is defined as:

$$\varepsilon_g^{(1)}(f) = (f \ominus b) \cup g, \tag{3.13}$$

where $\cup$ represents the union of sets, and the geodesic erosion of f with respect to g of size n is defined as:

$$\varepsilon_g^{(n)}(f) = \varepsilon_g^{(1)}\left[\varepsilon_g^{(n-1)}(f)\right], \tag{3.14}$$

where $\varepsilon_g^{(0)}(f) = f$. Figure 3.8 shows the geodesic erosion process of size 1.

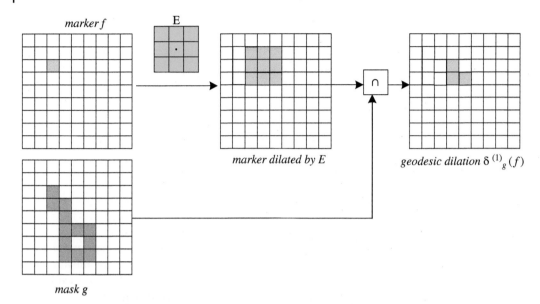

Figure 3.7 Schematic diagram of geodesic dilation.

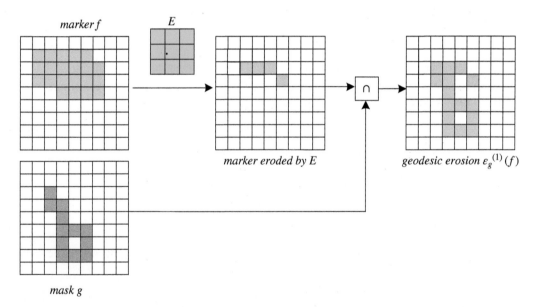

Figure 3.8 Schematic diagram of geodesic erosion.

Similarly, the morphological erosion reconstruction of the mask image g to the marker image f shown as $R_g^\varepsilon(f)$ is defined as the geodesic erosion of f to g, which keeps iterating until it reaches a stable state:

$$R_g^\varepsilon(f) = \varepsilon_g^{(k)}(f). \tag{3.15}$$

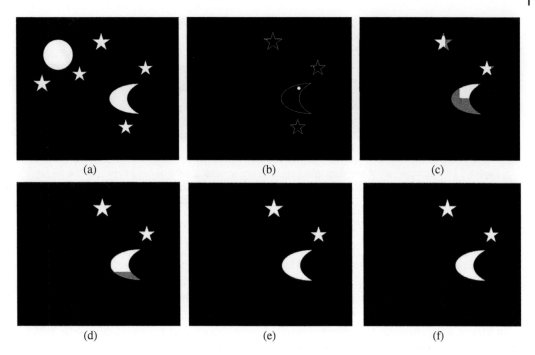

Figure 3.9 Geodesic erosion operation: (a) The original image. (b) The marker image. (c) The result after 50 iterations. (d) Intermediate results after 100 iterations. (e) Intermediate results after 200 iterations. (f) Final results after 300 iterations.

Figure 3.9a is the original binary image, and Figure 3.9b is the marker image. By performing iteration operations 50, 100, 200, and 300 times, respectively, it can be seen that the geodesic erosion gradually tends to reach stabilization.

3.3.2 Reconstruction of Opening Operations and Closing Operations

For dilation reconstruction, the mask image can be obtained by $g = \varepsilon(f)$, which meets the dilation reconstruction condition $g \leq f$. For erosion reconstruction, the marker image can be obtained by $g = \delta(f)$, which meets the erosion reconstruction condition $g \geq f$.

Based on the above two strategies, a more robust morphological closing reconstruction algorithm can be constructed. The calculation formula of the algorithm is:

$$R^C(g) = R^\varepsilon_{R^\delta_g(\varepsilon_E(g))}\left(\delta_E\left(R^\delta_g(\varepsilon(f))\right)\right), \tag{3.16}$$

where $R^C(f)$ represents the result of morphological closing reconstruction. Both ε and δ operations require the structuring element E, so the above formula can be expressed as:

$$R^C(g) = R^\varepsilon_{R^d_g(\varepsilon_E(g))}\left(\delta\left(R^d_g(\varepsilon_E(f))\right)\right), \tag{3.17}$$

where $\varepsilon_E(g)$ represents the erosion operation of the mask image g with the structuring element E, and E is generally a disk with a radius of r. By selecting the size of r in the structuring

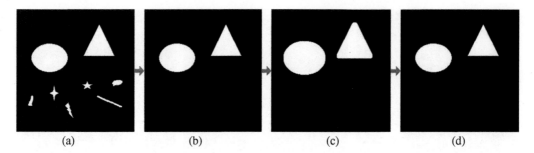

Figure 3.10 Comparison of morphological reconstruction results on a binary image: (a) The binary image. (b) Morphological dilation reconstruction. (c) Morphological erosion reconstruction. (d) Morphological closing reconstruction.

element E, different levels of noise or interference in the unprocessed image can be suppressed. When $r = 0$, $R^C(g) = g$ is equal to f. It indicates that the morphological closing reconstruction does not perform a filtering operation on g. Figure 3.10 shows the morphological reconstruction result when $r = 6$.

In Figure 3.10, although the morphological dilation reconstruction can significantly remove the interference information, it is restricted by its marker image, and the final reconstruction result can only show the shape of the erosion component. Due to the duality, morphological erosion reconstruction uses dilation constraints to enhance the missing component. Compared with the former, the morphological closing reconstruction not only compensates for their respective shortcomings but also effectively improves the final filtering effect through alternate operations.

According to the dual operation, the formula of morphological opening reconstruction is:

$$R^O(g) = R^\delta_{R^\varepsilon_g(\delta(g))}\left(\varepsilon\left(R^\varepsilon_g(\delta(g))\right)\right), \tag{3.18}$$

where $R^O(g)$ represents the result of morphological closing reconstruction. Both ε and δ operations require the structuring element E, so the above formula can also be expressed as:

$$R^O(g) = R^\delta_{R^\varepsilon_g(\delta_E(g))}\left(\varepsilon\left(R^\varepsilon_g(\delta_E(g))\right)\right), \tag{3.19}$$

where $\delta_E(g)$ means that the mask image g is expanded by the structuring element E, and E is generally a disk with a radius of r. Figure 3.11 shows the comparison between opening and reconstruction opening, and Figure 3.11a is the original image. Figure 3.11b shows the result of an erosion operation using linear structuring elements. Figure 3.11c is the result of a morphological opening operation using vertical fine structuring elements. Figure 3.11d shows the result of reconstruction using vertical fine structuring elements. In order to extract the text with long vertical lines in the picture, linear structuring elements are used to erode the original picture, and the result is shown in Figure 3.11b. Then, the reconstructed opening is performed on the corroded image, and the result is shown in Figure 3.11d. It can be seen that in Figure 3.11c, the vertical bar has been restored after the morphological opening, but the character does not include the rest except the vertical bar. Figure 3.11d shows that the characters with long vertical bars are completely restored using the reconstruction opening, and other letters have been removed.

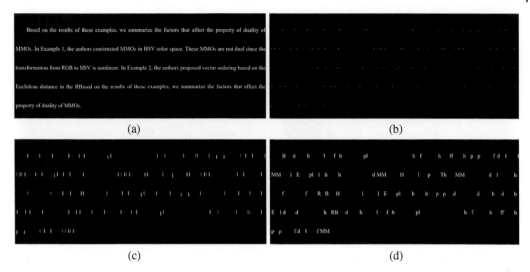

Figure 3.11 Geodesic opening operation: (a) Original image (test image size is 918 × 2018 pixels). (b) The image after vertical line erosion operation. (c) The results of operation calculation with vertical line. (d) The results of operation calculation by geodesic erosion using a vertical line.

3.4 Watershed Transform

The watershed transform can be classified as a region-based segmentation approach that has its origins in mathematical morphology. As one of the most important applications of morphological algorithms [21], the watershed transform has been an active research area, and it is considered the most effective yet simple operator in image segmentation since it was proposed in the 1980s [22].

3.4.1 Basic Concepts

In the watershed transform, an image is regarded as a topographic landscape with ridges and valleys. The elevation values of the landscape are typically defined by the gray values of the respective pixels or their gradient magnitude In Figure 3.12, based on such a 3D representation, the theoretical model of the algorithm is abstracted from the geomorphic features. The elevation of each terrain can be seen as the pixel value in the image, and each basin can be seen as the pixel grayscale minimum area in the image. Assuming that the entire model is immersed in water, and the water slowly expands outward from the lowest point of the model, a relatively high point will appear on the adjacent area of two catchment basins—the maximum point of the image. These maximum points separate the two catchment basins, thus constituting the watershed.

By simulating a flooding process, the watershed transform can be expressed as [23]:

$$Xh_{min} = T_{h_{min}}(I), \tag{3.20}$$

$$\forall h = [h_{min}, h_{max} - 1], \tag{3.21}$$

$$X_{h+1} = min_{h+1} \cup C_{X_h}(X_h \cap X_{h+1}). \tag{3.22}$$

Formula 3.18 shows the initial conditions of the flooding process, and T is the point with the smallest pixel gray value in image I. In Formula 3.18, h represents the range of altitude—the range

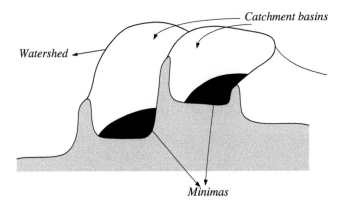

Figure 3.12 Schematic diagram of the watershed transform.

of the pixel value of the image. h_{min} is the minimum value of the pixel in image I, and h_{max} is the maximum value of the pixel. X_{h+1} is all the points with the pixel value of $h + 1$. min_{h+1} indicates that a new basin has been created at $(h + 1)$ elevation. And C_{Xh} indicates the basin where the X_h point is located [24].

By calculating the gradient of the image, the area with the smaller gradient value is regarded as a catchment basin, or a valley. The eight-connected domain method then collects all pixels in the catchment basin and numbers each catchment basin. The connecting part of the two catchment basins is called the watershed. Notably, the gradient is always higher at the edge of the segmentation target and lower in other places. Therefore, the watershed transform is usually applied to gradient images for segmentation tasks [25].

However, though the watershed transform is very sensitive to subtle gray changes and can accurately identify the edges of objects, in practice, with noisy mage data, a large number of small regions may arise, which is known as the "over-segmentation" problem, leading to over-segmentation.

3.4.2 Watershed Segmentation Algorithms

In practical applications, researchers have proposed a variety of algorithms to achieve watershed segmentation. These algorithms can be divided into three categories: watershed segmentation based on (1) distance transformation, (2) gradient, and (3) labeled images. We will introduce these three types of algorithms in detail [26].

1) **Watershed Segmentation Algorithm Based on Distance Transform**
    In binary image segmentation, a common tool used in conjunction with the watershed transform is the distance transformation. The distance transformation refers to the distance from each pixel to its nearest nonzero pixel in a binary image [27]. Notably, the watershed transform based on the distance transformation can be used not only in binary images but also in grayscale and multivariate images. The distance transform–based watershed segmentation process can be described as follow: First, the input original multivariable image is converted into grayscale to obtain a grayscale image, and then the grayscale image is converted into a binary image. Then, the distance from the catchment basin to the watershed is obtained by the distance

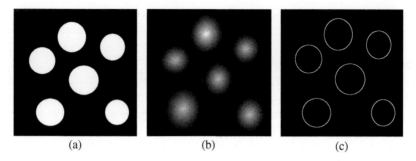

Figure 3.13 Experimental results on watershed transform based on distance variation: (a) Original image. (b) Distance transformation. (c) Watershed ridge of distance transformation.

transformation of the binary image [28]. Finally, the watershed segmentation is performed on the distance-transformed image to obtain the segmentation result.

The distance transformation of a binary image is a relatively simple concept. In Figure 3.13a, the distance-based watershed transform only needs to calculate the distance from each pixel to its nearest nonzero pixel and then find the watershed ridge. In Figure 3.13d, some objects are not properly separated, resulting in over-segmentation. This phenomenon is a common problem of segmentation algorithms based on the distance transform [29].

2) **Gradient-Based Watershed Segmentation Algorithm**

Because a binary image usually contains limited image information, in practice, grayscale images and color images are in more widespread use. However, when directly performed, the watershed segmentation algorithm based on the distance transform on the above images, the segmentation result is usually poor. Therefore, researchers prefer to use gradient-based watershed segmentation algorithms to obtain better image segmentation results for gradient images, which usually contain rich image contour information. However, gradient images are easily disturbed by noise and image details, so the watershed segmentation based on gradient images still faces serious over-segmentation problems [24].

The process of gradient-based watershed segmentation is shown in Figure 3.14. First, the original image is transformed into a gray image, and then the gradient image is determined according to a morphological gradient calculation method. Finally, the watershed transform is used to obtain the segmentation results. Compared with the watershed segmentation algorithm based

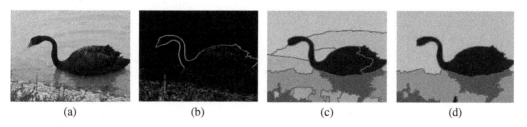

Figure 3.14 Experimental results on watershed transform based on gradient images: (a) Original image. (b) Gradient image. (c) Results of gradient-based watershed algorithm. (d) Results of gradient-based watershed algorithm after smoothing the gradient.

on the distance transform, this algorithm has better segmentation results for grayscale images. However, the final segmentation result still contains a large number of over-segmented regions [30].

3) **Marker-Based Watershed Segmentation Algorithm**

Figures 3.13 and 3.14 show that the above two types of watershed transforms have their own advantages for different types of image segmentation, but they both face the problem of over-segmentation. Over-segmentation is mainly caused by too many local minima in the process of watershed segmentation. Therefore, how to suppress the useless local minima is a problem worth exploring. To address this problem, researchers have proposed a watershed segmentation algorithm based on masker images. In fact, the mask image is a seed point image, in which each seed point corresponds to a region, and the selection result of the seed point will directly affect the final segmentation result. There are two main ways to generate marked images, namely internal marking and external marking. The former generates target markings, while the latter generates background markings [31].

The marker-based watershed segmentation algorithm is as follows: First, the original image is converted into a grayscale image, and then the gradient image of this grayscale image is calculated. Then, the internal and external marked images are calculated by means such as linear filtering, nonlinear filtering, and morphological processing. In the actual application process, different marked images-generating methods need to be selected according to the specific task [32].

In Figure 3.15, the mark-based watershed segmentation algorithm is used to mark the location of the local minimum area in the image with coins, and the result is an accurate mark of the middle position of the coins, as shown in Figure 3.15c, d. After getting the final internal and external tags, the final segmentation result is shown in Figure 3.15e.

Three commonly used watershed transforms are introduced above; all three suffer from the problem of over-segmentation, though, which is one of the most challenging issues facing the watershed transform. In order to solve this problem, researchers have put forward many solutions, and the most practical ideas can be divided into two types: (1) One is image preprocessing, that is, to suppress the noise, details, and other information in the image as much as possible before the watershed transform, so as to make the image as smooth as possible. This operation is conducive to obtaining a smoother gradient image and can effectively solve the over-segmentation problem. The commonly used image preprocessing strategies are image filtering and image reconstruction. However, image preprocessing may smooth the target contour of an image to some extent while suppressing the noise, resulting in the final calculated gradient image deviating from the actual target contour, leading to limited accuracy of the final segmentation contour. (2) The other is the gradient correction method. The calculation of the local

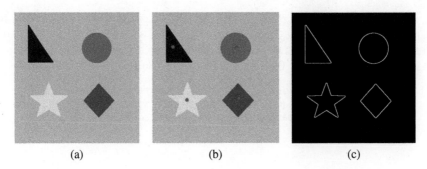

(a) (b) (c)

Figure 3.15 Experimental results on marker-based watershed transform: (a) Original diagram. (b) Internal marking of the image. (c) Image segmentation results.

minimum of the image depends on the gradient image. Therefore, correcting the gradient image is an effective means to solve the problem of over-segmentation. Common gradient correction methods include morphological gradient reconstruction, gradient threshold, and gradient optimization based on the morphological dual operation. These methods can keep the main gradient of the image from being smoothed and suppress gradients with lower gray values or smaller areas.

In summary, the image segmentation method based on the morphological watershed has many applications particularly in some special segmentation tasks. Although the current image segmentation methods are based on supervised deep learning methods, the morphological watershed transform can still be used as a supplementary method to further improve the final segmentation results [33]. Therefore, the fusion of morphological watershed segmentation and supervised deep learning is one of the future research directions in the field of image segmentation.

3.5 Multivariate Mathematical Morphology

3.5.1 Related Concepts

Before introducing multivariate mathematical morphology, we first recall some basic notions of classical mathematical morphology. A set L with a partial ordering $\leq$ is called a complete lattice if every subset H of L has a lowest upper bound (supremum) $\vee H$ and a greatest lower bound (infimum) $\wedge H$. Letting E and T represent nonempty sets, we denote $F(E, T)$ by the power set T^E, i.e. the functions mapping from E into T. If T is a complete lattice, then $F(E, T)$ is also a complete lattice. The operator $\psi: E \to T$ is increasing if $\forall f, g \in F(E, T)$ and $f \leq g \Rightarrow \psi(f) \leq \psi(g)$. ψ is extensive if $f \leq \psi(f)$, and is anti-extensive if $\psi(f) \leq f$. The operator $\psi: T \to T$ is called an operator on T. The simplest operator on T is the identity operator, which maps every element into itself and is denoted by *id*. $id(f) = f$, for $\forall f \in T$. An operator is idempotent if $\psi(\psi(f)) = \psi(f)$. It is a dilation if $\psi\left(\bigvee_{t \in b} f_{-t}\right) = \bigvee_{t \in b} \psi(f_{-t})$, where b is a structuringl element. Duality is erosion if $\psi\left(\bigwedge_{t \in b} f_{-t}\right) = \bigwedge_{t \in b} \psi(f_{-t})$. If $\varepsilon: E \to T$ and $d: T \to E$ are two operators, we define the pair (ε, δ) as an adjunction between T and E if $\delta(g) \leq f \Leftrightarrow g \leq \varepsilon(f), \forall f, g \in F(E, T)$. If (ε, δ) is an adjunction, then ε is an erosion operation and δ is a dilation operation. For every erosion $\varepsilon_E: E \to T$, there corresponds a unique dilation $\delta: T \to E$ such that (ε, δ) is an adjunction. Similarly, for every dilation δ, one can associate a unique erosion ε. We then say that ε_i and δ are adjoined to each other. According to the definition of ε and δ, $\varepsilon \leq id$ and $id \leq d$. $\delta\varepsilon$ and $\varepsilon\delta$ are the compositions of ε and δ, which form an opening on E and a closing on T, respectively. An opening operator is increasing, anti-extensive, and idempotent; a closing operator is increasing, extensive, and idempotent. With every complete lattice T, one can associate an opposite or dual complete lattice T' by reversing the partial ordering: $f \leq' g$ in T' if $g \leq f$ in T. This observation forms the basis of the duality principle, which states that every definition or statement concerning complete lattices has a dual counterpart. For example, the dual of opening is closing, and vice versa.

3.5.2 Duality of Grayscale Mathematical Morphology

Duality, a property allowing one operation to be expressed in terms of another, is one of the most important principles in mathematical morphology. In classical mathematical morphology theory, most operators occur in pairs of dual operators, such as erosion and dilation, or opening and closing [29, 34]. Such operators do not necessarily need to exist as opposites of each other. Moreover, some complex morphological transformations are also proposed in terms of dual operators, such as morphological gradient operators, reconstruction operators, the top-hat transformation, and self-dual morphological operators. Consequently, duality is an important factor for extending elementary morphological operators to color images. In the following, we first review some elementary concepts of mathematical morphology.

Definition 3.3 The complement of a grayscale image f, denoted by f^c, is defined for each pixel x as the maximum value t_{max} (e.g. t_{max} for an 8-bit grayscale image) of the unsigned data type used for storing the image minus the value of the image f at position x,

$$f^c(x) = t(x)_{max} - f(x). \tag{3.23}$$

Definition 3.4 The complement image of a color image f, $f \in (E, T^{RGB})$, is defined as the complement of each color component [35], i.e.

$$\begin{aligned} f^c &= (r^c, g^c, b^c) \\ &= (t_{max} - r + t_{min}, t_{max} - g + t_{min}, t_{max} - b + t_{min}) \end{aligned} \tag{3.24}$$

Definition 3.5 Any two image transformations ψ and ψ' are dual with respect to complementation if applying ψ to an image is equivalent to applying ψ' to the complement of the image and then complementing the resulting image,

$$\psi(f/\boldsymbol{f}) = (\psi'(f^c/\boldsymbol{f}^c))^c. \tag{3.25}$$

In general, given the dual operator ψ, we denote its dual operator by ψ^*. That is, $\psi^* = \psi'$ and $(\psi')^* = \psi$. Likewise, $\gamma^* = \phi$ and $\phi^* = \gamma$, while $\varepsilon^* = \delta$ and $\delta^* = \varepsilon$ (ε and δ denote erosion and dilation, respectively, while γ and ϕ denote opening and closing, respectively).

Generally, the erosion operation removes bright image structures, while the dilation operation has the same effect on dark image structures. Hence, as previously explained, the composition operators of morphological erosion and dilation are usually used to remove noise in images.

3.5.3 Ordering Relations

Partial ordering plays a key role in mathematical morphology since output values at a given pixel for fundamental operators require that the ordering of input image values falls within a user-defined neighborhood centered at this pixel. Typically, a set of grayscale values in their natural order forms a completely ordered set, and the maximum and minimum values for the set are uniquely defined. However, when there is no unambiguous means of defining the minimum and maximum values between two vectors in more than one dimension, the lack of a total ordering can only be partially answered by considering a suitable (less than total) sub-ordering. Therefore, a well-defined total ordering is necessary for mathematical morphological operators. Two examples are presented in the following section.

Example 3.1 A color morphological operator defined in HSV (hue, saturation, value) color space has been proposed in [30]. The vector ordering $\leq_{Louverdis}$ of two pixels $x_1(h_1, s_1, v_1)$ and $x_2(h_2, s_2, v_2)$ is defined as follows:

$$x_1(h_1, s_1, v_1) \leq_{Louverdis} x_2(h_2, s_2, v_2) \Leftrightarrow \begin{cases} v_1 < v_2 \text{ or} \\ v_1 = v_2 \text{ and } s_1 \geq s_2 \text{ or} \\ v_1 = v_2 \text{ and } s_1 = s_2 \text{ and } h_1 \leq h_2 \end{cases}$$

Based on the vector ordering $\leq_{Louverdis}$, Louverdis proposed the corresponding color erosion and dilation, which are denoted by $\vec{\varepsilon}_{iLouverdis}$ and $\overrightarrow{\delta d}_{Louverdis}$, respectively. They have some properties, such as being translation-invariant and monotonically increasing, but they do not satisfy duality.

Definition 3.6 The multivariate morphological erosion is given by

$$\vec{\varepsilon}_{ib}(f)(x) = \wedge_{VO} \{f(x+y)\}.$$ (3.26)
$$\substack{y \in b}$$

Definition 3.7 The multivariate morphological dilation is given by

$$\vec{\delta}_b(f)(x) = \vee_{VO} \{f(x-y)\}.$$ (3.27)
$$\substack{y \in b}$$

VO denotes the corresponding vector ordering. $\vec{\varepsilon}_{iLouverdis}$ and $\vec{\delta}_{Louverdis}$ are defined as

$$\left(\vec{\varepsilon}_{iLouverdis}\right)_b (f)(x) = \wedge_{Louverdis} \{f(x+y)\}, \text{ and}$$ (3.28)
$$\substack{y \in b}$$

$$\left(\vec{\delta}_{Louverdis}\right)_b (f)(x) = \vee_{Louverdis} \{f(x-y)\},$$ (3.29)
$$\substack{y \in b}$$

where f denotes a color image, and b is a structuring element.

To prove $\left(\vec{\varepsilon}_{iLouverdis}, \vec{\delta}_{Louverdis}\right)$ doesn't satisfy duality. we let $f = (r, g, b)$ be a color image, and $f \in (E, T^{RGB}), f^c = (r^c, g^c, b^c)$. The transformation from RGB to HSV color space is given as follows:

$$v = P_1(r, g, b) = \frac{max\{r, g, b\}}{255},$$ (3.30)

$$s = P_2(r, g, b) = 1 - \frac{min\{r, g, b\}}{max\{r, g, b\}},$$ (3.31)

$$h = P_3(r, g, b) = \begin{cases} arccos\left\{\dfrac{(r-g)+(r-b)}{2\sqrt{(r-g)^2+(r-b)^2+(g-b)^2}}\right\} & b \leq g \\ \\ 2\pi - arccos\left\{\dfrac{(r-g)+(r-b)}{2\sqrt{(r-g)^2+(r-b)^2+(g-b)^2}}\right\} & b > g \end{cases},$$ (3.32)

where P_1, P_2, and P_3 are functions of the three variables r, g, and b, respectively. $v^c = P'_1(r^c, g^c, b^c) = P'_1(tmaxmax_{max})$, $s^c = P'_2(r^c, g^c, b^c) = P'_2(tmaxmax_{max})$, and $h^c = P'_3(r^c, g^c, b^c) = P'_3(tmaxmax_{max})$. P'_1, P'_2, and P'_3 are also functions of the three variables r, g, and b, respectively. Then,

$$x_1(h_1, s_1, v_1) \leq_{Louverdis} x_2(h_2, s_2, v_2) \Leftrightarrow$$

$$x_1(P_1(r_1, g_1, b_1), P_2(r_1, g_1, b_1), P_3(r_1, g_1, b_1)) \leq_{Louverdis} x_2(P_1(r_2, g_2, b_2), P_2(r_2, g_2, b_2), P_3(r_2, g_2, b_2))$$

$$\Leftrightarrow x_1(r_1, g_1, b_1) \leq_{Louverdis} x_2(r_2, g_2, b_2)$$

$$\Leftrightarrow \begin{cases} P_1(r_1, g_1, b_1) < P_1(r_2, g_2, b_2) \, or \\ P_1(r_1, g_1, b_1) = P_1(r_2, g_2, b_2) \, and \, P_2(r_1, g_1, b_1) > P_2(r_2, g_2, b_2) \, or \\ P_1(r_1, g_1, b_1) = P_1(r_2, g_2, b_2) \, and \, P_2(r_1, g_1, b_1) = P_2(r_2, g_2, b_2) \, and \, P_3(r_1, g_1, b_1) \leq P_3(r_2, g_2, b_2) \end{cases}$$

Similarly,

$$x_2^c(h_2, s_2, v_2) \leq_{Louverdis} x_1^c(h_1, s_1, v_1) \Leftrightarrow$$

$$\begin{cases} P_1(r_2^c, g_2^c, b_2^c) < P_1(r_1^c, g_1^c, b_1^c) \, or \\ P_1(r_2^c, g_2^c, b_2^c) = P_1(r_1^c, g_1^c, b_1^c) \, and \, P_2(r_2^c, g_2^c, b_2^c) > P_2(r_1^c, g_1^c, b_1^c) \, or \\ P_1(r_2^c, g_2^c, b_2^c) = P_1(r_1^c, g_1^c, b_1^c) \, and \, P_2(r_2^c, g_2^c, b_2^c) = P_2(r_1^c, g_1^c, b_1^c) \, and \, P_3(r_2^c, g_2^c, b_2^c) \leq P_3(r_1^c, g_1^c, b_1^c) \end{cases}$$

As a result, we obtain $P_1(r_1, g_1, b_1) \leq P_1(r_2, g_2, b_2) \Leftrightarrow P_1(r_1^c, g_1^c, b_1^c) \geq P_1(r_2^c, g_2^c, b_2^c)$, but are unable to obtain $P_2(r_1, g_1, b_1) \leq P_2(r_2, g_2, b_2) \Leftrightarrow P_2(r_1^c, g_1^c, b_1^c) \geq P_2(r_2^c, g_2^c, b_2^c)$ and $P_3(r_1, g_1, b_1) \leq P_3(r_2, g_2, b_2) \Leftrightarrow P_3(r_1^c, g_1^c, b_1^c) \geq P_3(r_2^c, g_2^c, b_2^c)$ because they are uncertain. Therefore, $x_1(h_1, s_1, v_1) \leq_{Louverdis} x_2(h_2, s_2, v_2) \Leftrightarrow x_2^c(h_2, s_2, v_2) \leq_{Louverdis} x_1^c(h_1, s_1, v_1)$ is also uncertain, and

$$\vee_{Louverdis} \{t_{max} - f(x+y)\}_{y \in b} \neq t_{max} - \wedge_{Louverdis} \{f(x+y)\}_{y \in b}$$

$$\vee_{Louverdis} \{t_{max} - f(x+y)\}_{y \in b} \neq t_{max} - \wedge_{Louverdis} \{f(x+y)\}_{y \in b}$$

Furthermore,

$$\left(\vec{d}_{Louverdis}\right)_b (f^c) = \vee_{Louverdis} \{f^c(x+y)\}_{y \in b}$$

$$= \vee_{Louverdis} \{t - \{f(x+y)\}\}_{y \in b} \neq t_{max} - \wedge_{Louverdis} \{f(x+y)\}_{y \in b} \tag{3.33}$$

$$= t_{max} - \left(\vec{\varepsilon}_{Louverdis}\right)_b (f) = \left[\left(\vec{\varepsilon}_{iLouverdis}\right)_b (f)\right]^c$$

Therefore, the MMF proposed by Louverdis et al. [30] does not satisfy duality. The vector ordering depends on P_1, P_2, and P_3. $P_i(r_1, g_1, b_1) \leq P_i(r_2, g_2, b_2) \Leftrightarrow P_i(r_2^c, g_2^c, b_2^c) \leq P_i(r_1^c, g_1^c, b_1^c)$, $1 \leq i \leq 3$, is only correct for linear transformations of RGB color space, i.e. $r_1 \leq r_2 \Leftrightarrow r_2^c \leq r_1^c$ because $r_1 \leq r_2 \Leftrightarrow r_1 - t221_{21 \, max}^{cc} \, max \, max \, max$. Similarly, $g_1 \leq g_2 \Leftrightarrow g_2^c \leq g_1^c$ and $b_1 \leq b_2 \Leftrightarrow b_2^c \leq b_1^c$ are also obtained. From the analysis made above, $x_1 \leq_{Louverdis} x_2 \Leftrightarrow x_2^c \leq_{Louverdis} x_1^c$ is also not necessarily true. Similarly, we can prove that multivariate morphological frameworks based on vector orderings, such as $\leq_{Angulo \, 07}$, $\leq_{Angulo \, 10}$, $\leq_{Lei \, 13}$, and $\leq_{Lei \, 14}$, do not satisfy duality.

Example 3.2 A type of color morphological operators defined in RGB color space is proposed in [14]. In this approach, the vector ordering $\leq_{De\ Witte}$ of two pixels $x_1(r_1, g_1, b_1)$ and $x_2(r_2, g_2, b_2)$ is defined as follows:

$$x_1(r_1, g_1, b_1) \leq_{De\ Witte} x_2(r_2, g_2, b_2) \Leftrightarrow \begin{cases} d(x_1, Bl) < d(x_2, Bl)\ or \\ d(x_1, Bl) = d(x_2, Bl)\ and\ d(x_1, Wh) \geq d(x_2, Wh) \end{cases}. \quad (3.34)$$

However, vector ordering *De Witte* $\geq$ is defined as

$$x_1(r_1, g_1, b_1)_{De\ Witte} \geq x_2(r_2, g_2, b_2) \Leftrightarrow \begin{cases} d(x_1, Wh) < d(x_2, Wh)\ or \\ d(x_1, Wh) = d(x_2, Wh)\ and\ d(x_1, Bl) \geq d(x_2, Bl) \end{cases},$$

where d denotes the Euclidean distance, and

$$d(x, Bl) = \sqrt{(r_x - 0)^2 + (g_x - 0)^2 + (b_x - 0)^2}, \quad (3.35)$$

$$d(x, Wh) = \sqrt{(255 - r_x)^2 + (255 - g_x)^2 + (255 - b_x)^2}. \quad (3.36)$$

In the following, we use t_{min}, t_{max} to represent 0 and 255, respectively.

De Witte et al. also proposed the definitions of color erosion and dilation, denoted by $\vec{\varepsilon}_{iDe\ Witte}$ and $\vec{d}_{De\ Witte}$, respectively, in terms of the vector ordering ($\leq_{De\ Witte}$, $_{De\ Witte}\geq$). Unfortunately, $\vec{\varepsilon}_{iDe\ Witte}$ and $\vec{d}_{De\ Witte}$ do not satisfy the requirements of being translation-invariant and monotonically increasing.

Proof:

Let us study the duality of the morphological framework proposed by De Witte et al. Analogous to Eqs. 3.35–3.36, $d(x, Bl) = P_1(r_x, g_x, b_x)$ and $d(x, Wh) = P_2(r_x, g_x, b_x)$ are obtained, where P_1 and P_2 are functions of the three variables r, g, and b. Then,

$$x_1(r_1, g_1, b_1) \leq_{De\ Witte} x_2(r_2, g_2, b_2)$$
$$\Leftrightarrow \begin{cases} P_1(r_1, g_1, b_1) < P_1(r_2, g_2, b_2)\ or \\ P_1(r_1, g_1, b_1) = P_1(r_2, g_2, b_2)\ and\ P_2(r_1, g_1, b_1) \geq P_2(r_2, g_2, b_2) \end{cases}, \quad (3.37)$$

$$x_1(r_1, g_1, b_1)_{De\ Witte} \geq x_2(r_2, g_2, b_2)$$
$$\Leftrightarrow \begin{cases} P_2(r_1, g_1, b_1) < P_2(r_2, g_2, b_2)\ or \\ P_2(r_1, g_1, b_1) = P_2(r_2, g_2, b_2)\ and\ P_1(r_1, g_1, b_1) \geq P_1(r_2, g_2, b_2) \end{cases}, \quad (3.38)$$

$$x_1^c(r_1, g_1, b_1)_{De\ Witte} \geq x_2^c(r_2, g_2, b_2)$$
$$\Leftrightarrow \begin{cases} P_2(r_1^c, g_1^c, b_1^c) < P_2(r_2^c, g_2^c, b_2^c)\ or \\ P_2(r_1^c, g_1^c, b_1^c) = P_2(r_2^c, g_2^c, b_2^c)\ and\ P_1(r_1^c, g_1^c, b_1^c) \geq P_1(r_2^c, g_2^c, b_2^c) \end{cases}, \quad (3.39)$$

$$d(x_1^c, Bl) = \sqrt{(t_{max} - r_1)^2 + (t_{max} - g_1)^2 + (t_{max} - b_1)^2}, \text{ and}$$
$$= d(x_1, Wh)$$

$$d(x_1^c, Wh) = \sqrt{(t_{max} - (t_{max} - r)_1)^2 + (t_{max} - (t_{max} - g_1))^2 + (t_{max} - (t_{max} - b_1))^2}$$
$$= \sqrt{(r_1)^2 + (g_1)^2 + (b_1)^2} = d(x_1, Bl) \quad (3.40)$$

We then get

$$P_2\left(r_1^c, g_1^c, b_1^c\right) = \sqrt{\left(t_{\max} - r_1^c\right)^2 + \left(t_{\max} - g_1^c\right)^2 + \left(t_{\max} - b_1^c\right)^2},$$ (3.41)

and because $r_1^c = t1_{\max}$, $g_1^c = t1_{\max}$, $b_1^c = t1_{\max}$,

$$P_2\left(r_1^c, g_1^c, b_1^c\right) = \sqrt{\left(r_1\right)^2 + \left(g_1\right)^2 + \left(b_1\right)^2} = P_1(r_1, g_1, b_1), \text{ and}$$

$$\begin{cases} P_2\left(r_1^c, g_1^c, b_1^c\right) < P_2\left(r_2^c, g_2^c, b_2^c\right) \text{ or} \\ P_2\left(r_1^c, g_1^c, b_1^c\right) = P_2\left(r_2^c, g_2^c, b_2^c\right) \text{ and } P_1\left(r_1^c, g_1^c, b_1^c\right) \geq P_1\left(r_2^c, g_2^c, b_2^c\right) \end{cases}$$

$$\Leftrightarrow \begin{cases} P_1(r_1, g_1, b_1) < P_1(r_2, g_2, b_2) \text{ or} \\ P_1(r_1, g_1, b_1) = P_1(r_2, g_2, b_2) \text{ and } P_2(r_1, g_1, b_1) \geq P_2(r_2, g_2, b_2) \end{cases}, \text{ i.e.}$$

$$x_1^c(r_1, g_1, b_1)_{De\ Witte} \geq x_2^c(r_2, g_2, b_2) \Leftrightarrow x_1(r_1, g_1, b_1) \leq_{De\ Witte} x_2(r_2, g_2, b_2).$$ (3.42)

Then,

$$\left(\vec{d}_{De\ Witte}\right)_b(\boldsymbol{f}^c) = \vee_{De\ Witte} \{\boldsymbol{f}^c(\boldsymbol{x} + \boldsymbol{y})\}_{\boldsymbol{y} \in b}$$

$$= \vee_{De\ Witte} \{tc(\boldsymbol{x} + \boldsymbol{y})_{\max} \{\}\}_{\boldsymbol{y} \in b}$$

$$= t \wedge_{De\ Witte} \{\boldsymbol{f}(\boldsymbol{x} + \boldsymbol{y})\}_{\max} = t\left(\vec{\varepsilon}_{De\ Witte}\right)_b(\boldsymbol{f})_{\max}$$ (3.43)

$$= \left[\left(\vec{\varepsilon}_{iDe\ Witte}\right)_b(\boldsymbol{f})\right]^c.$$

Therefore,

$$x_1^c\left(r_1^c, g_1^c, b_1^c\right) \leq_{De\ Witte} x_2^c\left(r_2^c, g_2^c, b_2^c\right) \Leftrightarrow \begin{cases} d\left(x_1^c, Bl\right) < d\left(x_2^c, Bl\right) \text{ or} \\ d\left(x_1^c, Bl\right) = d\left(x_2^c, Bl\right) \text{ and } d\left(x_1^c, Wh\right) \geq d\left(x_2^c, Wh\right) \end{cases}$$

$$\Leftrightarrow \begin{cases} d(x_1, Wh) < d(x_2, Wh) \text{ or} \\ d(x_1, Wh) = d(x_2, Wh) \text{ and } d(x_1, Bl) \geq d(x_2, Bl) \end{cases}$$

$$\Leftrightarrow x_1(r_1, g_1, b_1)_{De\ Witte} \geq x_2(r_2, g_2, b_2).$$ (3.44)

Accordingly, the multivariate morphological operators in [31] satisfy duality. However, these operators are called pseudo morphological operators, since $\leq_{De\ Witte}$ and $De\ Witte\geq$ are two different vector orderings and are defined in two different lattices.

Based on the results of these examples, we summarize the factors that affect the property of duality of multivariate morphological operators (MMOs). The authors constructed MMOs in HSV color space. These MMOs are not dual since the transformation from RGB to HSV is nonlinear. The authors proposed vector ordering based on the Euclidean distance in the RGB color space. Although these MMOs are dual, they are defined in two different lattices because the corresponding vector ordering relies on two reference colors. Therefore, an accurate vector ordering should satisfy two conditions:

1) The vector ordering must be defined in a complete lattice.
2) There is a linear transformation system from RGB to other spaces in which multivariate dual morphological operators are defined.

3.5.4 Multivariate Dual Morphological Operators

The transformation from RGB to YUV color space is linear, i.e.

$$\vec{v}'(y, u, v) = T\left[\vec{v}(r, g, b)\right] \Leftrightarrow \begin{cases} y = T_y(r, g, b) \\ u = T_u(r, g, b), \\ v = T_v(r, g, b) \end{cases} \tag{3.45}$$

where $T\{T_y, T_u, T_v\}$ is a linear transformation system from RGB to YUV. Based on the two conditions mentioned above, we first propose an multivariate morphological filter (MMF) based on a color space transformation as follows:

$$\vec{v}'_1(y_1, u_1, v_1) \leq_{YUV} \vec{v}'_2(y_2, u_2, v_2) \Leftrightarrow \begin{cases} y_1 < y_2 \; or \\ y_1 = y_2 \; and \; u_1 < u_2 \; or \\ y_1 = y_2 \; and \; u_1 = u_2 \; and \; v_1 \leq v_2 \end{cases}. \tag{3.46}$$

First, YUV is a linear transformation of the RGB color space. Second, equation (3.46) is a vector ordering in a complete lattice $(T, \leq_{YUV})$. Finally, Y denotes the brightness in YUV color space, which is of a higher rank than color components U and V in terms of visual perception. Therefore, $(T, \leq_{YUV})$ satisfies the two conditions of an accurate vector ordering and can be used to define MMOs with duality.

Similarly, YIQ and YCbCr are the linear transformations of RGB color space and can also be used to define MMOs with duality in terms of $(T, \leq_{YIQ})$ and $(T, \leq_{YCbCr})$. However, how can we define the pairwise ranks between color components such as (U, V), (I, Q), and (Cb, Cr)? In fact, there is no specific ordering between U and V (I and Q, Cb and Cr). Thus, to propose a better vector ordering, we introduce principal component analysis (PCA) [36] into our system because a decided ordering exists among different PCA components, i.e. the first component is more important than the second, the second is more important than the third, and so on.

$$\vec{v}'_1(f^p_1, s^p_1, t^p_1) \leq_{PCA} \vec{v}'_2(f^p_2, s^p_2, t^p_2) \Leftrightarrow \begin{cases} f^p_1 < f^p_2 \; or \\ f^p_1 = f^p_2 \; and \; s^p_1 < s^p_2 \; or \\ f^p_1 = f^p_2 \; and \; s^p_1 = s^p_2 \; and \; t^p_1 \leq t^p_2 \end{cases}, \tag{3.47}$$

where f^p, s^p, and t^p denote the first, second, and third principal components, respectively. Since PCA is a popular dimensional reduction algorithm, it is easy to extend relation (3.47) to high-dimension data processing. (The signs of the projection vectors from PCA are arbitrary. Here, we only use PCA to reduce the dimensions. Each dimension of the reduced subspace has the same direction, and thus the signs do not influence the result.) Depending on relation (3.47), we define a linear transformation system based on PCA, where $\vec{v}(d_1, d_2, \cdots d_n)$ denotes the original high-dimension data, and $\vec{v}^r(D_1, D_2, \cdots D_n)$ denotes the transformed high-dimension data using a linear system $T\{T_1, T_2, \cdots, T_n\}$, $n \in R$, i.e.

$$\vec{v}^r(D_1, D_2, \cdots D_n) = T\left[\vec{v}(d_1, d_2, \cdots d_n)\right] \Leftrightarrow \begin{cases} D_1 = T_1(d_1, d_2, \cdots d_n) \\ D_2 = T_2(d_1, d_2, \cdots d_n) \\ \vdots \\ D_n = T_n(d_1, d_2, \cdots d_n) \end{cases}, \tag{3.48}$$

$$\vec{v}_1^r \leq {}_{PCA}\, \vec{v}_2^r \Leftrightarrow \begin{cases} T_1\!\left[\vec{v}_1\right] < T_1\!\left[\vec{v}_2\right] \ or \\ \quad T_1\!\left[\vec{v}_1\right] = T_1\!\left[\vec{v}_2\right]\ and\ T_2\!\left[\vec{v}_1\right] < T_2\!\left[\vec{v}_2\right]\ or \\ \qquad\qquad\qquad \vdots \\ T_1\!\left[\vec{v}_1\right] = T_1\!\left[\vec{v}_2\right]\ and\ T_2\!\left[\vec{v}_1\right] = T_2\!\left[\vec{v}_2\right] \cdots and\ T_n\!\left[\vec{v}_2\right] \leq T_n\!\left[\vec{v}_2\right] \end{cases} \tag{3.49}$$

However, the lexicographic ordering $\leq_{PCA}$ is so huge that when the value of n is large, it leads to a high computational cost. Fortunately, we can reduce the value of n by selecting the first m principal components that include the main information of the data. $m \ll n$ when the value of n is extremely large, for example, $n = 225$ for hyperspectral remote sensing images [37].

Based on the above analysis, the proposed vector ordering not only protects duality but also can be easily carried out. Finally, based on the different color spaces and PCA, we propose a unified representation form of the vector ordering ($\leq_{LT}$) that includes $\leq_{YUV}$, $\leq_{YIQ}$, $\leq_{YCrCb}$, and $\leq_{PCA}$, where $m = 3$ for $\leq_{YUV}$, $\leq_{YIQ}$, and $\leq_{YCbCr}$, and $3 \leq m \leq n$ for $\leq_{PCA}$.

Let $\mathbf{f}$ denote a color image, and $(T; \leq_{LT})$ be a completed lattice ordering. $\vee_{LT}$ and $\wedge_{LT}$ denote the supremum and infimum based on $\leq_{LT}$, respectively. We define the unified multivariate morphological erosion and dilation as follows [38]:

$$\left(\vec{\varepsilon}_{iLT}\right)_b(\mathbf{f}) = \underset{\substack{t \in b}}{\wedge_{LT}}(\mathbf{f}_{-t}), \tag{3.50}$$

$$\left(\vec{\delta}_{LT}\right)_b(\mathbf{f}) = \underset{\substack{t \in b}}{\vee_{LT}}(\mathbf{f}_{-t}), \tag{3.51}$$

where $\left(\vec{\varepsilon}_{iLT}, \vec{\delta}_{LT}\right)$ is an adjunction. We analyze the properties of $\vec{\varepsilon}_{iLT}$ and $\vec{d\delta}_{LT}$ in the following section.

Definition 3.8 The transformation ψ is invariant to translations if it commutes with image translations, i.e.

$$\psi \text{ is invariant to translations} \Leftrightarrow \forall f,\ \forall b,\ \psi(f_k) = [\psi(f)]_k.$$

Proposition 3.2 If $\vec{\varepsilon}_{iLT}$ and $\vec{\delta}_{LT}$ are a multivariate erosion and dilation based on vector ordering $\leq_T$, then [39]

$$\vec{\varepsilon}_{iLT} \text{ is invariant to translations} \Leftrightarrow \left[\left(\vec{\varepsilon}_{iLT}\right)_b(\mathbf{f})\right]_k + \mathbf{r} = \left(\vec{\varepsilon}_{iLT}\right)_b(\mathbf{f}_k + \mathbf{r}),\ \text{and}$$

$$\vec{d}_{LT} \text{ is invariant to translations} \Leftrightarrow \left[\left(\vec{d}_{LT}\right)_b(\mathbf{f})\right]_k + \mathbf{r} = \left(\vec{d}_{LT}\right)_b(\mathbf{f}_k + \mathbf{r}).$$

Proof:

$$\begin{aligned} \left(\vec{\varepsilon}_{iLT}\right)_b(\mathbf{f}_k + r)(x) &= \underset{\substack{y \in b}}{\inf}{}_T\{f_k(x+y) + r\} \\ &= \underset{\substack{y \in b}}{\wedge_T}\{f(x+y)\}_k + r \\ &= \left(\underset{\substack{y \in b}}{\wedge_T}\{f(x+y)\}\right)_k + r \\ &= \left[\left(\vec{\varepsilon}_{iLT}\right)_b(f)(x)\right]_k + r \end{aligned} \tag{3.52}$$

Similarly, $\left(\vec{d}_{LT}\right)_b(\mathbf{f}_k + r)(x) = \left[\left(\vec{d}_{LT}\right)_b(\mathbf{f})(x)\right]_k + \mathbf{r}.$

Definition 3.9 The transformation ψ is extensive if, for a multichannel image f, the transformed image is greater than or equal to the original image, i.e. if ψ is greater than or equal to the identity transformation id:

$$\psi \text{ is extensive} \Leftrightarrow i\delta \leq_{LT} \psi.$$

Definition 3.10 The transformation ψ is anti-extensive if, for a multichannel image f, the transformed image is less than or equal to the original image, i.e. if ψ is less than or equal to the identity transformation $i\delta$ [40]:

$$\psi \text{ is anti}-\text{extensive} \Leftrightarrow \psi \leq_{LT} id.$$

Proposition 3.3 If $\vec{\varepsilon}_{iLT}$ and $\vec{\delta}_{LT}$ are multivariate erosion and dilation based on vector ordering $\leq_{LT}$, then

$$\vec{\delta}_{LT} \text{ is extensive} \Leftrightarrow i\delta \leq_{LT} \vec{\delta}_{LT}, \text{ and}$$

$$\vec{\varepsilon}_{LT} \text{ is anti}-\text{extensive} \Leftrightarrow \vec{\varepsilon}_{LT} \leq_{LT} id.$$

Proof:

$$\left(\vec{\varepsilon}_{LT}\right)_b (f)(x) = \inf_{\substack{LT \\ y \in b}} \{f(x+y)\}$$
$$= \wedge_{LT} \{f(x+y)\} \quad .$$
$$\substack{y \in b}$$

$\forall x_i \in f, \exists \vec{e}(f)(x_i) \leq_{LT} x_i$, i.e. $\vec{\varepsilon}_{iLT} \leq_{LT} id$. Similarly, $id \leq_T \vec{\delta}_{LT}$.

Definition 3.11 The transformation ψ is increasing if it preserves the ordering relations between images [41], i.e.

$$\psi \text{ is increasing} \Leftrightarrow \forall f, g, f \leq_{LT} g \Rightarrow \psi(f) \leq_{LT} \psi(g).$$

Proposition 3.4 If $\vec{\varepsilon}_{iLT}$ and $\vec{d}_{LT}$ are multivariate erosion and dilation based on vector ordering $\leq_{LT}$, then

$$\vec{d}_{LT} \text{ is increasing} \Leftrightarrow f \leq_{LT} g \Rightarrow \vec{d}_{LT}(f) \leq_{LT} \vec{d}_{LT}(g), \text{ and}$$

$$\vec{\varepsilon}_{iLT} \text{ is increasing} \Leftrightarrow f \leq_{LT} g \Rightarrow \vec{\varepsilon}_{iLT}(f) \leq_{LT} \vec{\varepsilon}_{iLT}(g).$$

Proof:
If $\forall x_i \in f, \forall z_i \in g$, then $x_i \leq_{LT} z_i$ for $f \leq_{LT} g$, $x_i + y \leq_{LT} z_i + y$, so $\wedge_{LT} (x_i + y) \leq_{LT} \wedge_{LT} (z_i + y)$, i.e.

$\vec{\varepsilon}_{ib}(f) \leq_{LT} \vec{\varepsilon}_{ib}(g)$. Similarly, $f \leq_{LT} g \Rightarrow \vec{d}(f) \leq_{LT} \vec{d}(g)$.

Proposition 3.5 If $\vec{\varepsilon}_{iLT}$ and $\vec{\delta}_{LT}$ are multivariate erosion and dilation based on vector ordering $\leq_{LT}$, then

$$\vec{\varepsilon}_{iLT} \text{ and } \vec{d}_{LT} \text{ are dual forms with respect to complementation } C \Leftrightarrow \vec{\varepsilon}_{iLT} = C\vec{\delta}_{LT}C.$$

Proof:

$$\left(\vec{\delta}_{LT}\right)_b(f^c) = \sup_{\substack{T \\ y\in b}}\{f^c(x+y)\}$$
$$= \bigvee_{\substack{T \\ y\in b}}\{f^c(x+y)\}$$
$$= \bigvee_{\substack{T \\ y\in b}}\{t_{\max} -f(x+y)\}$$
$$= t_{\max} - \bigwedge_{\substack{T \\ y\in \check{b}}}\{f(x+y)\} \qquad (3.53)$$
$$= t_{\max} - \vec{\varepsilon}_b(f^c)$$
$$= \left[\left(\vec{\varepsilon}_{LT}\right)_b(f^c)\right]^c$$

The proposed vector ordering $\leq_{LT}$ is a framework in which classical grayscale morphological algorithms can be easily extended to multichannel images [42].

3.6 Discussion and Summary

In this chapter, we described the basic operations and applications of mathematical morphology, which is based on lattice theory and random geometry to study geometric structures. We defined the basic operators such as dilation and erosion and structuring elements of binary images and grayscale images in the second section. These operators are the basic components of mathematical morphology, and with these basic operations, a combined filter can be constructed for further improvement. In the third section, we described the basic knowledge of morphological reconstruction as well as the basic operations of morphological reconstruction by introducing geodesic erosion, geodesic dilation, and their applications. In the fourth section, we first described the simplest yet effective segmentation algorithm, namely the watershed transform. Then we introduced the commonly used watershed transforms based on distance, gradient, and control markers. The fifth section showed how to extend the mathematical morphology to the color space and then completed the basic operations such as the dilation and erosion in color space.

Nowadays, with the development of other related disciplines, mathematical morphology has cached up new developments and it can be combined with neural networks, genetic algorithms, and other algorithms to form new theories and application areas. Thus mathematical morphology is still an active research area with important theoretical and practical value.

References

1 Pesaresi, M. and Benediktsson, J.A. (2001). A new approach for the morphological segmentation of high-resolution satellite imagery. *IEEE Trans. Geosci. Remote Sens.* **39** (2): 309–320.

2 Kanungo, T. and Haralick, R.M. (1990). Character recognition using mathematical morphology[C]. In: *Proc. of the Fourth USPS Conference on Advanced Technology*, 973–986.

3 Gu, Y. (2003). Automated scanning electron microscope based mineral liberation analysis. *J. Miner. Mater. Character. Eng.* **2** (1): 33–41.

4 Lei, T., Jia, X., Liu, T. et al. (2019). Adaptive morphological reconstruction for seeded image segmentation. *IEEE Trans. Image Process.* **28** (11): 5510–5523.

5 Kamba, S.A., Ismail, M., Hussein-Al-Ali, S.H. et al. (2013). In vitro delivery and controlled release of doxorubicin for targeting osteosarcoma bone cancer. *Molecules* **18** (9): 10580–10598.

6 Gonzalez, R. C., Woods, R. E., and Masters, B. R. (2009). Digital image processing.

7 Lei, T., Zhang, Y., Wang, Y. et al. (2017). A conditionally invariant mathematical morphological framework for color images. *Inform. Sci.* **387**: 34–52.

8 Heijmans, H.J. (1997). Composing morphological filters. *IEEE Trans. Image Process.* **6** (5): 713–723.

9 Nam, C.S., Nijholt, A., and Lotte, F. (ed.) (2018). *Brain–Computer Interfaces Handbook: Technological and Theoretical Advances.* CRC Press.

10 Goutsias, J., Heijmans, H.J., and Sivakumar, K. (1995). Morphological operators for image sequences. *Computer Vision Image Understanding* **62** (3): 326–346.

11 Salembier, P. and Wilkinson, M.H. (2009). Connected operators. *IEEE Signal Process. Mag.* **26** (6): 136–157.

12 Serra, J. (1986). Introduction to mathematical morphology. *Computer Vision Graphics Image Process.* **35** (3): 283–305.

13 Soille, P. (2013). *Morphological Image Analysis: Principles and Applications.* Springer Science & Business Media.

14 Serra, J. (1979). Biomedical image analysis by mathematical morphology (author's transl). *Pathol. Biol.* **27** (4): 205–207.

15 Hÿtch, M. and Hawkes, P.W. (2020). *Morphological Image Operators.* Academic Press.

16 Santillán, I., Herrera-Navarro, A.M., Mendiola-Santibáñez, J.D., and Terol-Villalobos, I.R. (2010). Morphological connected filtering on viscous lattices. *J. Math. Imaging Vision* **36** (3): 254–269.

17 Wilkinson, M. (2009). Connected operators: a review of region-based morphological image processing techniques. *IEEE Signal Process. Mag.* **26** (6): 136–157.

18 Chen, J.J., Su, C.R., Grimson, W.E.L. et al. (2011). Object segmentation of database images by dual multiscale morphological reconstructions and retrieval applications. *IEEE Trans. Image Process.* **21** (2): 828–843.

19 Lei, T., Fan, Y., Zhang, C., and Wang, X. (2013). Vector mathematical morphological operators based on fuzzy extremum estimation. In: *2013 IEEE International Conference on Image Processing*, 3031–3034. IEEE.

20 Preston, K.E.N.D.A.L.L. (1971). Feature extraction by Golay hexagonal pattern transforms. *IEEE Trans. Comput.* **100** (9): 1007–1014.

21 Najman, L. and Schmitt, M. (1996). Geodesic saliency of watershed contours and hierarchical segmentation. *IEEE Trans. Pattern Anal. Mach. Intell.* **18** (12): 1163–1173.

22 Peleg, S. and Rosenfeld, A. (1981). A min-max medial axis transformation. *IEEE Trans. Pattern Anal. Mach. Intell.* **2**: 208–210.

23 Mangan, A.P. and Whitaker, R.T. (1999). Partitioning 3D surface meshes using watershed segmentation. *IEEE Trans. Vis. Comput. Graph.* **5** (4): 308–321.

24 Beucher, S. (1994). Watershed, hierarchical segmentation and waterfall algorithm. In: *Mathematical Morphology and its Applications to Image Processing*, 69–76. Dordrecht: Springer.

25 Acharjya, P.P. and Ghoshal, D. (2012). Watershed segmentation based on distance transform and edge detection techniques. *Int. J. Computer Appl.* **52** (13).

26 Zhao, Y., Liu, J., Li, H., and Li, G. (2008). Improved watershed transform for dowels image segmentation. In: *2008 7th World Congress on Intelligent Control and Automation*, 7644–7648. IEEE.

27 Hill, P.R., Canagarajah, C.N., and Bull, D.R. (2002). Texture gradient based watershed segmentation. In: *2002 IEEE International Conference on Acoustics, Speech, and Signal Processing*, vol. **4**, IV–3381. IEEE.

28 Levner, I. and Zhang, H. (2007). Classification-driven watershed segmentation. *IEEE Trans. Image Process.* **16** (5): 1437–1445.

29 Serra, J. (1998). Connectivity on complete lattices. *J. Math. Imaging Vision* **9** (3): 231–251.

30 Louverdis, G., Vardavoulia, M.I., Andreadis, I., and Tsalides, P. (2002). A new approach to morphological color image processing. *Pattern Recogn.* **35** (8): 1733–1741.

31 De Witte, V., Schulte, S., Nachtegael, M. et al. (2005). Vector morphological operators for colour images. In: *International Conference Image Analysis and Recognition*, 667–675. Berlin, Heidelberg: Springer.

32 Yu-Qian, Z., Wei-Hua, G., Zhen-Cheng, C. et al. (2006). Medical images edge detection based on mathematical morphology. In: *2005 IEEE engineering in medicine and biology 27th annual conference*, 6492–6495. IEEE.

33 Krinidis, S. and Chatzis, V. (2010). A robust fuzzy local information C-means clustering algorithm. *IEEE Trans. Image Process.* **19** (5): 1328–1337.

34 Lei, T. and Fan, Y.Y. (2011). Noise gradient reduction based on morphological dual operators. *IET Image Process.* **5** (1): 1–17.

35 Russ, J.C. (2006). *The Image Processing Handbook*. CRC Press.

36 Licciardi, G., Marpu, P.R., Chanussot, J., and Benediktsson, J.A. (2011). Linear versus nonlinear PCA for the classification of hyperspectral data based on the extended morphological profiles. *IEEE Geosci. Remote Sens. Lett.* **9** (3): 447–451.

37 Velasco-Forero, S. and Angulo, J. (2011). Supervised ordering in IR^{P}: application to morphological processing of hyperspectral images. *IEEE Trans. Image Process.* **20** (11): 3301–3308.

38 Lezoray, O. and Elmoataz, A. (2012). Nonlocal and multivariate mathematical morphology. In: *2012 19th IEEE International Conference on Image Processing*, 129–132. IEEE.

39 Batcher, K.E. (1980). Design of a massively parallel processor. *IEEE Trans. Comput.* **29** (09): 836–840.

40 Gerritsen, F.A. and Aardema, L.G. (1981). Design and use of DIP-1: a fast, flexible and dynamically microprogrammable pipelined image processor. *Pattern Recogn.* **14** (1–6): 319–330.

41 Sternberg, S.R. (1981). Parallel architectures for image processing. In: *Real-Time Parallel Computing*, 347–359. Boston, MA: Springer.

42 Brito-Loeza, C. and Chen, K. (2010). On high-order denoising models and fast algorithms for vector-valued images. *IEEE Trans. Image Process.* **19** (6): 1518–1527.

4

Neural Networks

4.1 Artificial Neural Networks

4.1.1 Overview

1) **Development of Artificial Neural Networks (ANNs)**

Before the neural network model was proposed, biomedical researchers made important contributions, especially the Nobel laureates Santiago Ramón y Cajal and Camillo Golgi in 1906. Camillo Golgi, an Italian scholar, first observed nerve cells with a microscope, while Ramón y Cajal, a Spanish scholar, first proposed that nerve cells (neurons) are the most basic unit of neural activity. Thanks to the rapid development of neuroscience, the neuroscientist Warren McCulloch and the computer scientist and logician Walter Pitts first proposed the M-P model of neurons in 1943, which is the first mathematical model of neurons in human history, thus accelerating the era of artificial intelligence.

In 1958, American scientist Frank Rosenblatt invented a binary linear classifier called perceptron. Perceptron is the first practical application of artificial neural networks, marking a new stage of development for ANNs. Because the perceptron is a single-layer feedforward neural network, it cannot deal with the linear inseparable problem. In 1983, the Canadian scientists Geoffrey Hinton and Terrence Sejnowski first proposed the concept of "hidden unit" and designed a Boltzmann machine containing a visible layer and a hidden layer. The increase of layers indicates that neural networks can deal with the complex nonlinear separable (linearly inseparable) problems. However, the parameter training algorithm has become an important bottleneck restricting the development of multilayer neural networks. In 1974, Paul Werbos, an American scholar, invented the famous back-propagation (BP) neural network learning algorithm with far-reaching influence in his doctoral thesis while studying for his doctoral degree at Harvard University. The proposal of the BP algorithm provides a practical solution for the learning and training of multilayer ANNs, which led to the rapid development of ANNs in the 1980s. However, due to the shallow structure of BP neural networks, they are easily overfitting and usually have slow parameter speed, thus, the development of ANNs was very slow in the 1990s and the early twenty-first century.

In 2006, benefiting from the improvement of computer performance and the emergence of massive data, Geoffrey Hinton and his student Ruslan Salakhutdinov started the era of deep learning [1] by realizing the first deep neural network. The representative achievements in the era of deep learning should belong to the rapid development of convolutional neural networks (CNN). In 1998, French scientist Yann LeCun first proposed the LeNet-5 [2] for character recognition and achieved good results in small-scale handwritten numeral recognition.

Image Segmentation: Principles, Techniques, and Applications, First Edition. Tao Lei and Asoke K. Nandi.
© 2023 John Wiley & Sons Ltd. Published 2023 by John Wiley & Sons Ltd.

In the twenty-first century, the emergence of massive multimedia data and the successful implementation of deep neural networks have paved the way for the rapid development of deep convolution neural networks. In 2012, Alex Krizhevsky and others proposed and implemented the deep convolution neural network AlexNet [3], accelerating the development of artificial intelligence.

2) **Basic Principles and Applications of ANNs**

Artificial neural networks [4], often referred to simply as neural networks, are mathematical models that aim to imitate the structure and function of biological neural networks. The typical neural network architecture is composed of an input layer, hidden layers, and an output layer, in which the hidden layers are located between the input layer and output layer. A layer in a neural network is usually composed of multiple neurons. The connection between different layers is realized by the full connection. The input layer is used to receive input data, and the number of neurons in the input layer is the same as the dimension of the input data. There can be one or more hidden layers, and the number of neurons in each layer is uncertain. The hidden layers are used to process the input data and carry out nonlinear mapping. Generally, a larger number of hidden layers indicate more significant nonlinearity and robustness of the neural network. The output layer is used to summarize and calculate the feature vectors generated by the last layer of the hidden layer and generate the output vector. The output layer for regression task usually contains only one neuron, and the number of neurons in the output layer for classification task is the same as the number of classes in the task.

An ANN mainly use a learning algorithm to update its internal neuron parameters, so it has the ability to learn from data. In addition, the introduction of a nonlinear activation function enables the artificial neural network to effectively fit the complex nonlinear relationship between input and output.

Applying a neural network to a data classification task usually has the following main steps: constructing data set and dividing it into training set and test set, designing network and initializing, training network, saving parameters, and outputting model. The specific flowchart is shown in Figure 4.1.

In the early days, neural networks were used for some simple data classification and regression tasks, especially for linear separable problems. Later, neural networks were used for complex problem modeling, especially for the classification and regression of high-dimensional complex data. Because the characteristics of complex distribution and high dimension of these data, it is difficult to achieve good results using traditional modeling methods. The essential reason lies in the limited representation ability of the models.

Neural networks have been widely used in all walks of life, especially in language understanding (NLP), object recognition, image understanding, target tracking, and so on. Taking the application of neural networks in image understanding as an example, The task usually is to let the computer automatically annotate images. How to use a neural network to complete the task? The general method is first to construct the image data set and annotation data set and then construct an encoder-decoder CNN. For example, VGG [5] is used as the pretraining network, the cross-entropy loss function is used as the objective function, and the stochastic gradient descent (SGD) learning

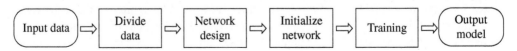

Figure 4.1 Neural network for data classification.

strategy is used to train the network until convergence. For the trained network, when we input an image, we can directly output its corresponding annotation information. A neural network for image semantic annotation is shown in Figure 4.2.

From the above application examples of neural networks, we can see that, compared with other mathematical modeling methods, the advantage of neural networks is that they don't need a lot of prior knowledge and artificial data features. ANNs have a simple model design and fast reasoning speed. The disadvantage is that its learning process has a high consumption of computing and memory resources. Thus the ANNs have the characteristics of black box and poor interpretability.

4.1.2 Neuron Model

1) **Neuron**

An artificial neuron, referred to just as neuron for short, is the basic unit of a neural network. It mainly simulates the structure and characteristics of biological neurons, receiving a group of input signals and generating outputs.

Biologists discovered the structure of biological neurons in the early twentieth century. A computing unit in the human brain is a simple brain neuron. In the neural system of the human brain, there are about 8.6×10^{10} neurons, which are connected by $10^{14} \sim 10^{15}$ synapses. A biological neuron usually has multiple dendrites and an axon. Dendrites are used to receive information, and axons are used to send information. When the accumulation of input signals obtained by neurons exceeds a certain threshold, it is in an excited state and generates electrical impulse. There are many terminals at the end of the axon, which can make connections (synapses) to the dendrites of other neurons and transmit electrical impulse signals to other neurons.

In 1943, neuroscientist Warren S. McCulloch and mathematician Walter Pitts proposed a very simple neuron model named MP neurons based on the structure of biological neurons. The neurons in modern neural networks have not changed much compared to the structure of MP neurons; the difference is that the activation function f in MP neurons is a step function of 0 or 1, while the activation function in modern artificial neurons is usually required to be a continuously derivable function.

In the calculation model of a neural network, the parameters of the model are equivalent to synapses. Then a input signal is transmitted to neurons as a linear combination of input data

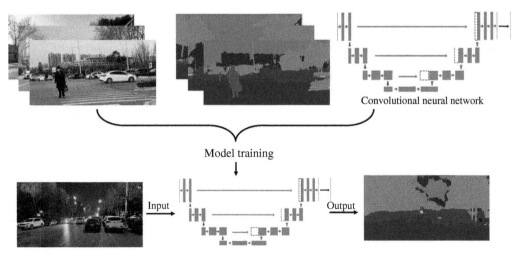

Convolutional neural network

Model training

Input   Output

Figure 4.2 Convolutional neural network used for image semantic annotation.

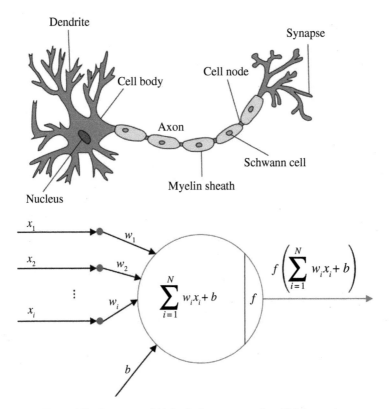

Figure 4.3 Structure of biological neurons and artificial neurons.

and model parameters, and then through the activation function, the model is finally obtained. The structures of biological neurons and artificial neurons are shown in Figure 4.3.

A neuron receives the output of N neurons from the last layer, which is considered as the input x_i. The connection between neurons is also called weight, denoted as w_i. Firstly, the neuron calculates the weighted sum of multiple input and adds the bias term b (for bias) with the weighted sum to obtain the superposition result z, which is expressed as:

$$z = \sum_{i=1}^{N} w_i x_i + b. \tag{4.1}$$

The superposition result z is sent to the activation function f_a, and the activated result a^r is taken as the output term of the neuron, i.e.

$$a^r = f_a\left(\sum_{i=1}^{N} w_i x_i + b\right). \tag{4.2}$$

Formula 4.2 gives the mathematical model or expression of an artificial neuron. Based on the combination of artificial neurons, we can build an artificial neural network.

2) **Activation Function**

In formula (4.2), if f_a is a linear function, then the artificial neural network can only perform an affine transformation on the input data, and the superposition of multiple affine transformations is still an affine transformation. Therefore, even if multiple hidden layers are introduced into the artificial neural network, its fitting ability to deal with complex input data cannot be enhanced. The solution to this problem is to introduce a nonlinear transformation into neurons: use nonlinear functions to process the summation results, and then take the processing results as the input of the next full connection layer. This nonlinear function is also called activation function. The activation functions are usually saturated or unsaturated, according to its saturation.

An activation function plays a very important role in artificial neurons. In order to enhance the representation and learning ability of a neural network, the activation function needs to have the following properties:

1) Continuity and differentiability (nondifferentiability at a few points is allowed); the differentiable activation function can directly use numerical optimization methods to learn network parameters.
2) Simplicity: The activation function and its derivative should be as simple as possible, which is beneficial for improving the efficiency of network computing.
3) Appropriate range: The range of the derivative function of the activation function should be limited in an suitable range, not too large or too small; otherwise, it would affect the efficiency and stability of training.

Several activation functions commonly used in neural networks are introduced below.

The sigmoid function and the hyperbolic tanh function are popular saturation activation functions.

1) **Sigmoid function:** The sigmoid function (also called logistic function) refers to a kind of S-shaped function, which is a saturation function at both ends. The sigmoid function maps variables between 0 and 1. Specifically, the larger negative numbers are converted closer to 0, and the larger positive numbers are converted closer to 1. The definition of the sigmoid function is shown in Eq. (4.3), its derivative is shown in Eq. (4.4), and the function image is shown in Figure 4.4a:

$$\sigma(x) = \frac{1}{1 + e^{-x}}, \tag{4.3}$$

$$\sigma'(x) = \sigma(x)(1 - \sigma(x)). \tag{4.4}$$

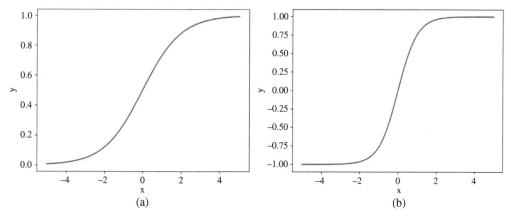

Figure 4.4 (a) Sigmoid function and (b) tanh function.

2) **Tanh function:** It is a double tangent sine curve. The tanh function is similar to the sigmoid function, converting the input value into an output between −1 and 1. The definition of the tanh function is shown in Eq. (4.5), its derivative is shown in Eq. (4.6), and the function image is shown in Figure 4.4b:

$$f_a(x) = \tanh(x) = \frac{e^x - e^{-x}}{e^x + e^{-x}},\tag{4.5}$$

$$f_a'(x) = 1 - (f_a(x))^2.\tag{4.6}$$

The saturation activation function can limit the activation value to a certain range and slow down the gradient explosion phenomenon to a certain extent. Because of its normalization characteristics, the sigmoid function is typically used as the activation function of the output layer in binary classification tasks or regression tasks; compared with the sigmoid function, due to the characteristics of zero mean, the tanh function is used in the hidden layer(s) is more conducive to the parameter updating of neural networks. It can be seen from Eq. (4.4) and (4.6) that the derivatives of both the sigmoid function and tanh function contain the function itself, and the back-propagation derivation process is relatively simple, but the index needs to be calculated to solve the function itself, so it still has a large amount of calculation. In addition, when the input value is too large or too small, the derivative of the saturation activation function approaches 0, which will lead to the disappearance of the gradient during the back-propagation process, and the chain product effect will further amplify this phenomenon in the shallow layer of networks. This means that it is difficult to effectively update the shallow parameters of networks during the training process, which will eventually lead to difficulty in the convergence of networks, even training failure. Therefore, a saturated activation function is less used in deep network taining, while an unsaturated activation function is more popular.

Rectified linear unit (ReLU) and its variants are mainstream unsaturated activation functions.

3) **ReLU function:** It is also known as the rectifier function, and is very popular in neural networks. This activation function simply retains the result greater than 0 and performs zero-setting on the result less than 0. Its definition is shown in Eq. (4.7), its derivative is shown in Eq. (4.8), and the function image is shown in Figure 4.5:

$$f_a(x) = ReLU(x) = \begin{cases} x, x \geq 0 \\ 0, x < 0 \end{cases}\tag{4.7}$$

$$f_a'(x) = \begin{cases} 1, x \geq 0 \\ 0, x < 0 \end{cases}\tag{4.8}$$

Compared with the derivatives of the sigmoid and tanh activation functions, the derivative calculation of the ReLU function is very simple, in which only the operations of addition, multiplication, and comparison are needed, and using a threshold filter can get the results, with no need to carry out other complex operations. Therefore, the ReLU function is more efficient in calculation in Figure 4.5, it can be seen that the wider activation boundary of ReLU can effectively alleviate the gradient disappearance problem and improve the efficiency of the network. However, the ReLU neuron is more likely to appear "dead" during the training process. If the parameters are updated improperly, the ReLU neuron in the first hidden layer may not be activated on all training data, then the gradient value of the parameters of the neuron will always be 0, which means it will never be activated in the future training process. Moreover,

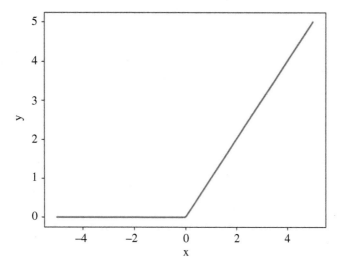

Figure 4.5 The ReLU function.

the one-sided inhibition of ReLU can block the negative gradient propagation, which makes the network have a certain sparse expression ability. However, when the learning rate is large, it may cause irreversible death of a certain number of neurons in the network, which seriously affects the network performance, as shown in Eq. (4.9).

$$\omega_{i+1} = \omega_i + \eta \Delta \omega x_i. \tag{4.9}$$

If $x_i < 0$, then $\eta \Delta \omega x_i \ll 0$. If the output is 0 after activating the function, the neuron output will always be 0 in subsequent iterations. In order to improve the ReLU activation function, scholars have proposed a large number of improved ReLU functions, such as parametric ReLU (PReLU), leaky ReLU, and exponential linear unit (ELU), RReLU. We will introduce several variants of ReLU in detail latter.

4) **Leaky ReLU**
Leaky ReLU. It is also called ReLU function with leakage. This activation function improves the weakness of ReLU function in training. When input $x < 0$, this function gives all negative values a nonzero slope a ($a \neq 1$, usually $a = 0.01$). Therefore, even neurons are inactive, there can also be a nonzero gradient to update parameters to avoid neurons being dead. Leaky ReLU is defined by Eq. (4.10). The function image is shown in Figure 4.6a:

$$LeakyReLU = \begin{cases} x, x > 0 \\ ax, x \leq 0 \end{cases}, \tag{4.10}$$

where a is a small constant, usually taken as 0.01.

RReLU is a random ReLU function, where a will randomly extract a value from the uniform distribution $U(l, u)$: $a \sim U(l, u)$, $l < u$, $l, u \in [0, 1)$. The definition of RReLU is shown in Eq. (4.11), and the image of the function is shown in Figure 4.6b.

$$RReLU(x) = \begin{cases} x, x \geq 0 \\ ax, x < 0 \end{cases} \tag{4.11}$$

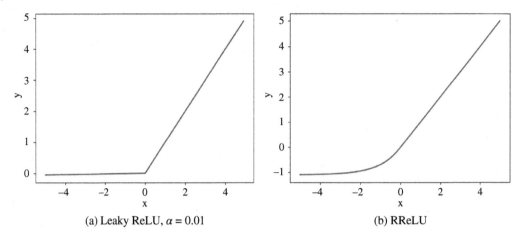

(a) Leaky ReLU, $\alpha = 0.01$ (b) RReLU

Figure 4.6 (a) The Leaky ReLU function, a = 0.01 and (b) The RReLU function.

5) **Parametric ReLU (PReLU)**

PReLU is called ReLU function with parameters. These parameters are learnable during the training process and different neurons can have different parameter values. The definition of PReLU is shown in Eq. (4.12):

$$PReLU(x) = \begin{cases} x, x > 0 \\ ax, x \leq 0 \end{cases}.$$ (4.12)

As can be seen from formula (4.11), when $x < 0$, a is the slope of the function, so PReLU is an unsaturated function. If $x = 0$, the PReLU function degenerates to the ReLU function. If x is a small constant, the PReLU can be regarded as a ReLU with leakage. PReLU can allow different neurons to have different parameter values, or a group of neurons share a parameter.

6) **Exponential Linear Unit (ELU)**

ELU is an approximately zero centered nonlinear function, where a is a manually specified hyperparameter and $a \geq 0$. When $x \leq 0$, it can adjust the output mean value to be near 0. The definition of ELU is shown in Eq. (4.13), and the function is shown in Figure 4.7a:

$$ELU = \begin{cases} x, x > 0 \\ a(e^x - 1), x \leq 0 \end{cases}.$$ (4.13)

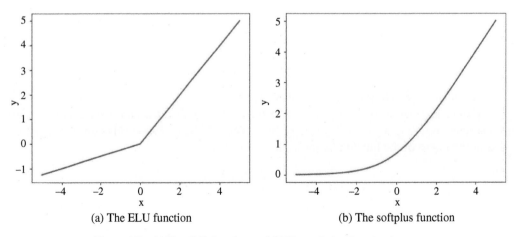

(a) The ELU function   (b) The softplus function

Figure 4.7 (a) The ELU function and (b) The softplus function image.

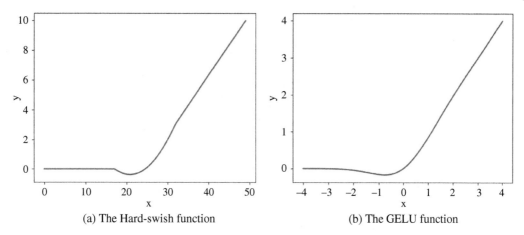

(a) The Hard-swish function   (b) The GELU function

Figure 4.8 (a) The Hard-swish function and (b) The GELU activation function.

7) **Softplus Function**
   Softpls can be regarded as a smooth version of ReLU function. The definition of the softplus function is shown in Eq. (4.14), and the function is shown in Figure 4.7b:

$$Softplus(x) = log(1 + exp(x)) \tag{4.14}$$

   The derivative of the softplus function is the sigmoid function. Although the softplus function also has the characteristics of unilateral inhibition and wide excitation boundary, it does not have the characteristic of sparse activation.

8) **Swish Function**
   This function is a self-gating activation function. The swish function is defined by Eq. (4.15):

$$swish = x\sigma(\beta x), \tag{4.15}$$

   where $\sigma(\cdot)$ is the sigmoid function, and β is a learnable parameter or a fixed hyperparameter. $\sigma(\cdot) \in (0, 1)$ can be regarded as a soft gating mechanism. When $\sigma(\beta x)$ is close to 1, the gate is in the "open" state, and the output of the activation function is similar to x itself; when $\sigma(\beta x)$ is close to 0, the state of the gate is "off," and the output of the activation function is close to 0. The hard swish function is shown in Figure 4.8a.

9) **GELU Function**
   It is also called Gauss error linear element and is a kind of activation function that adjusts the output value through a gate mechanism; it is similar to the swish function. The GELU definition is shown in Eq. (4.16), and the function is shown in Figure 4.8b.

$$GELU(x) = xp(X \leq x). \tag{4.16}$$

10) **Maxout Unit**
   It is also a piecewise linear function. The input of the sigmoid type function, ReLU and other activation functions is the net input z of neurons, which is a scalar. The input of the maxout unit is all the original outputs of the previous layer neurons, which is the vector $x = [x1, x2, \cdots, xD]$.
   Each maxout cell has K weight vectors $w_k \in R^D$ and offset $b_k(1 \leq k \leq K)$. For input x, one can get K net inputs $z_k, 1 \leq k \leq K$.

$$z_k = w_k^T x + b_k, \tag{4.17}$$

   where $w_k = [w_{k,1}, \cdots, w_{k,D}]^T$ is the k-th weight vector.

The nonlinear function of maxout element is defined as:

$$mxout(x) = \max_{k \in [1, k]} (z_k). \tag{4.18}$$

In fact, the maxout model is also a new type of activation function. In the feedforward neural network, the maxout is the maximum value of this layer. In the convolutional neural networks, a maxout feature map can be obtained by taking the maximum value from multiple feature maps. Maxout's fitting ability is very strong. It can fit any convex function.

In the process of practical applications of neural network, we usually normalize the input before using the ReLU activation function. The purpose of normalization is to limit the input of the activation function has a value from 0 to 1, so as to ensure that the output of ReLU is not 0. This normalization usually normalizes the data to a normal distribution.

4.1.3 Single-Layer Perceptron and Linear Network

4.1.3.1 Single Layer Perceptron

Perceptron is an early neural network model proposed by American scholar Frank Rosenblatt in 1957. The concept of learning was first introduced into the perceptron, which simulated the learning function of human brain to a certain extent in mathematics based on symbol processing, and it attracted extensive attention.

A single layer perceptron is the simplest neural network. It includes an input layer and an output layer, and the input layer and the output layer are directly connected. Different from the earliest MP model, its neuron weight is variable, so it can be learned through certain rules. The problem of linear separability can be solved quickly and reliably.

The single-layer perceptron is composed of a linear combiner and a binary threshold element. Each component of the input vector is multiplied by the weight and then is summed in the linear combiner to get a scalar result. The output of the linear combiner is the input of the binary threshold element, and the obtained linear combination result is transmitted from the hidden layer to the output layer through a binary threshold element. In fact, this step performs a symbolic function. The binary threshold element is usually a rising function whose typical purpose is to map nonnegative input values to 1 and negative input values to −1 or 0.

Consider a two-class pattern classification problem: the input is an N-dimensional vector $x = [x_1, x_2, ..., x_N]$, where each component is linked with a weight ω_i. The output of the linear layer is summed to get a scalar value:

$$v = \sum_{i=1}^{N} x_i \omega_i. \tag{4.19}$$

Then, the obtained v value is judged in the binary threshold element to generate a binary output:

$$y = \begin{cases} 1 & v \geq 0 \\ -1 & v < 0 \end{cases} \tag{4.20}$$

A single layer perceptron can divide the input data into two categories: l_1 and l_2. When $y = 1$, it is considered to enter $x = [x_1, x_2, ..., x_N]$ of l_1, when $y = -1$, it is considered as input $x = [x_1, x_2, ..., x_N]$ of l_2. In practical applications, in addition to the input N-dimensional vector, there is an external offset with a constant value of 1 and a weight of b. The structure is shown in Figure 4.9.

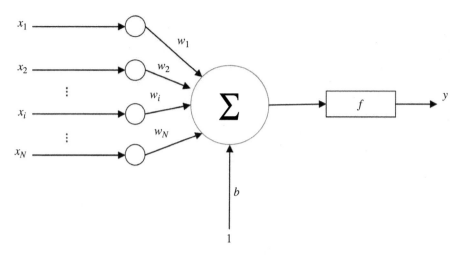

Figure 4.9 Single layer perceptron structure.

In this way, the output $\mathbf{y}$ can be expressed as:

$$y = sgn\left(\omega^{\mathrm{T}}\mathbf{x}\right) \tag{4.21}$$

According to Figure 4.9, the hyperplane of single-layer perceptron is determined by the following formula:

$$\sum_{i=1}^{N} \omega_i x_i + \mathbf{b} = 0. \tag{4.22}$$

From the above formula, when the dimension $N = 2$, the input vector can be expressed as a point in the plane rectangular coordinate system. The classification hyperplane is a straight line:

$$\omega_1 x_1 + \omega_2 x_2 + b = 0. \tag{4.23}$$

Suppose there are three points, which are divided into two classes. The first class includes points (3, 0) and (4, −1), and the second class includes points (0, 2.5). The selected weight is $\omega_1 = 2$, $\omega_2 = 3$, $b = 1$, so the coordinate points on the plane can be divided into two classes by a straight line.

4.1.3.2 Perceptron Learning Algorithm

In the previous example, the selected weight $\omega_1 = 2$, $\omega_2 = 3$, $b = 1$ can well separate the data. In practical applications, it is necessary to use a computer to automatically learn and obtain the correct weight according to the training data. Usually, the learning algorithm of error correction learning rules is adopted. The learning algorithm steps are as follows:

Step1: Define variables and parameters.
- $x(n) = [+1, x_1(n), x_2(n), ..., x_N(n)]^{\mathrm{T}}$, $N + 1$-dimensional input vector;
- $\omega(n) = [b(n), \omega_1(n), \omega_2(n), ..., \omega_N(n)]^{\mathrm{T}}$, $N + 1$-dimensional weight value vector;
- $b(n) =$ bias;
- $y(n) =$ actual output;
- $d(n) =$ expected output;
- $\eta =$ the learning rate parameter, which is a constant that is larger than 0 and smaller than 1.

Step 2: Initialization. $n = 0$, and set the weight vector ω to random or all zero values.

Step 3: Activation. Input training samples. For each training sample, $x(n) = [+1, x_1(n), x_2(n), ..., x_N(n)]^T$, specify its expected output d, i.e. if $x \epsilon l_1$, $d = 1$, if $x \epsilon l_2$, $d = -1$.

Step 4: Calculate the real output.

$$y = sgn\left(\omega^T \mathbf{x}\right),$$

where *sgn* is a symbolic function.

Step 5: Update the weight vector.

$$\omega_{i+1} = \omega_i + (y - d)\mathbf{x}_i.$$

If the perceptron predicts the training sample correctly, that is, $d = y$, the perceptron will not change; otherwise, the weight will be adjusted according to the degree of error.

Step 6: If the convergence condition is met, the algorithm ends; if not, i will increase by 1: $i = i + 1$ and go to step 3 to continue.

Thus, what are the convergence conditions? When the weights of a perceptron can realize accurate classification for samples, the algorithm converges, and the error of the network is zero. In calculation, the convergence conditions can generally be:

- The error is less than a preset smaller value ε. Namely,

$$|d(n) - y(n)| < \varepsilon.$$

- The weight change between the two iterations is very small, that is,

$$|\omega(n + 1) - \omega(n)| < \varepsilon.$$

- Set the maximum number of iterations M, and the algorithm stops iterating after M iterations.

In practical applications, updating weights usually requires the introduction of a new parameter η, namely,

$$\omega_{i+1} = \omega_i + \eta(y - d)\mathbf{x}_i, \tag{4.24}$$

where $\eta \in (0, 1)$ is called the learning rate, The value of η determines the influence of the error on the weight, Notably, the value of η cannot be too large or too small,

- η should not be too large, so as to provide a relatively stable weight estimation for the input vector and avoid the problem of network convergence difficulty.
- η should not be too small, so that the weight can change in real time according to the input vector $\mathbf{x}$, reflecting the correction effect of error on the weight, and avoiding too slow convergence.

In practical applications, for the learning rate η, the selection of variable parameters is often more effective than fixed parameters. That is to say, firstly using a large learning step and then using a small learning step. The change of error value or weight is used to control η during the iterative process. Generally, the change of error value or weight will gradually decrease, and the corresponding learning rate should also gradually decrease; or preset the decrease law of learning rate, such as exponential decline. Using the above method, by controlling η, the change of parameters can realize more effective learning and make the network converge faster.

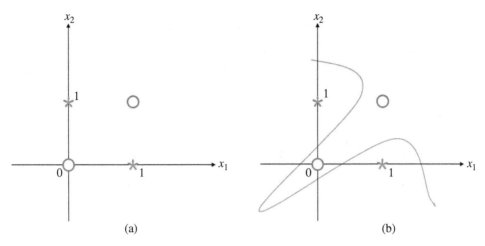

Figure 4.10 Linear inseparable. (a) ○ and ∗ indicate the output from the XOR gate. (b) Using a curve to ○ and ∗.

The single-layer perceptron is not suitable for all binary classification problems. It is only available to linearly separable problems; that is, through learning and adjusting the weights, we can finally find the appropriate decision surface and realize the correct classification. For linearly inseparable problems, the learning algorithm of a single-layer perceptron does not converge and cannot achieve correct classification, as shown in Figure 4.10.

In Figure 4.10(a), ○ and ∗ cannot be distinguished by a single straight line, but it can be realized if we remove the limitation of linearity and replace it by a curve, as shown in Figure 4.10(b).

4.1.3.3 Linear Neural Network

The most typical example of linear neural network is adaptive linear element, which was proposed by Widrow and Hoff in the late 1950s. It is mainly used for pattern association, signal filtering, prediction, model recognition, and control through linear approximation of a functional formula.

A linear neural network and a perceptron network have similar structures but different transfer function, but the neuron transfer function is different. The structure of a linear neural network is shown in Figure 4.11.

As shown in Figure 4.11, in addition to generating binary output, a linear neural network can also generate analog output—that is, using a linear transfer function, the output can be any value.

The first difference between a linear neural network and a perceptron is that the perceptron can only output two possible values, while the output of a linear neural network can take a range of values, and its transfer function is a linear function. The linear neural network adopts the Widrow-Hoff learning rule by utilizing the linear mean squared error (LMSE) algorithm to adjust the weight and bias of the network. A linear neural network is very similar to a perceptron in structure, but the neuron activation function is different. Here, the *purelin* activation function is used for model training, which can get a better effect. The *sign* activation function is adopted when outputting binary result.

The second difference between a linear network and a perceptron is that a perceptron can only solve the classification problem, while a linear neural network can realize the regression problem. As shown in Figure 4.12, points (1,3), (2,2), (3,3), (4,1), and (4,2) are passed through a linear network, and the linear network finally uses a straight line to fit all these points.

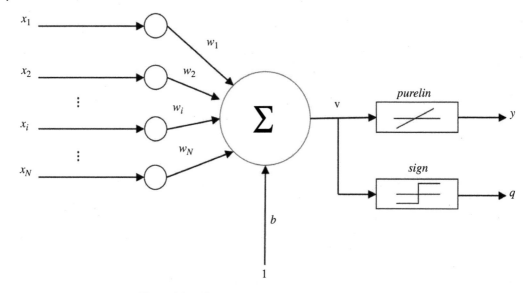

Figure 4.11 The structure of a linear neural network.

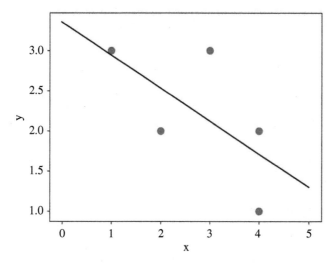

Figure 4.12 Using a linear neural network to realize regression problem.

4.2 Convolutional Neural Network

4.2.1 Convolution and its Application in Images

4.2.1.1 Definition

Convolution is an important operation in mathematical analysis. It is the result of the sum of two variables multiplied in a certain range. In mathematics (especially in functional analysis), convolution is a mathematical operator that generates the third function $(f_a * g)$ through the mathematical operation of two functions (f_a and g), which represents, in continuous domain, the integral of the overlapping part of function value product of function f_a and g after flipping and translation to the overlapping length.

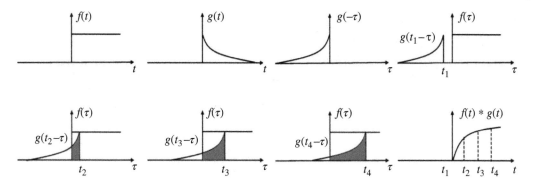

Figure 4.13   The definition of convolution.

Simple definition: Let $f_a(x)$ and $g(x)$ be two integrable functions on R for integration:

$$\int_{-\infty}^{\infty} f_a(\tau)g(x-\tau)d\tau \tag{4.25}$$

It can be proved that the above integral exists for almost all real numbers x. In this way, with different values of x, this integral defines a new function $h(x)$, called the convolution of functions f_a and g, denoted as $h(x) = (f_a * g)(x)$. The definition of convolution is shown in Figure 4.13.

Convolution has been widely used in acoustics, spectroscopy, signal processing, image processing, and computer vision. In signal processing and image processing, one-dimensional convolution and two-dimensional convolution are mainly used.

4.2.1.2   One-Dimensional Convolution in Discrete Domain

One-dimensional convolution is often used in the field of signal processing to calculate the delay accumulation of signals. Let $x[t]$ be the input signal, $h[t]$ be the pulse signal, and t, k be integers; then, the convolution of $x[t]$ and $h[t]$ is defined as:

$$y[t] = x[t] * h[t] = \sum_{k=-\infty}^{\infty} x[k]h[t-k]. \tag{4.26}$$

The $h[t]$ is called a filter or a convolution kernel. The convolution of signal sequence $x[t]$ and filter $h[t]$ is defined as: $y = h * x$, where $*$ represents the convolution operation. In general, the length of the filter is much less than the length of the signal sequence.

Generally, different filters can be designed to extract different features of signal sequences. For example, when the filter is $h = \left[\frac{1}{K}, ..., \frac{1}{K}\right]$, the convolution operation is equivalent to the simple moving average of the signal sequence (the window size is K); when the filter is $h = [1, -2, 1]$, the second-order differentiation of the signal sequence can be approximately realized, that is

$$x''(t) = x(t+1) + x(t-1) - 2x(t). \tag{4.27}$$

Figure 4.14 shows an example of one-dimensional convolution of two filters. It can be seen that the two filters extract different features of the input sequence, respectively. The filter (a) $h = \left[\frac{1}{2}, \frac{1}{4}, \frac{1}{2}\right]$ can detect the low-frequency information in the signal sequence, while the filter (b) $h = [1, -2, 1]$ can detect the high-frequency information in the signal sequence.

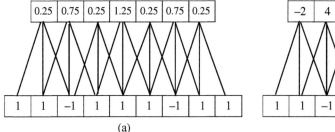 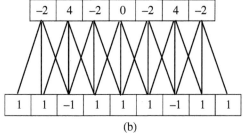

(a)   (b)

Figure 4.14 one-dimensional convolution example. (a) employs the filter $h = [1/2,1/3,1/2]$ and (b) employs the filter $h = [1,-2,1]$

Let $y(t) = h(t) * x(t)$. If a is any real or complex constant, the convolution has the following properties:

1) Commutative law: $x(t) * h(t) = h(t) * x(t)$
2) Distribution law: $x(t) * [g(t) + h(t)] = x(t) * g(t) + x(t) * h(t)$
3) Associative law: $[x(t) * g(t)] * h(t) = x(t) * [g(t) * h(t)]$
4) Combination law of number multiplication: $a[x(t) * h(t)] = [ax(t)] * h(t) = x(t) * ah(t)$, where a is a scalar.
5) Translation properties: $x(t - t_1) * h(t - t_2) = y(t - t_1 - t_2)$
6) Differential characteristics: $y'(t) = x'(t) * h(t) = x(t) * h'(t)$
7) Integral characteristic: $y^{(-1)}(t) = x^{(-1)}(t) * h(t) = x(t) * h^{(-1)}(t)$
8) Equivalent characteristic: $y(t) = x^{(-1)}(t) * h'(t) = x'(t) * h^{(-1)}(t)$

4.2.1.3 Two-Dimensional Convolution in Discrete Domain

In image processing, an image is input into a neural network in the form of a two-dimensional matrix, so we need a two-dimensional convolution. Given an image $X \in R^{M \times N}$ and a filter $H \in R^{U \times V}$, generally $U \ll M$, $V \ll N$, and its convolution is:

$$y_{ij} = \sum_{u=1}^{U} \sum_{v=1}^{V} h_{uv} x_{i-u+1,j-v+1}. \tag{4.28}$$

The two-dimensional convolution of input information X and filter H is defined as:

$$Y = H * X. \tag{4.29}$$

An example of two-dimensional convolution is shown in Figure 4.15. Two-dimensional convolution is an extension of one-dimensional convolution, and their principle is similar. First, a

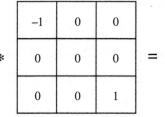

Figure 4.15 Two-dimensional convolution example.

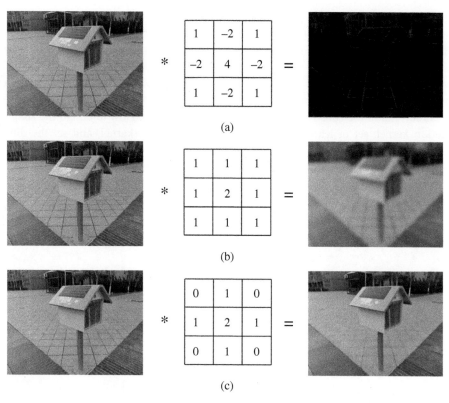

Figure 4.16 Several common filters in image processing. (a) An ideal high-pass filter. (b) A Gaussian smoothing filter. (c) A morphological corrosion operation.

convolution kernel is translated on an image, and then we perform the multiplication operation of the convolution kernel at the corresponding window in the image. Finally, a sum operation is performed and the result is considered as the convolution result at the position in the image.

In image processing, convolution is often used as an effective method of feature extraction. The result of an image after convolution is called feature map. Figure 4.16 shows several filters commonly used in image processing and their corresponding feature mapping. Figure 4.16(a) shows an ideal high-pass filter, which can be used to extract and enhance the edge of the image; Figure 4.16(b) shows a Gaussian smoothing filter, which can be used to smooth the noise in an image.

Convolution has been widely used in digital image processing, such as image filtering, image edge detection, and image enhancement. Convolution is very popular in image processing due to two reasons. Firstly, convolution can extract local features of images and then carry out various operations on images; secondly, the natural instinct of the convolution operation is convenient for the conversion of time domain and frequency domain in image processing. In a convolution neural network, using a large number of convolution kernels to extract the deep features of images is of great significance for advanced tasks such as image classification and semantic understanding.

In traditional image processing and analysis, we usually give a fixed convolution kernel for image feature extraction. This artificial feature engineering can lead to the extracted image features being independent of the classes, which makes the extracted features not conducive to subsequent image analysis, such as target detection, classification and recognition. The deep convolution neural network is superior to the traditional artificial feature engineering because the convolution kernel is learned from the data itself, and the number of convolution kernels is huge. The obtained image features that can be obtained can effectively represent the high-level semantic information of the image itself, especially for advanced image analysis (target detection, classification and recognition). Figure 4.17 shows an example of the result of feature extraction using different convolution.

Obviously, convolution plays a very important role in the field of image processing. The convolution operation directly determines the quality of feature extraction, and the core of the convolution operation lies in the construction of the convolution kernel. How to design the shape of the convolution kernel, set or learn the parameters of convolution kernel, and set the dimension and number of the convolution kernel will affect the final feature extraction result. Therefore, convolution is one of the key research contents in image engineering.

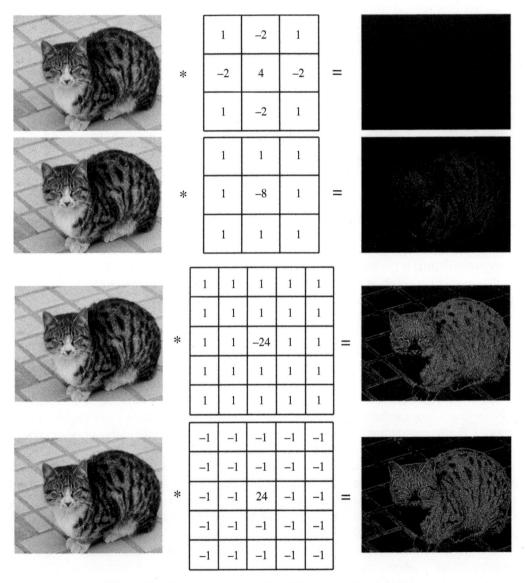

Figure 4.17 Feature extraction using different convolution kernels.

4.2.1.4 Extended Convolution Operation

1) Point Convolution

Point convolution [6] is also called 1×1 convolution. Because the smallest window is used, 1×1 convolution loses the function of convolution layer on adjacent elements in a feature map. In fact, 1×1 convolution is mainly calculated in the channel dimension. Figure 4.18 shows that if we use 3 channels and 2 output channels 1×1 cross correlation calculation of convolution kernel. It is worth noting that the input and output have the same height and width. Each element in the output comes from the weighted accumulation between different channels of elements at the same position of feature maps. Suppose we take the channel dimension as the feature dimension and the elements at the high and wide dimensions as the data samples; then the function of 1×1 convolution layer is equivalent to that of full connection layer.

2) Channel Convolution

   1) Single-Channel Convolution

   In deep learning, convolution is a process in which elements are multiplied first and then added. For an image with one channel, the convolution is shown in Figure 4.19. The filter is a 3×3 matrix, whose elements are [[0,1,2], [2,2,0], [0,1,2]]. The filter slides from left to right at the input and performs element multiplication and addition at each position. The final output is a 3×3 matrix.

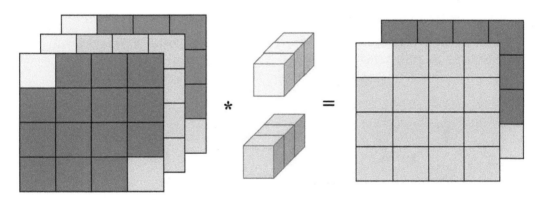

Figure 4.18 Point convolution.

3 ×0	3 ×1	2 ×2	1	0
0 ×2	0 ×2	1 ×0	3	1
3 ×0	1 ×1	2 ×2	2	3
2	0	0	2	2
2	0	0	0	1

*

0	1	2
2	2	0
0	1	2

=

12	12	17
10	17	19
9	6	74

Figure 4.19 Single-channel convolution.

2) Multichannel Convolution

In CNNs, multi-channel convolution means that multiple filters are employed at a convolutional layer. The convolution network layer is usually composed of multiple channels (hundreds of convolution kernels), and each channel extracts the abstract features of different aspects of the previous layer.

Multichannel convolution is shown in Figure 4.20, in which different convolution kernels are applied to input channels of the previous layers to generate an output channel. We repeat this process for all convolution kernels to generate multiple channels. Each of these channels is then added together to form a single output channel.

The input layer is a $5 \times 5 \times 3$ matrix with three channels. The filter is a $3 \times 3 \times 3$ matrix. Firstly, each kernel in the filter is applied to the three channels of the input layer and added. Then, a cubic convolution is performed to produce three sizes of 3×3 channels.

In the first step of multichannel 2D convolution, each kernel in the filter is applied to three channels of the input layer.

In the second step of multi-channel 2D convolution, the three channels are added together (element by element) to form a final single channel.

3) 3D Convolution

A 3D filter can move in three directions (image height, width, channel). At each position, multiplication and addition are performed pixel-by-pixel to get a number. Since the filter can slide in 3D space, the output numbers are arranged in 3D space, and the output of 3D convolution is 3D data. An example of 3D convolution is shown in Figure 4.21.

Similar to the spatial relationship of objects in a 2D convolution, a 3D convolution [7] can describe the spatial relationship of objects in 3D space. This 3D relationship has important applications, such as 3D segmentation/reconstruction in biomedical images, CT, and MRI.

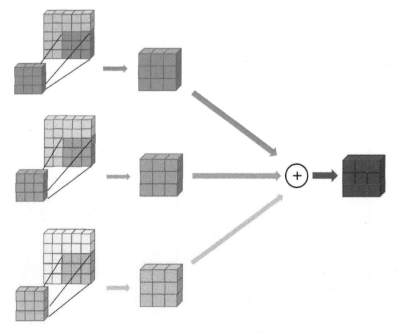

Figure 4.20 Multichannel convolution.

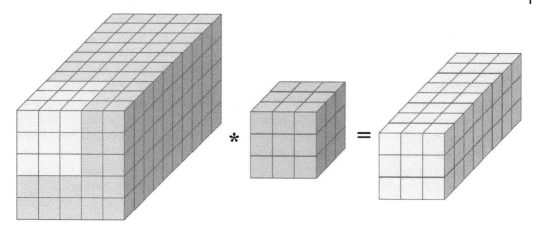

Figure 4.21   3D convolution.

4.2.2   Convolutional Network Architecture and Parameter Learning

4.2.2.1   Convolutional Network Architecture

A CNN is a kind of feedforward neural network with depth structure including convolution calculation. It is one of the representative algorithms of deep learning. A CNN has the ability of representation learning and can classify the input information according to its hierarchical structure. Therefore, it is also called "translation-invariant artificial neural network."

A CNN is essentially a multilayer perceptron. The key reason for its success lies in the ways of local connection and weight sharing. On the one hand, it reduces the number of weights to achieve network optimization. On the other hand, there is a risk of overfitting. The advantages of convolution networks are more obvious for multidimensional image processing. Because an image can be directly used as the input of the network, the complex process of feature extraction and data reconstruction in traditional recognition algorithms is avoided. It has many advantages in two-dimensional image processing, such as automatically extracting image features, including color, texture, shape, and image topology. It also has good robustness and computational efficiency in dealing with two-dimensional images, especially for the applications such as recognizing displacement, scaling, and other forms of distortion invariance.

Convolution neural network is a multilayer supervised learning neural network. The convolution layers and pooling layers in the hidden layers are the core modules to realize the feature extraction function of a CNN. The network model adopts the gradient descent method to minimize the loss function, adjusts the weight parameters layer by layer, and improves the accuracy of the network through frequent iterative training. The low-level structure of a CNN is usually composed of a convolution layer and a pooling layer alternately. The high-level structure is the full connection layer, which corresponds to the hidden layer and logistic regression classifier of traditional multilayer perceptron. The input of the first full connection layer is the feature image obtained by feature extraction from the convolution layer and sub-sampling layer. The last output layer is a classifier, which can classify the input feature map by logistic regression, softmax regression, support vector machine, and so forth.

The structure of a CNN includes a convolution layer, a pooling layer, and a full connection layer. Each layer has multiple feature maps, each feature map extracts an input feature through a convolution filter, and each layer has multiple neurons.

A traditional neural network requires the input data to be a set of high-dimensional vectors. Obviously, the spatial structure information of the image will be ignored when processing a digital image based on the traditional neural network, which limits the application of neural network in digital image processing. Because digital image processing mainly depends on the convolution operation, how to introduce convolution into neural network is the key to solve this problem. Yann LeCun first proposed the first-generation CNN, LeNet, in 1998 and successfully applied it to postal code number recognition. LeNet has a far-reaching impact on the CNN in the era of deep learning. Its structure is shown in Figure 4.22. The network realizes the extraction of complex features by stacking a large number of convolution layers, effectively reduces the resolution of the feature map by using pooling operation, so as to increase the receptive field range of the convolution kernel, and finally realizes the classification of the feature map by using a full connection layer.

Obviously, the characteristic of a CNN is that the neurons of each layer are only connected with some neurons of the previous layer. It is a locally connected, weight-sharing artificial neural network. This kind of network has obvious advantages for image-based data analysis and recognition. Therefore, CNNs have become an important cornerstone for deep learning to achieve revolutionary results in the field of computer vision. At present, most CNNs used in image classification tasks adopt the structural paradigm of LeNet.

Local connection means that the nodes of the convolution layer are only connected with some nodes of the previous layer, which is only used to learn local features. The idea of local perceptual structure comes from the cortical structure of animal vision. Only some neurons play a key role during the process of perceiving external objects. In computer vision, it is a wildly acceptable view that, the correlation between pixels is also related to the distance between pixels. The correlation between neighborhood pixels is strong, but the correlation is weak when one pixels is far away from the pixels. It can be seen that the local correlation theory is also applicable to the field of image processing in computer vision. Therefore, local perception uses some neurons to receive image information and then enhances image information by synthesizing all image information. This local connection (as shown in Figure 4.23(b)) greatly reduces the number of parameters, speeds up the learning rate, and reduces the possibility of overfitting to a certain extent.

Weight sharing means that the convolution kernel h as a parameter is the same for all neurons. As shown in Figure 4.23(c), the weights on all connections of the same color are the same. Weight sharing can be understood as a convolution kernel that captures only local features from the input data. Therefore, if we want to extract multiple features, we need to use multiple different convolution kernels.

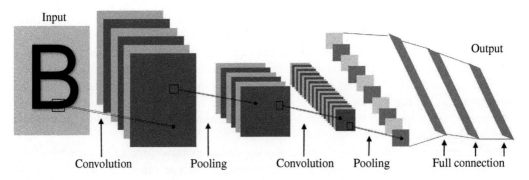

Figure 4.22 LeNet structure diagram.

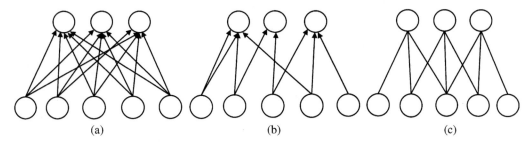

Figure 4.23 Structure diagram of full connection, local connection, and weight sharing. (a) Full connection. (b) Local connection. (c) Shared weights.

4.2.2.2 Convolution Layer

The function of the convolution layer is to extract the features of the input data. It contains multiple convolution kernels, and each element of the convolution kernel corresponds to a weight coefficient and an offset, which is similar to the neuron of a feedforward neural network. Each neuron in the convolution layer is connected with multiple kernel in the area close to the previous layer. The size of the area depends on the size of the convolution nucleus, which is called "receptive field" and its meaning can be similar to the receptive field of visual cortical cells. When the convolution kernel works, it will scan the input features regularly, multiply and sum the input features in the receptive field, and superimpose the offset.

The receptive field is usually determined by the size of the convolution kernel. Generally, a too small receptive field will make it difficult to obtain the rich context information of the image, and a too large receptive field will lead to too much parameters and lose the significance of local feature extraction. Therefore, the size of the convolution kernel is usually set based on experience and task requirements.

In convolution neural network, for a typical convolution layer, the relationship between the input tensor size $W_1 \times H_1 \times C_1$ and output characteristic diagram tensor dimension $W_2 \times H_2 \times C_2$ is determined by four hyperparameters, namely, the number of convolution kernels K, the space size F of convolution kernels, the sliding step size S^c of convolution kernels, and the filling size P of input tensors. Their physical meanings are as follows:

1) Number of convolution kernels K: A convolution layer usually contains 2^n convolution kernels, which are used to extract feature maps of the input tensor. Generally, the more convolution kernels corresponds to the stronger feature representation ability. The number of convolution kernels determines the number of channels of the output feature maps.
2) Convolution kernel space size F: The hyperparameter represents the width or height of the convolution kernel. Generally, the width and height of the convolution kernel are the same, and they are odd pixels, where the size of the convolution kernel is $F_1 \times F_1 \times C_1$.
3) Convolution kernel sliding step S^c: The sliding step determines the moving range of the convolution kernel in the two dimensions of the width and height of the input tensor. In order to ensure that the convolution kernel can traverse all positions of the input tensor, the value of the sliding step is usually set not to be greater than the spatial size of the convolution kernel.
4) Filling size P of the tensor: The hyperparameter determines the width of filling zero value around the input tensor. It is typically used in combination with the convolution kernel space size F and the sliding step size S^c to flexibly control the space size of the output feature maps.

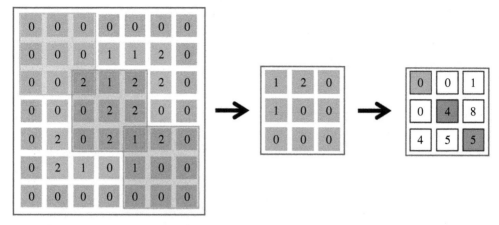

The input feature map   The convolution kernel   The output feature map

Figure 4.24 Image convolution.

The width W_2, height H_2, and number of channels C_2 of the output feature maps are calculated by the following three formulas:

$$W_2 = \frac{(W_1 - F + 2P)}{S^c} + 1, \tag{4.30}$$

$$H_2 = \frac{(H_1 - F + 2P)}{S^c} + 1, \tag{4.31}$$

$$C_2 = K. \tag{4.32}$$

A typical convolution process is shown in Figure 4.24. Input tensor parameter: $W_1 = H_1 = 5$, convolution kernel $F = 3$, sliding step of convolution kernel $S^c = 2$, filling size of input tensor $P = 1$. By performing convolution operation, output feature map parameters are: $W_2 = H_2 = 3$.

4.2.2.3 Pooling Layer

A pooling layer usually appears between several continuous convolution layers. A pooling layer converges the local information of the feature map to achieve the effect of feature dimension reduction. On the one hand, it reduces the redundant components and improves the robustness of the feature maps; on the other hand, it reduces the spatial size of the feature maps, controls the number of parameters and calculation lost of CNN, and can avoid the occurrence of model overfitting. The process of pooling is similar to convolution, that is, a pooling kernel traverses the whole input tensor from top to bottom and from left to right and acts with the input tensor to produce the pooled feature maps. Generally, the size of the pooling window is even, and the sliding mode without overlap and separation is adopted. In practice, the setting of $F = 2$ and $S^c = 2$ is popular.

Max pooling and mean pooling are the two most commonly used pooling methods.

In the cases of maximum pooling, the pooling kernel selects the largest element from its overlapping input tensor and records it to the corresponding position of the output feature maps. The characteristic of maximum pooling is that it can effectively suppress the estimated mean shift caused by network parameter error, so it can better retain texture information. For a region

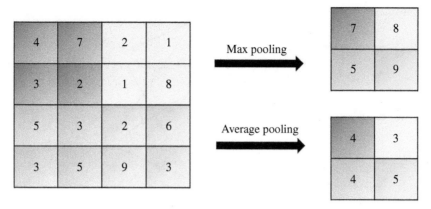

Figure 4.25 Max pooling operation and average pooling.

$R_{m,n}^d$, the maximum element value of all neurons in this region is selected as the representation of this region, i.e.

$$y_{m,n}^d = \max_{i \in R_{m,n}^d} x_i. \tag{4.33}$$

However, the average pooling process is to calculate the average value of the input tensor elements that overlap with the pooling kernel, i.e.

$$y_{m,n}^d = \frac{1}{\left| R_{m,n}^d \right|} \sum_{i \in R_{m,n}^d} x_i \tag{4.34}$$

Its characteristic is that it can suppress the large variance of the estimated value caused by the limited size of the pooling kernel area and better retain the background information. Figure 4.25 shows a 2×2 maximum pooling kernel and a 2×2 average pooling kernel in a 4×4 pooling process on the input tensor.

A typical pooling layer divides each feature map into a non-overlapping area of size 2×2, and then use the maximum pooling method for downsampling.

4.2.2.4 Full Connection Layer

The full connection layer is the basis of the artificial neural network as described in the previous chapter, which is used as a feature classifier in CNNs. In the early CNNs such as AlexNet and VGG, the feature map obtained by the feature extractor is first expanded into a one-dimensional tensor, then sent to several full connection layers for nonlinear mapping. As a result, the tensor is converted into a one-dimensional tensor with the same length as the number of categories of classification tasks. Finally, the activation is performed by the softmax layer to obtain the probability value corresponding to each category, so as to achieve the purpose of image classification. However, on the one hand, the full connection layer introduces a large number of trainable parameters for CNNs, which increases the training difficulty and overfitting risk of the network. On the other hand, the full connection layer requires its input tensor to have the same length, which means that the size of the input image of CNN must also be consistent; otherwise, the forward propagation process will not be completed. The structure diagram of fully connected neural network is shown in Figure 4.26.

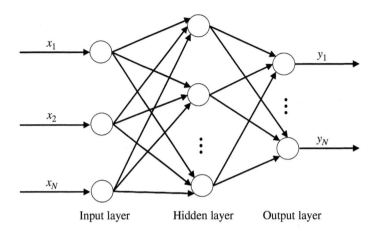

x_1

y_1

x_2

y_N

x_N

Input layer   Hidden layer   Output layer

Figure 4.26 Structure diagram of fully connected neural network.

ResNet [8] and subsequent networks adopt a more flexible way, and its feature classifier is composed of global average pooling layer and point rolling layer. Let the size of the feature map output by the feature extractor be $W_o \times H_o \times C_o$. The global average pooling layer is responsible for converting the feature map size to $1 \times 1 \times C_o$. That is, the number of channels remains unchanged, but the space size is reduced to only unit width and height. The point convolution layer is responsible for nonlinear mapping of the transformed feature map, and the size of the feature map is changed from $1 \times 1 \times C_o$ to $1 \times 1 \times N$, where N is the number of categories, and the probability distribution on each category is finally activated by the softmax layer. The global average pooling operation is independent of the spatial size of the feature map, so the forward propagation is no longer limited by the size of the input image, which not only expands the range of available images but also reduces the workload of image preprocessing. The point convolution layer makes full use of the characteristics of local connection of convolution operation, greatly reduces the trainable parameters of feature classifier compared with the full connection layer, reduces the network scale, and effectively avoids the occurrence of overfitting.

4.2.2.5 Parameter Learning

In CNNs, the parameters are the weight and bias existing in convolution kernels. Similar to fully connected feedforward networks, convolutional networks can also learn parameters through an error back-propagation algorithm.

In the fully connected feedforward neural network, the gradient mainly passes through the error term of each layer *err*. Back propagation is carried out, and the gradient of each layer parameter is further calculated.

In convolution neural networks, there are mainly two neural layers with different functions: the convolution layer and the pooling layer. The parameters are the convolution kernel and bias, so only the gradient of the parameters in the convolution layer needs to be calculated.

Without loss of generality, layer l is a convolution layer, and the input characteristics of layer $(l-1)$ are mapped as $\chi^{(l-1)} \epsilon R^{M \times N \times D}$, the net input $Z^{(l)} \epsilon R^{M' \times N' \times P}$ of the feature map of layer l is obtained through a convolution calculation, and the number $p(1 \leq p \leq P)$ feature mapping net input of layer l is:

$$Z^{(l,p)} = \sum_{d=1}^{D} F^{(l,p,d)} \otimes X^{(l-1,d)} + b^{(l,p)} \tag{4.35}$$

where $F^{(l,p,d)}$ and $b^{(l,p)}$ are convolution kernels and biases. There are $P \times D$ convolution kernels and P offsets in layer l, and the chain rule can be used to calculate their gradients respectively.

Supposing $Y = F \otimes X$, where $X \in R^{M \times N}$, $F \in R^{U \times V}$, $Y \in R^{(M-U+1) \times (N-V+1)}$, and the function $f_a(Y) \in R$ is a scalar function. According to the chain rule, the partial derivative of $f_a(Y)$ with respect to Y is the convolution of X and $\dfrac{\partial f_a(Y)}{\partial Y}$. $\dfrac{\partial f_a(Y)}{\partial F} = \dfrac{\partial f_a(Y)}{\partial Y} \otimes X$ and the net input formula of the above feature mapping, the partial derivative of the loss function $\mathcal{L}$ about the convolution kernel $F^{(l,p,d)}$ of layer l is

$$\frac{\partial \mathcal{L}}{\partial F^{(l,p,d)}} = \frac{\partial \mathcal{L}}{\partial Z^{(l,p)}} \otimes X^{(l-1,d)} = err^{(l,p)} \otimes X^{(l-1,d)} \tag{4.36}$$

Note that in the formula above, $err^{(l,p)} = \dfrac{\partial \mathcal{L}}{\partial Z^{(l,p)}}$ is the partial derivative of the net input $Z^{(l,p)}$ of the loss function with respect to the P-th feature map of layer l. Similarly, it can be obtained that the partial derivative of the loss function with respect to the P-th offset $b^{(l,p)}$ of the l-th layer is

$$\frac{\partial \mathcal{L}}{\partial b^{(l,p)}} = \sum_{i,j} \left[err^{(l,p)} \right]_{i,j} \tag{4.37}$$

In convolution networks, the gradient of each layer's parameters depends on the error term of its layer $err^{(l,p)}$, $\otimes$ represents cross-correlation operation.

4.2.2.6 Back-Propagation Algorithm

The calculation of error terms in the convolution layer and pool layer is different, so the error terms are calculated respectively.

1) The pooling layer: when layer $(l + 1)$ is a pooling layer, because the pooling layer is a downsampling operation, the error term err of each neuron in layer $(l + 1)$ corresponds to a region of the corresponding feature map of layer l. Each neuron in the p-th feature map of layer l has an edge connected to a neuron in the p-th feature map of layer $(l + 1)$. According to the chain rule, the error term $err^{(l,p)}$ of an feature mapping of layer l can be obtained only by upsampling the error term $err^{(l+1,p)}$ of the corresponding feature mapping of layer $(l + 1)$ (same as the size of layer l), and then multiplying the partial derivative of the activation value of the feature mapping of layer l element by element.

The specific derivation process of the error term $err^{(l,p)}$ of the p feature map of the l layer is as follows:

$$err^{(l,p)} \triangleq \frac{\partial \mathcal{L}}{\partial Z^{(l,p)}} \tag{4.38}$$

$$= \frac{\partial X^{(l,p)}}{\partial Z^{(l,p)}} \frac{\partial Z^{(l+1,p)}}{\partial X^{(l,p)}} \frac{\partial \mathcal{L}}{\partial Z^{(l+1,p)}} \tag{4.39}$$

$$= f_l'\left(Z^{(l,p)} \right) \odot \mathrm{up}\left(\delta^{(l+1,p)} \right), \tag{4.40}$$

where $f_l'(\cdot)$ is the derivative of the activation function used in layer l, and $\mathrm{up}(\bullet)$ is the upsampling function, which is just opposite to the downsampling operation used in the pooling layer. If the downsampling is maximum pooling, the error term $err^{(l+1,p)}$ will be directly transferred to the neuron corresponding to the maximum value in the corresponding area of the upper layer, and the error term of other neurons in the area is set to 0. If the downsampling is average pooling, the error term $err^{(l+1,p)}$ will be evenly distributed to all neurons in the corresponding area of the previous layer.

2) Convolution layer: when layer $(l+1)$ is a convolution layer, we assume the network input of feature mapping is $Z^{(l+1)} \epsilon R^{M' \times N' \times P}$, where the p-th $(1 \leq p \leq P)$ feature map is the network input

$$Z^{(l+1,p)} = \sum_{d=1}^{D} F^{(l+1,p,d)} \otimes X^{(l,d)} + b^{(l+1,p)}, \tag{4.41}$$

where both $F^{(l+1,p,d)}$ and $b^{(l+1,p)}$ are convolution kernels and offsets of layer $(l+1)$, and there are $P \times D$ convolution kernels and P offsets in layer $(l+1)$.

The error term $\delta^{(l,d)}$ corresponds to the d-th feature mapping of the l-th layer. The specific derivation process is as follows:

$$err^{(l,d)} \triangleq \frac{\partial \mathcal{L}}{\partial Z^{(l,d)}}, \tag{4.42}$$

$$= \frac{\partial X^{(l,d)}}{\partial Z^{(l,d)}} \frac{\partial \mathcal{L}}{\partial X^{(l,d)}}, \tag{4.43}$$

$$= f_l'\left(Z^{(l,d)}\right) \odot \sum_{P=1}^{P} \left(\text{rot180}\left(W^{(l+1,p,d)} \widetilde{\otimes} \frac{\partial \mathcal{L}}{\partial Z^{(l+1,p)}} \right) \right), \tag{4.44}$$

$$= f_l'\left(Z^{(l,d)}\right) \odot \sum_{P=1}^{P} \left(\text{rot180}\left(W^{(l+1,p,d)} \widetilde{\otimes} err^{(l+1,p)} \right) \right), \tag{4.45}$$

where $\widetilde{\otimes}$ is used for wide convolution, and rot180$(\cdot)$ represents a rotation of $180°$.

4.3 Graph Convolutional Network

4.3.1 Overview

In recent years, deep learning technology has gradually become a research hotspot and mainstream development direction in the field of artificial intelligence. It is mainly used in non-Euclidean data processing with regular distribution of high-dimensional features and has made remarkable achievements in the fields of image processing, speech recognition, and semantic understanding.

The concept of graph originated from the famous Königsberg Seven Bridges problem in the eighteenth century. By the middle of the twentieth century, research on matroid theory, hypergraph theory, and extreme graph theory developed vigorously, making graph theory an important mathematical research before the birth of electronic computing. As a relational data structure, the application of graphs in deep learning has received more and more attention in recent years. The evolution of graphs is divided into three stages: mathematical origin, computing application, and neural network extension. With the advent of computers and the development of the era of machine learning, graphs have been widely used as an important data structure that can effectively and abstractly express the entities in information and data and the relationships between entities. Graph databases have effectively solved many traditional database problems such as modeling defects as well as slow calculation speed in a large amount of complex data. Graph databases have also become a hot research direction. Graph structures can connect structured data points through edges and connect data nodes of different types and structures according to the relationship between data. Therefore, graph structures have been are widely used in data storage, retrieval, and calculation.

Based on graph structure data, the knowledge graph can accurately describe the relationship between entities in the real world through the semantic relationship between points and edges, including research directions such as knowledge extraction, knowledge reasoning, and knowledge graph visualization. Graph computing has the characteristics of large data scale, low locality, and high computing performance, and its algorithms are mainly divided into three categories: path search algorithms, centrality algorithms, and community discovery algorithms. They are time-sensitive on large-scale data with complex relationships and have strong and accurate performance on important applications in the fields of social networking, anti-fraud system, and user recommendation.

The research of graph neural networks [9] mainly focuses on the propagation and aggregation of adjacent node information. As an emerging technology in the field of artificial intelligence in recent years, graph neural networks have opened up a new space for processing complex graph structure data. With the help of artificial intelligence technologies such as deep learning and reinforcement learning, graph neural networks can quickly mine topological information and complex features in graph structures and have solved many major problems in the fields of computer vision, recommendation systems, and knowledge graphs. Therefore, the combination of graph neural networks and future convolutional neural networks is an important way to solve network optimization problems, enhance network reliability, and improve network resource utilization. In response to other problems of graph neural networks, researchers have given many solutions. With more in-depth research and exploration in the field of graph neural networks, the field of artificial intelligence will be further expanded.

4.3.2 Convolutional Network over Spectral Domains

In the field of signal processing, the Fourier transform of function convolution is the convolution of function Fourier transform, namely:

$$F\{f^*g\} = F\{f\} \cdot F\{g\} = \hat{f} \cdot \hat{g}, \tag{4.46}$$

where $F\{f\}$ represents the corresponding spectral domain signal $\hat{f}$ obtained by the Fourier transform of f.

Through the inverse Fourier transform F^{-1}, the convolution form can be obtained as follows:

$$f^*g = F^{-1}\{F\{f\} \cdot F\{g\}\}. \tag{4.47}$$

Given a graph G with n nodes, if the Laplace matrix L can be equally decomposed as $U\Lambda U^T$, then for the graph signal x, its graph-theoretic Fourier transform is $F(x) = U^T x$, and its inverse graph-theoretic Fourier transform is $F^{-1}(x) = Ux$. Put it into formula (4.47), and the convolution operation of graph signal x with a filter g can be obtained:

$$x^*g = U(U^T x \odot U^T g) \tag{4.48}$$

where $\odot$ represents the Hadamard Product. According to the above formula, we can take the whole $U^T g$ as a convolution kernel θ, then there is:

$$x^*f = U(U^T x \odot \theta) = U(\theta \odot U^T x) = Ug_\theta U^T x \tag{4.49}$$

where, g_θ is a diagonal matrix whose diagonal element is θ:

$$g_\theta = diag(\theta) = \begin{pmatrix} \theta_1 & \cdots & \\ \vdots & \ddots & \vdots \\ & \cdots & \theta_n \end{pmatrix} \tag{4.50}$$

For Formula 4.49, we can view it as a three-step transformation of a graph signal x:

1) Take the graph signal x in the spatial domain for the graph theory Fourier transform, and get $F(x) = U^T x$.
2) The convolution kernel g_θ is defined in the spectral domain, and the spectral signal is transformed to get $g_\theta U^T x$.
3) The spectral domain signal is transformed into the spatial domain signal $F^{-1}(g_\theta U^T x) = U g_\theta U^T x$ by graph theory inverse Fourier transform.

Finally, a concise form of graph convolution can be obtained:

$$g * x = U g U^T x = U \begin{pmatrix} \theta_1 & \cdots & \\ \vdots & \ddots & \vdots \\ & \cdots & \theta_n \end{pmatrix} U^T x. \tag{4.51}$$

It is also necessary to extend the definition of graph convolution above from an n-dimensional graph signal x to an $(n \times d)$-dimensional graph node feature matrix X. If the state of the node at layer l is X^l and its dimension is $n \times d_l$, then the state of the node can be updated as:

$$x_j^{l+1} = \sigma\left(U \sum_{i=1}^{d_l} F_{i,j}^l U^T x_i^l\right), j = 1, \cdots, d_{l+1}, \tag{4.52}$$

where x_i^l is the i-th column of matrix X^l, that is, the i-th dimensional graph signal; $F_{i,j}^l$ corresponds to the convolution kernel of the l-layer and i-dimension graph signal (x_i^l), i.e. g_θ in Eq. (4.51). If the node states at the next level have dimension $n \times d_{l+1}$, then the number of convolution kernels at this level is $d_l \times d_{l+1}$.

If $F_{i,j}^l$ is a learnable parameter, then we get the early spectral domain graph convolution network. The grid data represented by images can be regarded as a special graph structure, so the graph convolution can also be used on the grid data. In this case, $U F_{i,j}^l U^T$ corresponds to a convolution kernel in CNNs.

Although this early model points the way for graph convolution over spectral domains, it still leaves much to be desired:

1) The above graph convolution process needs to calculate the eigenvector of the graph Laplacian matrix. The computational complexity of this operation is $O(n^3)$, where n is the number of nodes. When the graph is large, the computational cost generated by this operation is high. In addition, UF needs to be calculated in each forward propagation process $U F_{i,j}^l U^T$; this kind of matrix operation also has a high computational cost.
2) Each layer of the network needs $n \times d_l \times d_{l+1}$ parameters to define the convolution kernel. When the graph is large, the number of parameters of the network is huge.
3) This kind of spectral domain convolution method has no clear meaning in spatial domain and can not be clearly localized to a fixed point.

4.3.3 Chebyshev Network

Defferrard et al. [10] proposed a new spectral domain graph convolutional network, which has the advantages of fast localization and low complexity. Because the network uses Chebyshev polynomial expansion approximation, the network thus is also called Chebyshev network (ChebyNet).

From the perspective of graph signal analysis, we hope that g in formula 4.51 has good localization characteristics, that is, it affects nodes in a smaller area around the graph node, so we can define g as a Laplacian matrix. The function $g_\theta(L)$, because the Laplacian matrix is applied once, is equivalent to spreading the information to the adjacent points with a distance of 1 on the graph.

The result obtained after the signal x is filtered by this filter, and the output result is:

$$y = g_\theta(L)x = g_\theta(U\Lambda U^T)x = Ug_\theta(\Lambda)U^T x. \tag{4.53}$$

That is, the convolution kernel g_θ can be regarded as the function $g_\theta(\Lambda)$ of the eigenvalue Λ of the Laplacian matrix. Usually, you can choose a polynomial convolution kernel:

$$g_\theta(\Lambda) = \sum_{k=0}^{K} \theta_k \Lambda^k, \tag{4.54}$$

where the parameter θ_k is the coefficient of the polynomial. It can be seen from Eq. (4.54) that only $K+1$ parameters ($K \ll n$) are needed at this time, and obviously the number of parameters has been greatly reduced. In addition, it can be seen from Eq. (4.53) that this is equivalent to the definition of $g_\theta(L) = \sum_{k=0}^{K} \theta_k L^k$, that is, the information can be propagated at most K steps at each node, thus making the graph. The product has the characteristics of localization.

ChebyNet proposes a further acceleration scheme on this basis, approximating $g_\theta(\Lambda)$ as the K-order truncation of the Chebyshev polynomial:

$$g_\theta(\Lambda) = \sum_{k=0}^{K} \theta_k T_k(\widetilde{\Lambda}), \tag{4.55}$$

where T_k is a Chebyshev polynomial of order k, and $\widetilde{\Lambda} = 2\Lambda_n/\lambda_{max} - I_n$ is a diagonal matrix whose main purpose is to map the diagonal matrix of eigenvalues to the interval $[-1,1]$. The Chebyshev polynomial is used because it can be solved recursively:

$$T_k(x) = 2xT_{k-1}(x) - T_{k-2}(x) \tag{4.56}$$

Starting from the initial value $T_0 = 1$, $T_1 = x$, using the above recursive formula, the value of k-th order T_k can be easily obtained.

To avoid eigenvalue decomposition, we write Eq. (4.53) back as a function of L:

$$y = U \sum_{k=0}^{K} \theta_k T_k(\widetilde{\Lambda}) U^T x = \sum_{k=0}^{k} \theta_k T_k(\widetilde{L})x, \tag{4.57}$$

where $\widetilde{L} = 2L/\lambda_{max} - I_n$. The Eq. (4.54) is actually the K-degree polynomial of the Laplace matrix, so it still retains K-localization. In practical applications, the symmetric normalized Laplacian matrix L^{sys} is typically used to replace the original L.

4.3.4 Graph Convolutional Network

Thomas Kipf et al. [11] proposed the most classic graph convolutional network based on the Chebyshev network. Thomas Kipf et al. limited the polynomial convolution kernel in the Chebyshev network to order 1. Consequently, Eq. (4.57) is approximately a linear function of $\widetilde{L}$, so the amount of calculation is greatly reduced. Obviously, by doing so, the node can only be affected by the neighboring nodes of the first order around it. However, we only need to superimpose K such graph convolutional layers to extend the influence of the node to K-order adjacent nodes. In fact, the superimposed multilayer 1st-order graph convolution makes the dependence of nodes on the K-order adjacent nodes more flexible, and good results have also been achieved in the experiment.

Next, starting from Eq. (4.57), the graph convolutional network is derived. We'll take the symmetric normalized version of the Laplace matrix. Since the maximum eigenvalue of the Laplacian matrix can be approximately taken as $\lambda \approx 2$, the first-order graph convolution can be written as:

$$
\begin{aligned}
y = g_\theta(L^{sym})x &\approx \theta_0 T_0(\tilde{L})x + \theta_1 T_1(\tilde{L})x \\
&= \theta_0 x + \theta_1(L^{sym} - I_n)x \\
&= \theta_0 x + \theta_1 D^{-\frac{1}{2}} A D^{-\frac{1}{2}} x.
\end{aligned} \tag{4.58}
$$

In order to further reduce the number of parameters and prevent the occurrence of overfitting, take $\theta' = \theta_0 = -\theta_1$, so the above formula can be written as:

$$
y = \theta'\left(I_n + D^{-\frac{1}{2}} A D^{-\frac{1}{2}}\right)x. \tag{4.59}
$$

Observe the matrix $I_n + D^{-\frac{1}{2}} A D^{-\frac{1}{2}}$, and its eigenvalue range is [0,2]. If we perform multiple iterations, it may cause numerical instability, such as gradient vanishing or gradient explosion. In order to alleviate this kind of problem, we need to normalize again so that its characteristic value range becomes [0,1]. Here we define $\tilde{A} = A + I_n$, and the diagonal matrix $\tilde{D}$ has $\tilde{D}_{ii} = \sum_j \tilde{A}_{ij}$, so the normalized matrix becomes:

$$
I_n + D^{-\frac{1}{2}} A D^{-\frac{1}{2}} \rightarrow \tilde{D}^{-\frac{1}{2}} \tilde{A} \tilde{D}^{-\frac{1}{2}}. \tag{4.60}
$$

Now, our convolution operation becomes $\theta' \tilde{D}^{-\frac{1}{2}} \tilde{A} \tilde{D}^{-\frac{1}{2}} x$. Extend the graph signal to $X \in \mathbb{R}^{n \times c}$ (equivalent to n nodes, each node has c-dimensional attributes, and X is the initial attribute matrix of all nodes):

$$
Z = \theta' \tilde{D}^{-\frac{1}{2}} \tilde{A} \tilde{D}^{-\frac{1}{2}} X\Theta, \tag{4.61}
$$

where $\Theta \in \mathbb{R}^{c \times d}$ is the parameter matrix, and $Z \in \mathbb{R}^{n \times d}$ is the output from graph convolution.

In practical applications, it is usually possible to superimpose a multilayer graph convolution to obtain a graph convolution network, as shown in Figure 4.27. We use H^l to represent the node vector of the l-th layer and W^l to represent the parameters of the corresponding layer. We define $\hat{A} = \tilde{D}^{-\frac{1}{2}} \tilde{A} \tilde{D}^{-\frac{1}{2}}$, and then each layer graph convolution can be formally defined as:

$$
H^{(l+1)} = f(H^l, A) = \sigma(\hat{A} H^l W^l). \tag{4.62}
$$

As shown below, we employ a commonly used two-layer graph convolutional network to explain how the graph convolutional network classifies nodes semi-supervised. Suppose we have a graph $G = \{V, \varepsilon\}$ with n nodes, the node attribute matrix in the graph is $X \in \mathbb{R}^{n \times d}$, the adjacency matrix is M_A, and each node in the graph can be divided into m categories. We use the following method to predict the label of a node:

$$
\hat{Y} = f(X, M_A) = \text{softmax}\left(\hat{M}_A \text{ReLU}\left(\hat{M}_A X W^0\right) W^1\right). \tag{4.63}
$$

First, we input the node attribute matrix X and the adjacency matrix M_A of the entire graph. Through a two-layer graph convolution network, we get the node embedding matrix $Z = \hat{M}_A \text{ReLU}\left(\hat{M}_A X W^0\right) W^1$ and then use the softmax function to output the predicted classification

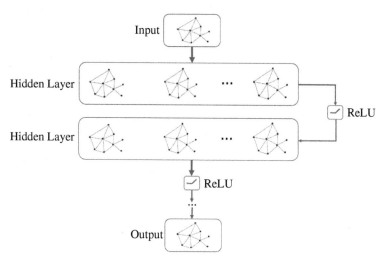

Figure 4.27 Graph convolutional network.

results and finally compare the difference between the predicted result $\hat{Y}$ and the real label Y on the node V_{train} of the training set. Then we calculate the cross-entropy between them and use the result as the loss function:

$$L = -\sum_{l=0}^{m-1}\sum_{i\in V_{train}} Y_{li}\ln\hat{Y}_{li}. \tag{4.64}$$

Applying the SGD method for training, the weight of the network can be obtained.

4.4 Discussion and Summary

Neural networks have been developed for more than 60 years since the model was proposed in the 1950s. At present, they have been widely used in various fields and have achieved great success, especially in speech recognition, computer vision, industrial inspection, and national defense.

This chapter focuses on artificial neural networks, convolution neural networks, and graph neural networks, deeply analyzing the basic principles of each type of neural network algorithm and their advantages and disadvantages. Although artificial neural networks can model complex nonlinear relationships, they have limited applications due to a shallow network structure caused by full connection. Deep CNNs are easily constructed since they employ the weight-sharing strategy, and they can achieve high-precision prediction by learning the deep features of data. However, CNNs are only suitable for structured data such as images, voice, and text. Graphic neural networks can model unstructured data by building relationships between different nodes, but they have limited applications and can only work on a shallow network. In summary, each type of network has different application scenarios because of its own characteristics.

Neural networks are brilliant in various fields, and especially in recent years, deep convolution neural networks have been widely used in the field of computer vision. However, there are still many challenges in neural networks. Firstly, the interpretability of neural network is a problem. Neural networks can model complex nonlinear relationships, but it is difficult to effectively explain these excellent results, which limits the application and further development of neural networks in some respects. Secondly, the layer depth of a neural network is directly proportional to the representation ability of the network. Therefore, a strong network means that the model has a huge number of parameters. How to solve the contradiction between the feature representation ability of the model and the number of parameters is a research hotspot in the future. Finally, the training process of neural networks is often complex, which usually requires a lot of computing power to complete the network training. Therefore, excellent optimization algorithms are important for improving neural networks in the future.

References

1 LeCun, Y., Bengio, Y., and Hinton, G. (2015). Deep learning. *Nature* **521** (7553): 436–444.

2 LeCun, Y. et al. (1998). Gradient-based learning applied to document recognition. *Proc. IEEE* **86** (11): 2278–2324.

3 Krizhevsky, A., Sutskever, I., and Hinton, G.E. (2012). Imagenet classification with deep convolutional neural networks. *Adv. Neural Inform. Process. Syst.* **25**: 1097–1105.

4 Michalski, R.S. and Carbonell, J.G. (2013). *Machine Learning: An Artificial Intelligence Approach* (ed. T.M. Mitchell). Springer Science & Business Media.

5 Simonyan, K. and Zisserman, A. (2014). Very deep convolutional networks for large-scale image recognition. arXiv preprint *arXiv:1409.1556*.

6 Chollet, F. (2017). Xception: Deep learning with depthwise separable convolutions. In: *Proceedings of the IEEE Conference on Computer Vision and Pattern Recognition*, 1251–1258.

7 Ji, S., Xu, W., Yang, M., and Yu, K. (2012). 3D convolutional neural networks for human action recognition. *IEEE Trans. Pattern Anal. Mach. Intell.* **35** (1): 221–231.

8 He, K., Zhang, X., Ren, S., and Sun, J. (2016). Deep residual learning for image recognition. In: *Proceedings of the IEEE Conference on Computer Vision and Pattern Recognition*, 770–778.

9 Scarselli, F., Gori, M., Tsoi, A.C. et al. (2008). The graph neural network model. *IEEE Trans. Neural Netw.* **20** (1): 61–80.

10 Defferrard, M., Bresson, X., and Vandergheynst, P. (2016). Convolutional neural networks on graphs with fast localized spectral filtering. *Adv. Neural Inform. Process. Syst.* **29**: 3844–3852.

11 Kipf, T.N. and Welling, M. (2016). Semi-supervised classification with graph convolutional networks. arXiv preprint *arXiv:1609.02907*.

Part II

Methods

5

Fast and Robust Image Segmentation Using Clustering

As the fuzzy c-means clustering (FCM) algorithm is sensitive to noise, local spatial information is often introduced to an objective function to improve the robustness of the FCM algorithm for image segmentation. However, the introduction of local spatial information often leads to a high computational complexity, arising out of iterative calculations of distances between pixels within local spatial neighbors and clustering centers. To address this issue, an improved FCM algorithm based on morphological reconstruction and membership filtering, namely FRFCM that is significantly faster and more robust than FCM, is presented in this chapter. This is based on the recently published article by Lei et al. [1]. First, the local spatial information of images is incorporated into FRFCM by introducing a morphological reconstruction operation to guarantee noise immunity and image detail preservation. Second, the modification of membership partition, based on the distance between pixels within local spatial neighbors and clustering centers is replaced by local membership filtering that depends only on the spatial neighbors of membership partition.

Compared with state-of-the-art algorithms, the presented FRFCM algorithm is simpler and significantly faster since it is unnecessary to compute distances between pixels within local spatial neighbors and clustering centers. In addition, it is efficient for noisy image segmentation because membership filtering is able to improve the membership partition matrix efficiently. Experiments performed on synthetic and real-world images demonstrate that the presented algorithm not only achieves better results but also requires less time than the state-of-the-art algorithms for image segmentation.

5.1 Introduction

Image segmentation aims to partition an image into several nonoverlapping regions that are consistent according to the requirements of different applications, and it is always one of the most challenging tasks in image understanding and computer vision due to the variety and complexity of images. Even though numerous approaches [2] of image segmentation have been proposed, none of them are sufficiently robust and efficient for a large number of different images. The technologies of image segmentation include clustering [3], region growth [4], watershed transform [5], active contour model [6], mean shift [7], graph cut [8], spectral clustering [9], Markov random field [10], and neural networks [11].

Among those technologies mentioned above, clustering is one of the most popular methods used for image segmentation because of its effectiveness and rapidity. The aim of clustering is to partition a set into some clusters so that members of the same cluster are similar and members of different clusters are dissimilar. Generally, clustering methods can be categorized into hierarchical model,

Image Segmentation: Principles, Techniques, and Applications, First Edition. Tao Lei and Asoke K. Nandi.
© 2023 John Wiley & Sons Ltd. Published 2023 by John Wiley & Sons Ltd.

graph theory, decomposing a density function, and minimizing an objective function. In this chapter, we will focus on image segmentation based on clustering methods by minimization of an objective function.

As conventional k-means clustering is crisp or hard, it leads to poor results for image segmentation. Based on fuzzy set theory, FCM had been proposed by Bezdek [12]. FCM is superior to hard clustering as it has more tolerance to ambiguity and retains more original image information. Even though FCM is efficient for images with simple texture and background, it fails to segment images with complex texture and background or images corrupted by noise because it only considers gray-level information without considering the image spatial information. To address the problem, one of the most popular ideas is to incorporate the local spatial information in an objective function to improve the segmentation effect. Motivated by this idea, Ahmed et al. [13] proposed the FCM algorithm with spatial constraints (FCM_S), where the objective function of the classical FCM algorithm is modified in order to take into account of the intensity inhomogeneity and to allow the labeling of a pixel to be influenced by the labels in its immediate neighborhood. However, FCM_S is time-consuming because the spatial neighborhoods term is computed in each iteration. To reduce the execution time of FCM_S, Chen and Zhang [14] employed average filtering and median filtering to obtain the spatial neighborhood information in advance. The two proposed variants, FCM_S1 and FCM_S2, have a lower computational cost than FCM_S since both mean-filtered images and median-filtered images can be computed before the start of the iterative stage. However, it is difficult to ascertain the type of noise and intensity before using FCM_S1 or FCM_S2.

Enhanced FCM (EnFCM) [15] is an excellent algorithm from the viewpoint of its low computational time; it performs clustering based on gray-level histograms instead of pixels of a summed image. The computational time is low because the number of gray levels in an image is generally much smaller than the number of its pixels. However, the segmentation result produced by EnFCM is only comparable to that produced by FCM_S. To improve the segmentation results obtained by EnFCM, Cai et al. [16] proposed the fast generalized FCM algorithm (FGFCM), which introduced a new factor as a local similarity measure aiming to guarantee both noise immunity and detail preservation for image segmentation and meanwhile removes the empirically adjusted parameter that is required in EnFCM and finally performs clustering on gray-level histograms. Although FGFCM is able to improve the robustness and computational efficiency of FCM to some extent, they require more parameters than EnFCM.

To develop new FCM algorithms that are free from any parameter selection, Krinidis and Chatzis [17] proposed a robust fuzzy local information c-means clustering algorithm (FLICM) by replacing the parameter employed by EnFCM with a novel fuzzy factor that is incorporated into an objective function to guarantee noise immunity and image detail preservation. Although the algorithm FLICM overcomes the problem of parameter selection and promotes image segmentation performance, the fixed spatial distance is not robust for different local information of images. Gong et al. [18] utilized a variable local coefficient instead of the fixed spatial distance and proposed a variant of the FLICM algorithm (RFLICM) that is able to exploit more local context information in images. Furthermore, by introducing a kernel metric to FLICM and employing a trade-off weighted fuzzy factor to adaptively control the local spatial relationship, Gong et al. [19] proposed a novel FCM with local information and kernel metric (KWFLICM) to enhance the robustness of FLICM to noise and outliers. Similar to FLICM, KWFLICM is also free of any parameter selection. However, KWFLICM has a higher computational complexity than FLICM. In fact, the parameter selection depends on image patches and local statistics.

Image patches have been successfully used in nonlocal denoising [20] and texture feature extraction [21], and a higher classification accuracy can be obtained by using the similarity measurement based on patch. Therefore, patch-based denoising methods, where the nonlocal spatial information is introduced in an objective function by utilizing a variant parameter that is adaptive according to noise level for each pixel of images [19] are extended to FCM to overcome the problem of parameter selection to improve the robustness to noise. However, it is well known that patch-based nonlocal filtering and parameter estimation have a very high computational complexity. To reduce the running time of FLICM and KWFLICM, Zhao et al. [22] proposed the neighborhood weighted FCM algorithm (NWFCM), which replaces the Euclidean distance in the objective function of FCM with a neighborhood weighted distance obtained by patch distance. Even though the NWFCM is faster than FLICM and KWFLICM, it is still time-consuming because of the calculation of patch distance and parameter selection. To overcome this shortcoming, Guo et al. [23] proposed an adaptive FCM algorithm based on noise detection (NDFCM), where the trade-off parameter is tuned automatically by measuring local variance of gray levels. Despite the fact that NDFCM employs more parameters, it is fast since image filtering is executed before the start of iterations.

Following the work mentioned above, this chapter presents a significantly fast and robust algorithm for image segmentation. The presented algorithm [1] can achieve good segmentation results for a variety of images with a low computational cost yet achieve a high segmentation precision.

The main findings of the chapter can be summarized as follows.

1) FRFCM employs morphological reconstruction (MR) [24, 25], to smooth images in order to improve the noise immunity and image detail preservation simultaneously, which removes the difficulty of having to choose different filters suitable for different types of noise in existing improved FCM algorithms. Therefore, the presented FRFCM is more robust than these algorithms for images corrupted by different types of noise.

2) FRFCM modifies membership partition by using a faster membership filtering instead of the slower distance computation between pixels within local spatial neighbors and their clustering centers, which leads to a low computational complexity. Therefore, the presented FRFCM is faster than other improved FCM algorithms.

The rest of this chapter is organized as follows. In Section 5.2, we provide the related work. In Sections 5.3 and 5.4, we describe local spatial information integration to FCM and a membership filtering for FCM. Finally, we present the discussion and summary in Section 5.5.

5.2 Related Work

To improve the drawback that the FCM algorithm is sensitive to noise, most algorithms try to incorporate local spatial information, such as FLICM, KWFLICM, and NWFCM. However, high computational complexity is a problem for them. In fact, the introduction of local spatial information is similar to image filtering in advance. Thus, local spatial information of an image can be calculated before applying the FCM algorithm, which will efficiently reduce computational complexity, such as FCM_S1 and FCM_S2. Besides, if the membership is modified through the use of the relationship of the neighborhood pixels but the objective function is not modified, then the corresponding algorithm will be simple and fast [26].

Motivated by this, in this chapter, we improve the FCM algorithm in two ways: one is to introduce local spatial information using a new method with a low computational complexity, and the other is to modify pixels' membership without depending on the calculation of distance between pixels within local spatial neighbors and clustering centers. The presented algorithm on image segmentation can be implemented efficiently with a small computational cost.

5.2.1 Objective Function of FCM Based on Neighborhood Information

By introducing local spatial information to an objective function of FCM algorithms, the improved FCM algorithms are insensitive to noise and show better performance for image segmentation. Generally, the modified objective function of these algorithms is given as follows:

$$J_m = \sum_{i=1}^{N} \sum_{k=1}^{c} v_{ki}^m \parallel x_i - pv_k \parallel^2 + \sum_{i=1}^{N} \sum_{k=1}^{c} G_{ki}, \tag{5.1}$$

where $g = \{x_1, x_2, \cdots, x_N\}$ represents a grayscale image, x_i is the gray value of the i-th pixel, pv_k represents the prototype value of the k-th cluster, v_{ki} denotes the fuzzy membership value of the i-th pixel with respect to cluster k, $U = [v_{ki}]^{c \times N}$ represents membership partition matrix, N is the total number of pixels in the image f, and c is the number of clusters. The parameter m is a weighting exponent on each fuzzy membership that determines the amount of fuzziness of the resulting classification. The fuzzy factor G_{ki} is used to control the influence of neighborhood pixels on the central pixel. Different G_{ki} usually leads to variant clustering algorithms, such as FCM_S, FCM_S1, FCM_S2, FLICM, KWFLICM, and NWFCM. From these algorithms, we found that the form of G_{ki} directly decides the computational complexity. For example, in FCM_S, the G_{ki} is defined as

$$G_{ki} = \frac{\alpha}{N_R} v_{ki}^m \sum_{r \in N_i} \parallel x_r - pv_k \parallel^2, \tag{5.2}$$

where α is a parameter that is used to control the effect of the neighbors term, N_R is the cardinality of v_{ki}, x_r denotes the neighbor x_i, and N_i is the set of neighbors within a window around x_i.

For FLICM, the G_{ki} is defined as

$$G_{ki} = \sum_{\substack{r \in N_i \\ i \neq r}} \frac{1}{d_{ir} + 1} (1 - u_{kr})^m \parallel x_r - pv_k \parallel^2, \tag{5.3}$$

where d_{ir} represents the spatial Euclidean distance between pixels x_i and x_r. It is obvious that G_{ki} is more complex than that in FCM_S, and thus FLICM has a higher computational complexity than FCM_S. In FCM_S1 and FCM_S2, G_{ki} is defined as

$$G_{ki} = \alpha v_{ki}^m \parallel \hat{x}_i - pv_k \parallel^2, \tag{5.4}$$

where $\hat{x}_i$ is a mean value or median value of neighboring pixels lying within a window around x_i. G_{ki} in FCM_S1 and FCM_S2 has a more simplified form than FCM_S, and the clustering time can be reduced because $\sum_{r \in N_i} \parallel x_r - pv_k \parallel^2 / N_R$ is replaced by $\alpha \parallel \hat{x}_i - pv_k \parallel^2$.

Although FCM_S1 and FCM_S2 simplify the neighbors term in the objective function of FCM_S and present excellent performance for image segmentation, it is difficult to ascertain the noise type that is required to choose a suitable filter (mean or median filter). FCM_S2 is able to obtain good segmentation results for images corrupted by salt-and-pepper noise, but it is incapable of doing so for images corrupted by Gaussian noise. FCM_S1 produces worse results compared with FCM_S2. In practical applications, we expect to obtain a robust $\hat{x}$ in which different types of noise are efficiently removed while image details are preserved. Similarly, we introduce MR to FCM because MR is not only able to obtain a good result, but it also requires a short running time. Therefore, in this chapter we introduce MR to FCM to address the drawback produced by conventional filters. MR uses a marker image to reconstruct the original image to obtain a better image, which is favorable to image segmentation based on clustering. Similar to FCM_S1 and FCM_S2, the reconstructed image will be computed in advance, and thus the computational complexity of the presented algorithm is low. We will present the computation of the reconstructed image in detail in Section 5.3.

5.2.2 Membership Correction Based on Local Distance

In the FCM algorithm, according to the definition of the objective function and the constraint that $\sum_{k=1}^{c} u_{ki} = 1$ for each pixel x_i, and using the Lagrange multiplier method, the calculations of membership partition matrix and the clustering centers is given as follows:

$$v_{ki} = \frac{\|x_i - v_k\|^{-2/(m-1)}}{\sum_{j=1}^{c} \|x_i - v_j\|^{-2/(m-1)}}, \tag{5.5}$$

$$v_k = \frac{\sum_{i=1}^{N} u_{ki}^m x_i}{\sum_{i=1}^{N} u_{ki}^m}. \tag{5.6}$$

According to (5.5), it is easy and fast to compute v_{ki} by using a vector operation for the FCM algorithm. However, it is complex and slow to compute v_{ki}, shown in (5.7) for an improved FCM algorithm, such as FLICM and KWFLICM, because a vector operation cannot be used in the computation of G_{ki} in (5.3):

$$u_{ki} = \frac{\left(\|x_i - v_k\|^2 + G_{ki}\right)^{-1/(m-1)}}{\sum_{j=1}^{c} \left(\|x_i - v_j\|^2 + G_{ji}\right)^{-1/m-1}}. \tag{5.7}$$

Therefore, a multiple loop program is employed by FLICM and KWFLICM, which causes high computational complexity. On the one hand, the introduction of G_{ki} in (5.7) is able to improve the robustness of FCM to noise image segmentation, but on the other hand, the introduction of G_{ki} causes a high computational cost. Clearly, there is a contradiction or trade-off between improving the robustness and reducing the computational complexity simultaneously for FCM. We found that if G_{ki} can be computed in advance, the contradiction will disappear because the v_{ki} in (5.7) can be computed by using a vector operation without multiple loops.

In this chapter, we introduce membership filtering to FCM to address the contradiction mentioned above. First, because a reconstructed image is computed in advance, we perform clustering on the gray-level histogram of an image reconstructed by MR. After obtaining a fuzzy membership partition matrix, we use membership filtering to modify the membership partition matrix to avoid the computation of distance between pixels within local spatial neighbors and clustering centers. We illustrate the presented method in detail in Section 5.3.

5.3 Local Spatial Information Integration to FCM

We employ MR to replace mean or median filters due to its robustness to noise. MR is able to efficiently suppress different noise without considering noise type. Moreover, the MR algorithm is fast, as parallel algorithms exist for the implementation of MR. Motivated by EnFCM, we perform clustering on the gray-level histogram of an image reconstructed by MR to obtain a fuzzy membership matrix via an iteration operation. Finally, a filter is employed to modify the membership partition matrix. Using this method, we can obtain a good segmentation result requiring less time.

5.3.1 Fast and Robust FCM Based on Histogram

Similar to EnFCM, the clustering of the presented FRFCM [1] is performed on the gray-level histogram, and thus the objective function can be written as

$$J_m = \sum_{g_v=1}^{\tau} \sum_{k=1}^{c} \gamma_{g_v} v_{kg_v}^m \| \xi_{g_v} - pv_k \|^2,$$ (5.8)

where v_{kg_v} represents the fuzzy membership of gray value g_v with respect to cluster k, and

$$\sum_{g_v=1}^{\tau} \gamma_{g_v} = N,$$ (5.9)

where ξ is an image reconstructed by MR, and ξ_{g_v} is a gray level, $1 \le g_v \le \tau$, τ denotes the number of the gray levels contained in ξ, which is generally much smaller than N. ξ is defined as follows:

$$\xi = R^C(f_o),$$ (5.10)

where R^C denotes morphological closing reconstruction, and f_o represents an original image.

Utilizing the Lagrange multiplier technique, the aforementioned optimization problem can be converted to an unconstrained optimization problem that minimizes the following objective function:

$$\tilde{J}_m = \sum_{g_v=1}^{\tau} \sum_{k=1}^{c} \gamma_{g_v} u_{kg_v}^m \| \xi_{g_v} - pv_k \|^2 - \lambda_L \left(\sum_{k=1}^{c} u_{kg_v} - 1 \right),$$ (5.11)

where λ_L is a Lagrange multiplier. Therefore, the problem of the minimization of the objective function is converted to finding the saddle point of the above Lagrange function and taking the derivatives of the Lagrangian $\tilde{J}_m$ with respect to the parameters, i.e. u_{kg_v} and pv_k.

By minimizing the objective function (5.8), we obtain the corresponding solution as follows:

$$u_{kl} = \frac{\| \xi_l - pv_k \|^{-2/(m-1)}}{\sum_{j=1}^{c} (\| \xi_l - v_j \|)^{-2/(m-1)}},$$ (5.12)

$$pv_k = \frac{\sum_{i=1}^{\tau} \gamma_l u_{kl}^m \xi_l}{\sum_{i=1}^{\tau} \gamma_l u_{kl}^m}.$$ (5.13)

According to (5.12), a membership partition matrix $U = \left[u_{kg_v} \right]^{c \times \tau}$ is obtained. To obtain a stable U, (5.12) and (5.13) are repeatedly implemented until $max|U^{(b)} - U^{(b+1)}| < \eta$, where η is a minimal error threshold. Because $u_{kg_v}^{(b)}$ is a fuzzy membership of gray value g_v with respect to cluster k, a new membership partition matrix $U' = [v_{ki}]^{c \times N}$ that corresponds to the original image f_o is obtained, i.e.

$$v_{ki} = u_{kg_v}^{(b)}, \quad \text{if } x_i = \xi_{g_v}$$ (5.14)

To obtain a better membership partition matrix and to speed up the convergence of our algorithm, we modify v_{ki} by using a membership filtering. Considering the trade-off between performance of membership filtering and the speed of algorithms, we employ a median filter in this chapter as follows:

$$U'' = med(U'),$$ (5.15)

where med represents median filtering.

Based on the analysis mentioned above, the presented algorithm FRFCM can be summarized as follows:

Step 1: Set the cluster prototypes value c, fuzzification parameter m, the size of filtering window w, and the minimal error threshold η.

Step 2: Compute the new image f_n using (5.10), and then compute the histogram of f_n.

Step 3: Initialize randomly the membership partition matrix $U^{(0)}$.

Step 4: Set the loop counter $b = 0$.

Step 5: Update the clustering centers using (5.13).

Step 6: Update the membership partition matrix $U^{(b+1)}$ using (5.12).

Step 7: If $max\{U^{(b)} - U^{(b+1)}\} < \eta$, then stop; otherwise, set $b = b+1$ and go to Step 5.

Step 8: Implement median filtering on membership partition matrix U' using (5.15).

5.3.2 Morphological Reconstruction

For the FCM algorithm, the rate of convergence is always decided by the distribution characteristics of data. If the distribution characteristic of data is favorable to clustering, the corresponding number of iterations is small; otherwise, the number of iterations is large. FCM is sensitive to noise because the distribution characteristics of data is always affected by noise corruption, which causes two problems. One is that the result obtained by the FCM algorithm is poor for noisy image segmentation; the other is that the number of iterations of FCM is larger for an image corrupted by noise than an image uncorrupted by noise. It is well known that the distribution characteristic of data can be described by a histogram. If the histogram is uniform, it is difficult to achieve a good and fast image segmentation. On the contrary, it is easy if the histogram has several apparent peaks. Figure 5.1 shows an example.

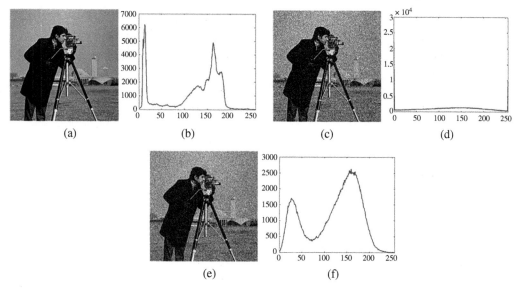

Figure 5.1 Comparison of distribution characteristics of data for noisy image and filtered image [1] / with permission of IEEE. (a) Original image "cameraman" (image size: 512×512). (b) Histogram of the original image. (c) Image corrupted by Gaussian noise (the mean value is zero, and the variance is 5%) (d) Histogram of (c). (e) image filtered by a mean filter (3×3). (f) Histogram of (e).

Table 5.1 Comparison of numbers of iterations for the three different images (the numbers represent the averages of repeating 100 times) [1] / with permission of IEEE.

	Original image (Figure 5.1a)	Noisy image (Figure 5.1a)	Filtered image (Figure 5.1c)
Numbers of iterations with standard deviations	21.06 ± 1.91	38.46 ± 7.51	24.56 ± 1.48

Figure 5.1 shows that the histogram of the original image has two obvious peaks while the histogram of the image corrupted by Gaussian noise has no obvious peaks except the extrema (0 and 255). There are two obvious peaks in Figure 5.1f, similar to the original Figure 5.1b, demonstrating a mean filter is efficient for the removal of Gaussian noise. We implemented the FCM algorithm on three images: original image, the image corrupted by Gaussian noise (the mean value is zero, and the variance is 5%), and the image filtered by a mean filter (the size of the filtering window is 3×3). Table 5.1 shows the comparison of numbers of iterations of FCM for the three images ($c = 2$).

Table 5.1 shows that the number of iterations of FCM is the smallest for the original image and it is the largest for the noisy image. Mean filter is efficient for the optimization of data distribution because the number of iterations is reduced.

In this chapter, we introduce MR to the FCM algorithm to optimize the distribution characteristic of data before applying clustering. MR is able to preserve object contour and remove noise without knowing the noise type in advance [27], which is useful for optimizing the distribution characteristic of data.

There are two basic MR operations, morphological dilation and erosion reconstruction [28]. Morphological dilation reconstruction is denoted by $R_g^\delta(f)$ and is defined as:

$$R_g^\delta(f) = \delta_g^{(k)}(f), \tag{5.16}$$

where g is the original image, f is a marker image with $f \le g$, δ represents the dilation operation, $\delta_g^{(1)}(f) = \delta(f) \wedge g$, $\delta_f^{(k)}(g) = \delta(\delta^{(k-1)}(f)) \wedge g$, and $\wedge$ stands for the pointwise minimum.

By duality, morphological erosion reconstruction is denoted by $R_g^\varepsilon(f)$ and is defined as:

$$R_g^\varepsilon(f) = \varepsilon_g^{(k)}(f), \tag{5.17}$$

where $f \ge g$, ε represents the erosion operation, $\varepsilon_g^{(1)}(f) = \varepsilon(f) \vee g$, $\varepsilon_f^{(k)}(g) = \varepsilon(\varepsilon^{(k-1)}(f)) \vee g$, and $\vee$ stands for the pointwise maximum.

The reconstruction result of an image depends on the selection of marker images and mask images [29]. Generally, if the original image is used as a mask image, then the transformation of the original image is considered as the marker image. In practical applications, $f = \varepsilon(g)$ meets the condition $f \le g$ for dilation reconstructions, and $f = \delta(g)$ meets the condition $f \ge g$ for erosion reconstructions. Thus, $f = \varepsilon(g)$ and $f = \delta(g)$ are always used to obtain a marker image due to simplicity and effectiveness.

Based on the composition of erosion and dilation reconstruction, some reconstruction operators with stronger filtering capability can be obtained, such as morphological opening and closing reconstruction. Because morphological closing reconstruction, denoted by R^C, is more suitable for texture detail smoothing, we employ R^C to modify the original image. R^C is defined as follows:

$$R^C(g) = R_{R_g^\delta(\varepsilon(g))}^\varepsilon \left(\delta \left(R_g^\delta(\varepsilon(g)) \right) \right). \tag{5.18}$$

In [20], the modified image $\xi = (\xi_i)_{i=1}^{N}$ is defined as

$$\xi_i = \frac{1}{1+\alpha}(x_i + \alpha \hat{x}_i). \tag{5.19}$$

According to (5.19), $x_i \in f$ and $\hat{x}_i \in R^C(g)$, where $R^C(g)$ denotes a reconstructed image obtained by R^C. To obtain a marker image, a structuring element E including center element is required for ε or δ, i.e. $\varepsilon_E(g) \le g$ and $\delta_E(g) \ge g$. Then, R^C is rewritten as

$$R^C(g) = R_{R_E^\delta(\varepsilon_E(g))}^\varepsilon \left(\delta_E \left(R_g^\delta(\varepsilon_E(g)) \right) \right). \tag{5.20}$$

For example, a disk with radius r can be considered as E. If $r = 0$, then $R^C(g) = g$; else, g will be smoothed to a different degree according to the change of r. Therefore, the effect of α is similar to r. And thus, we can replace ξ with $R^C(g)$, and the parameter α will be removed, which solves the problem of noise estimation because MR is able to remove different noises efficiently.

To show the effect of MR for different type of noise removal in images, Figure 5.2 shows comparative results generated by a mean filter, a median filter, and R^C. The original image is Figure 5.1a, and the size of the filtering window employed by the mean and the median filters is 3×3. For consistency, the structuring element, in this case, is also a square of size 3×3 ($r = 1$).

Figure 5.2c, e, and g show filtering results of an image corrupted by Gaussian noise by using the mean filter, the median filter, and R^C, respectively. It is clear that R^C is efficient for Gaussian noise removal. Similarly, Figure 5.2d, f, and h show filtering results of image corrupted by salt-and-pepper noise by using the mean filter, the median filter, and R^C, respectively. It is also clear that R^C is efficient for salt-and-pepper noise removal. Therefore, on the one hand, MR removes the difficulty of choosing filters for images corrupted by noise; on the other hand, MR integrates spatial information into FCM to achieve a better image segmentation. Compared with mean filtering and median filtering, Figure 5.2 shows that MR is able to optimize data distribution without considering noise type. Moreover, MR can obtain better results for image filtering than mean and median filters, which is important for subsequent clustering and image segmentation.

5.4 Membership Filtering for FCM

According to the above results in Section 5.3, we found that the introduction of local spatial information is useful and efficient for improving the FCM algorithm. However, the computation of distance between pixels within local spatial neighbors and clustering centers does introduce a high computational complexity, such as FCM_S. Although some improved algorithms, such as FCM_S1 and FCM_S2, reduce computational complexity by computing spatial neighborhood information in advance, these algorithms need to ascertain the noise type before applying an image filter. To exploit spatial neighborhood information during the iteration process of clustering, FLICM and KWFLICM compute the distance between the neighbors of pixels and clustering centers in each iteration. Although FLICM and KWFLICM produce good segmentation results for noisy images, they have a high computational complexity.

In [22], FLICM will equal to FCM, if G_{ki} is removed. For this, the idea is to replace G_{ki} in a simple way where the computation of distance between pixels within local spatial neighbors and pv_k is unnecessary. Motivated by the idea, membership filtering is introduced. We will replace the contribution of G_{ki} with the spatial neighborhood information of membership partition. To further

Figure 5.2 Comparison of noise removal using different methods [1] / with permission of IEEE. (a) Image corrupted by Gaussian noise (the mean value is zero, and the variance is 5%). (b) Image corrupted by salt-and-pepper noise (the noise density is 20%). (c) Filtered result using mean filtering for (a). (d) Filtered result using mean filtering for (b). (e) Filtered result using median filtering for (a). (f) Filtered result using median filtering for (b). (g) Filtered result using MR for (a). (h) Filtered result using MR for (b).

analyze the contribution of G_{ki}, Figure 5.3 shows the effect of spatial neighborhood information on membership partition.

FCM and FLICM are used to segment Figure 5.3b. Figure 5.3c, d show the membership partition provided by FCM and FLICM, respectively, when the number of iterations is 10. Figure 5.3c shows that some pixels marked with red color will be misclassified because the original image is corrupted by Gaussian noise. By introducing local spatial information into FLICM, the misclassified pixel will be corrected as shown in Figure 5.3d (the corrected pixels are marked with blue color). For a pixel (the gray value is 115) in Figure 5.3b, we obtained three fuzzy memberships (0.01, 0.98, 0.01) of the pixel shown in Figure 5.3c by using FCM, which clearly indicates that the pixel belongs to the second cluster according to FCM. However, in reality, it belongs the third cluster (the gray value is 170)

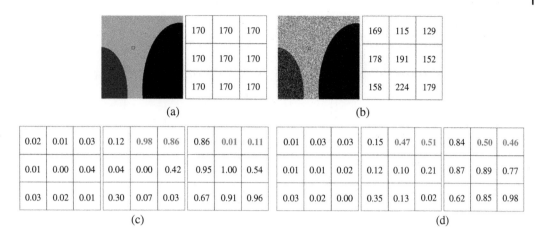

Figure 5.3 Comparison of membership partition from FCM and FLICM ($c = 3$, and the iteration step is 10) [1] / with permission of IEEE. (a) Original synthetic image included three gray levels (0, 85, 170). (b) Image corrupted by Gaussian noise (the mean value is zero, and the variance is 3%). (c) Membership partition using FCM. (d) Membership partition using FLICM.

according to the ground truth. In Figure 5.3d, we obtained new fuzzy memberships (0.03, 0.47, 0.50) of the pixel by using FLICM, which shows that the pixel belongs the third cluster because 0.5 is the maximal membership. Even though pixels corrupted with Gaussian noise are classified accurately by FLICM, the maximal membership value of pixels is small. And thus, FLICM has a slow speed of convergence.

According to (5.2) and (5.7), u_{ki} depends on the distance $\|x_i - pv_k\|$ and $\|x_r - pv_k\|$. But in fact, $\|x_r - pv_k\|$ is a repetitive or redundant computation since it can be obtained according to $\|x_i - pv_k\|$. It is the same as KWFLICM. It is clear that we can use a membership filter to correct the misclassified pixels, i.e. it is unnecessary to compute the distance between the neighbors of pixels and clustering centers. According to G_{ki} show in (5.3), the modified membership partition is considered as

$$u'_{ki} = v_{ki} + \sum_{\substack{r \in N_i \\ i \neq r}} \frac{1}{d_{ir} + 1} u_{kr}, \tag{5.21}$$

where d_{ir} represents the Euclidean distance between v_{ki} and u_{kr}, and u_{kr} is the neighbors of v_{ki}. The factor $1/(d_{ir}+1)$ reflects the spatial structure information of the membership partition.

Because FLICM is sensitive to salt-and-pepper noise, it is inefficient to use (5.21) to remove the noise. In this chapter, we use a median filter to modify the membership partition as shown in (5.15), In fact, it can be demonstrated that the introduction of local spatial information is similar to a membership filter for improving segmentation results. However, membership filtering does not require to compute the distance between the pixels within local spatial neighbors and clustering centers. Therefore, the corresponding computational complexity of improved algorithms based on membership filtering is lower than other algorithms, such as NWFCM, FLICM, and KWFLICM. For the membership partition obtained by FCM in each iteration shown in Figure 5.3, a median filter is used to modify the membership partition, and Figure 5.4 shows results (the result is normalized, and the filtering window is the same as the structuring element E).

Figure 5.4 shows membership filtering has a capability of correcting misclassified pixels. Moreover, it provides a better membership partition than FLICM. Therefore, it is a good idea to use

0.01	0.02	0.02	0.04	0.04	0.07	0.95	0.94	0.91
0.01	0.01	0.05	0.04	0.04	0.14	0.95	0.95	0.81
0.02	0.01	0.01	0.04	0.04	0.05	0.94	0.95	0.94

Figure 5.4 Membership partition using FCM based on membership filter for Figure 5.3 (b) (the iteration step is 10) [1] / with permission of IEEE.

Table 5.2 Comparison of clustering centers produced by different algorithms (the data represents average results of repeating 100 times, the iteration step is 10, and three gray levels of ground truth Figure 5.3a is (0, 85, 170)). The best values are highlighted [1] / with permission of IEEE.

Method	FCM	FLICM	MF-FCM	FRFCM
Values of clustering centers	(11.4, 106.9, 192.0)	(18.9, 117.3, 179.1)	(18.9, 91.2, 169.7)	**(5.6, 85.4, 169.1)**
MSE	33.07	38.51	19.89	**5.69**

membership filtering instead of the introduction of a fuzzy factor G_{ki}. Also, the FCM algorithm based on membership filtering (MFFCM) provides better clustering centers than FCM, as shown in Table 5.2. Therefore, the objective function of the FCM algorithm based on membership filtering converges quickly. However, if membership filtering is implemented in each iteration, the corresponding algorithm will be complex and inefficient. To improve the computational efficiency of MFFCM further, membership filtering is just implemented once on the final membership partition matrix.

Based on the analysis above, MR is used to optimize data distribution, and then we implement the FCM algorithm on the histogram of the reconstructed image. Finally, we use a median filter to modify the membership partition. The presented FRFCM is implemented on Figure 5.3b, and Table 5.2 shows the comparison of values of clustering centers produced by FCM, FLICM, MFFCM, and FRFCM, where the mean square error (MSE) of clustering centers is used to evaluate the performance of different algorithms.

In Table 5.2, 11.4, 18.9, 18.9, and 5.6 are values of the first clustering center obtained by FCM, FLICM, MFFCM, and FRFCM, respectively. The value 33.07 is the MSE between the FCM results of (11.4, 106.9, 192.0) and the ground truth of (0, 85, 170). It is clear that the value 5.6 from FRFCM is the closest value to 0 that is the first clustering center from the ground truth. Consequently, Table 5.2 demonstrates that FRFCM provides the best clustering centers after 10 iterations. Based on the analysis mentioned above, we conclude that the presented FRFCM has following advantages:

1) Similar to FLICM and KWFLICM, it is free parameters except the size of filtering window.
2) It has a low computational complexity because the redundant computation of distance is unnecessary.
3) It is able to provide good results for image segmentation because of the introduction of MR and membership filtering, and thus spatial information is efficiently exploited.

5.5 Discussion and Summary

To estimate the effectiveness and efficiency of the presented FRFCM [1], synthetic noise images, real images including a medical image and an aurora image, and color images are tested in our experiments. Nine state-of-the-art clustering algorithms, FCM [12], FCM_S1 [14], FCM_S2 [14], EnFCM [15], FGFCM [16], FLICM [17], KWFLICM [19], NWFCM [22], and NDFCM [23], are employed in these experiments to compare with the presented FRFCM [1]. These algorithms have different advantages. FCM, FCM_S1, FCM_S2, EnFCM, FGFCM, and NDFCM have a low computational complexity. FLICM, KWFLICM, and NWFCM have a strong capability of noise removal, while FLICM and KWFLICM do not require parameter values to be set.

In the following experiments, a fixed 3×3 window is used in all the algorithms except FCM for fair comparison. The weighting exponent is set to $m = 2$, $\eta = 10^{-5}$. In addition, according to FCM_S1, FCM_S2, and EnFCM, α is used to control the effect of the neighbors term; experientially, $\alpha = 3.8$. In FGFCM and NDFCM, the spatial scale factor and the gray-level scale factor are $\lambda_s = 3$ and $\lambda_g = 5$, respectively. Besides, a new scale factor λ_a equals to 3is used for the NDFCM [23]. For NWFCM, λ_g equals to 5. Except m, η, and the number of the cluster prototypes, there is no other parameters for FLICM and KWFLICM. For our FMFFCM, the mask image is the original image, and a square structuring element of size 3×3 is used to obtain a marker image. In addition, a median filter is used for fuzzy membership filtering, and the filtering window is also 3×3.

5.5.1 Results on Synthetic Images

In this section, two synthetic images with size 256×256 are used in the experiment. The first image includes three classes (three intensity values: 0, 85, and 170, respectively), and the second image includes four classes (four intensity value: 0, 85, 170, and 255, respectively). The two synthetic images are shown in Figure 5.5a and 5.6a, respectively. These images are corrupted by Gaussian, salt-and-pepper, and uniform noise, respectively, and these corrupted images are utilized for testing the efficiency and robustness of the above algorithms. Figure 5.5c-l and 5.6c-l show segmentation results obtained by different algorithms.

In addition, a performance index, the optimal segmentation accuracy (SA), and a quantitative index Score (S) [13] are used to assess the denoising performance of different algorithms, where SA is defined as the sum of the correctly classified pixels divided by the sum of the total number of the pixels:

$$SA = \sum_{k=1}^{c} \frac{A_k \cap C_k}{\sum_{j=1}^{c} C_j}, \tag{5.22}$$

and S is defined as the degree of equality between pixel sets A_k and the ground truth C_k,

$$S = \sum_{k=1}^{c} \frac{A_k \cap C_k}{A_k \cup C_k}, \tag{5.23}$$

where c is the number of the cluster prototypes, A_k denotes the set of pixels belonging to the k-th class found by the algorithm, and C_k denotes the set of pixels belonging to the class in the ground truth. All the algorithms are repeatedly run 100 times on synthetic images corrupted by different noises. Tables 5.3 and 5.4 give the average segmentation accuracy and the scores results of the repeated experiments for the 10 algorithms.

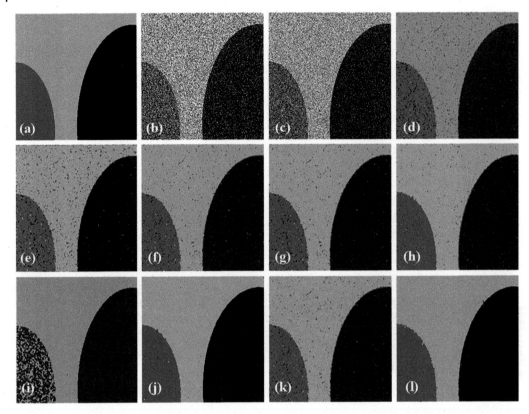

Figure 5.5 Comparison of segmentation results on the first symmetric image [1] / with permission of IEEE. (a) Original image. (b) Noisy image (Gaussian noise with zero mean and 5% variance). (c) FCM result. (d) FCM_S1 result. (e) FCM_S2 result. (f) EnFCM result. (g) FGFCM result. (h) FLICM result. (i) NWFCM result. (j) KWFLICM result. (k) NDFCM result. (l) FRFCM result.

In Figure 5.5, the FCM algorithm does not overcome its sensitivity to noise. FCM_S1 and FCM_S2 are able to reduce the effect of noise on segmentation results due to the introduction of local spatial information. EnFCM, FGFCM, and NDFCM improve segmentation results to some extent, the segmented images have better visual effect than FCM, FCM_S1, and FCM_S2. Although NWFCM obtains a better visual effect for the first and the third classes (the clustering centers are 0 and 170), it causes a poor effect on the second class (the clustering center is 85). FLICM and KWFLICM are superior to FGFCM depending on Figure 5.5h, j. Figure 5.5l shows that the presented FRFCM obtains a better segmentation result than other algorithms.

Figure 5.6 shows that FCM_S1 obtains a poor segmentation result, which is close to FCM because mean filters employed by FCM_S1 are incapable of removing salt-and-pepper noise. But FCM_S2 obtains a good segmentation result because median filters employed by FCM_S2 are able to efficiently remove salt-and-pepper noise. NWFCM provides a good segmentation result for images corrupted by salt-and-pepper noise because a weight function incorporating both patch structure information and the local statistics is introduced in distance measurement between pixels. FLICM and KWFLICM are sensitive to salt-and-pepper noise, which leads to poor results; even KWFLICM obtains a wrong result, as shown in Figure 5.6j. FGFCM is superior to EnFCM because FGFCM introduces a new factor as a local (both spatial and gray) similarity measure aiming to guarantee both noise immunity and detail preservation and meanwhile removes the empirically adjusted

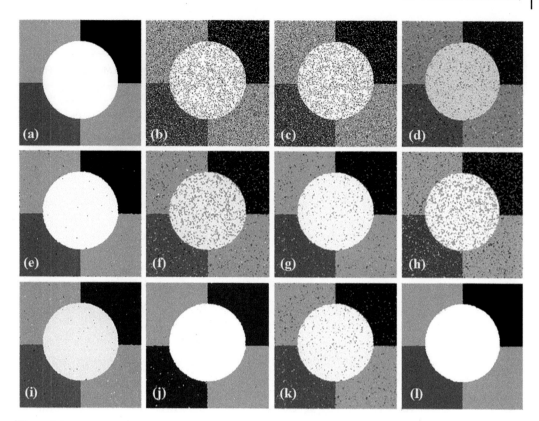

Figure 5.6 Comparison of segmentation results on the second symmetric image [1] / with permission of IEEE. (a) Original image. (b) Noisy image (salt-and-pepper, the noise intensity is 20%). (c) FCM result. (d) FCM_S1 result. (e) FCM_S2 result. (f) EnFCM result. (g) FGFCM result. (h) FLICM result. (i) NWFCM result. (j) KWFLICM result. (k) NDFCM result. (l) FRFCM result.

parameter α for image segmentation. FRFCM has superior noise immunity and detail preservation, and it provides better segmentation results than other algorithms due to the introduction of MR and membership filtering.

From Tables 5.3 and 5.4, we can see that the segmentation accuracy of FRFCM is consistently higher than other algorithms for synthetic images containing different noises. It is obvious that FRFCM is much more robust to different noise than other algorithms. KWFLICM is sensitive to salt-and-pepper and uniform noise when the noise level is high. FCM_S1 is efficient for images corrupted by Gaussian noise, but FCM_S2 is efficient for images corrupted by salt-and-pepper and uniform noise. NWFCM is able to provide good segmentation results for image corrupted by salt-and-pepper and uniform noise, but it is sensitive to Gaussian noise. Both NDFCM and FGFCM are robust to different noises, and they have close performance, according to Tables 5.3 and 5.4.

5.5.2 Results on Real Images

Image segmentation plays a key role in medical diagnosis support systems. It is always difficult to segment a medical image because of the complexity of medical images such as noise, blur, and intensity nonuniformity. To demonstrate the superiority of the presented FRFCM, a liver CT image

Table 5.3 Segmentation accuracy (SA%) of 10 algorithms on the first synthetic image with different noise. The best values are highlighted [1] / with permission of IEEE.

Noise	FCM	FCM_S1	FCM_S2	EnFCM	FGFCM	FLICM	NWFCM	KWFLICM	NDFCM	FRFCM
Gaussian 3%	73.70	98.47	98.23	98.86	98.88	99.1	91.91	**99.78**	98.81	**99.78**
Gaussian 5%	67.18	94.53	94.2	97.22	97.18	98.17	89.59	99.52	97.06	**99.64**
Gaussian 10%	59.67	80.53	80.31	88.17	87.46	90.85	89.61	95.45	87.49	**99.04**
Gaussian 15%	56.47	76.39	75.25	80.94	80.46	79.36	88.35	85.46	80.6	86.61
Salt-and-pepper 10%	94.28	94.85	99.80	95.70	98.65	93.37	99.91	**99.97**	98.60	99.94
Salt-and-pepper 20%	83.42	88.31	99.60	86.84	95.27	82.01	99.61	86.88	95.17	**99.86**
Salt-and-pepper 30%	77.05	78.42	98.89	83.00	89.29	71.51	98.62	99.31	87.75	**99.78**
Uniform 10%	93.64	97.05	99.85	98.08	98.88	96.53	99.89	**99.96**	98.81	99.93
Uniform 20%	87.14	93.09	99.28	94.82	96.64	90.22	99.42	**99.91**	96.43	99.88
Uniform 30%	80.6	88.3	97.43	90.25	92.58	80.72	97.97	**99.81**	92.19	99.79

Table 5.4 Comparison scores (S%) of the 10 algorithms on the second synthetic image with different noises. The best values are highlighted [1] / with permission of IEEE.

Noise	FCM	FCM_S1	FCM_S2	EnFCM	FGFCM	FLICM	NWFCM	KWFLICM	NDFCM	FRFCM
Gaussian 3%	56.69	95.47	95.34	96.64	96.67	97.73	95.31	99.32	96.48	**99.49**
Gaussian 5%	39.57	75.27	75.14	82.89	82.53	88.50	59.97	94.67	82.42	**98.73**
Gaussian 10%	36.61	67.92	68.82	77.04	76.70	82.44	55.63	89. 79	76.51	**97.97**
Gaussian 15%	36.31	67.34	68.55	76.98	76.65	81.93	54.81	89. 47	76.52	**96.41**
Salt-and-pepper 10%	85.65	89.23	99.72	89.78	96.90	84.66	99.77	45.51	96.81	**99.87**
Salt-and-pepper 20%	73.56	77.68	99.06	77.55	90.32	69.15	98.92	46.37	89.48	**99.67**
Salt-and-pepper 30%	62.33	61.53	96.98	67.86	77.8	53.32	95.36	41.88	76.89	**99.36**
Uniform 10%	86.15	92.79	99.56	95.09	97.12	79.59	99.60	44.67	96.91	**99.84**
Uniform 20%	73.96	84.50	97.93	88.17	91.43	77.99	98.44	52.6	90.85	**99.58**
Uniform 30%	63.25	74.70	93.13	78.68	82.23	63.33	94.81	55.98	81.56	**99.28**

(256×256) including a tumor is considered as a test image in this section. Figure 5.7 shows segmentation results of the tumor produced by different algorithms with $c = 5$.

In Figure 5.7a, the tumor is marked by a blue square. Our aim is to segment the tumor region from the liver CT image. It is clear that our algorithm shows an excellent performance for the detection of a tumor. Figure 5.7 shows that FCM, FCM_S1, FCM_S2, FLICM, NDFCM, and FRFCM are able to segment the tumor accurately, as shown in Figure 5.7b–d, g, j, and k, and EnFCM, FGFCM, NWFCM, and KWFLICM fail to segment the tumor, as shown in Figure 5.7e, f, h, i. Compared with the result from FCM, segmentation results from FCM_S1 and FCM_S2 are better, and the segmentation result from FRFCM provides a better visual effect for the tumor area. The segmentation result

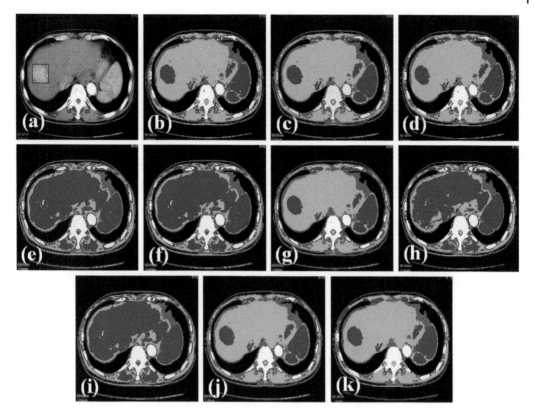

Figure 5.7   Comparison of segmentation results on the liver CT image [1] / with permission of IEEE. (a) Original image. (b) FCM result. (c) FCM_S1 result. (d) FCM_S2 result. (e) EnFCM result. (f) FGFCM result. (g) FLICM result. (h) NWFCM result. (i) KWFLICM result. (j) NDFCM result. (k) FMFFCM result.

can be used in 3D reconstruction of the tumor. And then, by computing the volume of the tumor, a doctor will make a correct diagnosis depending on the variation of the tumor volume.

Aurora is formed when solar wind collides with charged particles. It carries important information that reflects the invisible coupling between atmospheric layers. The analysis on aurora images is significant for research on space physics, such as climate change, global warming, and electromagnetic wave interference, [30]. Auroral oval segmentation is a key step in aurora image analysis, and it remains a challenging topic because of random noise, low contrast, and dayglow contamination in ultraviolet images. To extend the application of FRFCM in specified image segmentation, an aurora image (228×200) shown in Figure 5.8a is considered as a test image. Figure 5.8b–k show the comparison of segmentation results on auroral oval provided by different algorithms with $c = 3$.

As can be seen from Figure 5.8b, FCM is sensitive to noise. FCM_S1 and FCM_S2 improve the segmentation result obtained by FCM, but they are unable to segment aurora oval efficiently, as shown in Figure 5.8c, d. EnFCM and FGFCM fail to obtain segmentation results of aurora oval, as shown in Figure 5.8e, f. NWFCM and NDFCM are sensitive to noise, leading to poor segmentation results, as shown in Figure 5.8h, j. FLICM and KWFLICM provide better segmentation results than other algorithms, as shown in Figure 5.8g, i. However, they are time-consuming. The presented FRFCM achieves aurora oval segmentation, as shown in Figure 5.8k, with low computation time, and yet the segmentation result is better than for other algorithms.

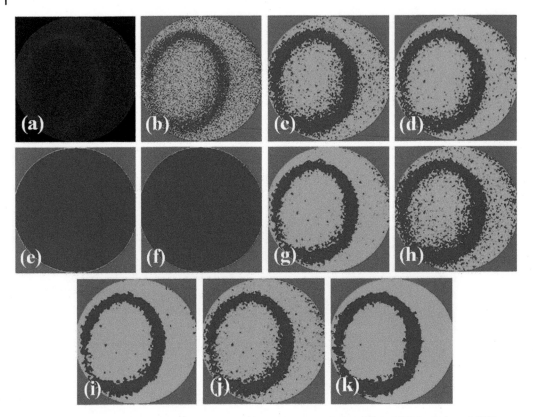

Figure 5.8 Comparison of segmentation results on the aurora image [1] / with permission of IEEE. (a) Original image. (b) FCM result. (c) FCM_S1 result. (d) FCM_S2 result. (e) EnFCM result. (f) FGFCM result. (g) FLICM result. (h) NWFCM result. (i) KWFLICM result. (j) NDFCM result. (k) FRFCM result.

5.5.3 Results on Color Images

Most of the improved FCM algorithms are only efficient for gray image segmentation, for it is difficult to obtain local spatial information of color images. However, FCM is able to segment color image with a shorter time, as local spatial information is neglected in FCM. It is easy to extend FCM_S1, FCM_S2, EnFCM, and NDFCM to color image segmentation because image filtering is performed on each channel of color images, respectively. Euclidean distance of pixels (3D vector) is employed in FLICM, KWFLICM, NWFCM, and FGFCM for color image segmentation, where the local spatial information is computed in each iteration of FLICM and KWFLICM. Thus, FLICM and KWFLICM have a very high computational complexity for color image segmentation. For EnFCM, FGFCM, and FMFFCM, the clustering is performed on pixels but not the gray-level histogram because it is difficult and complex to obtain the histogram of a color image. In addition, multivariate MR [31] is used in FRFCM to optimize data distribution, and the other steps are similar to gray image segmentation using FRFCM.

In this experiment, the tested images are chosen from the Berkeley Segmentation Dataset (BSDS500), which includes 500 images [32]. The selection of all parameters is the same as that for gray image segmentation except r in FRFCM ($r = 3$). We conducted experiments and applied these algorithms on BSDS500, and Figures 5.9 and 5.10 show the segmentation results. From Figure 5.9, we can see that all the algorithms fail to segment the color image except FRFCM. Figure 5.9k shows that FRFCM obtains excellent segmentation results without changing any

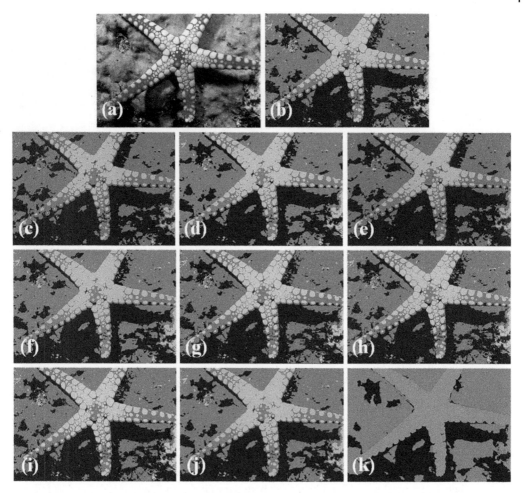

Figure 5.9 Comparison of segmentation results on color image "12003" from BSDS500 (c = 3) [1] / with permission of IEEE. (a) Original image. (b) FCM result. (c) FCM_S1 result. (d) FCM_S2 result. (e) EnFCM result. (f) FGFCM result. (g) FLICM result. (h) NWFCM result. (i) KWFLICM result. (j) NDFCM result. (k) FRFCM result.

parameters. Furthermore, to demonstrate the superiority of FRFCM, we implemented FRFCM on the data set of BSDS500, and some selected segmentation results are shown in Figure 5.10. Figure 5.10 shows that the segmentation results of different images have accurate contours, and we can obtain good object segmentation results using FRFCM, which is simple and significantly fast. It is clear that FRFCM provides excellent segmentation results for color images.

In this chapter, four performance measures, probabilistic rand index (PRI) [33], the boundary displacement error (BDE) [34], the covering (CV) [32], and the variation of information (VI) [32] are used to quantitatively evaluate segmentations obtained by different algorithms against the ground truth segmentation.

The *PRI* is a similarity measure that counts the fraction of pairs of pixels whose labels are consistent between the computed segmentation, S', and the corresponding ground truth segmentation, S. *PRI* can be calculated as follows:

$$PRI(S, S') = 1 - \left(\sum_i \left(\sum_j p_{ij} \right)^2 - 2 \sum_j \left(\sum_j p_{ij} \right)^2 + 2 \sum \sum p_{ij}^2 \right) / N, \qquad (5.24)$$

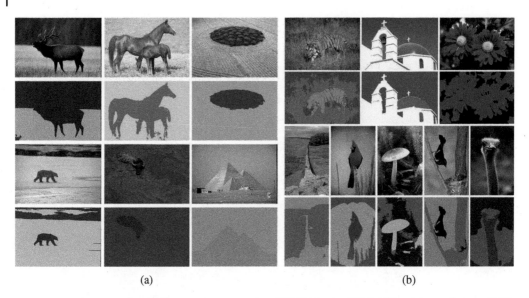

(a)   (b)

Figure 5.10 Segmentation results on color images from BSDS500 using FRFCM [1] / with permission of IEEE. (a) $c = 2$. (b) $c = 3$.

where p_{ij} is the number of pixels in the ith cluster of S and the jth cluster of S', and N is the total number of pixels of the image.

The *BDE* is an error measure that is used to measure the average displacement error of boundary pixels between two segmentations, and it is defined as

$$BDE(S, S') = \left(\left(\sum_i^{N_1} d(p_i, S) \right) / N_1 + \left(\sum_i^{N_2} d(p_i, S') \right) / N_2 \right) / 2, \tag{5.25}$$

where N_1 and N_2 denote the total number of points in the boundary sets S' and S, respectively. d is a distance between a pixel p_i in S' and its closest boundary pixel p in S, and it is defined as follows:

$$d(p_i, S) = \min_{p \in S} \| p_i - p \| . \tag{5.26}$$

The *CV* is an overlap measure that can be also used to evaluate the segmentation effect. It is defined as

$$CV(S \to S') = \left(\sum_{R \in S} |R| \cdot \max_{R' \in S'} O(R, R') \right) / N, \tag{5.27}$$

where $O(R, R') = |R \cap R'| / |R \cup R'|$ denotes the overlap between two regions R and R'.

The *VI* is a similarity measure that is always used to measure the distance between two segmentations in terms of their average conditional entropy and is given by

$$VI(S, S') = H(S) + H(S') - 2I(S, S'), \tag{5.28}$$

where H and I represent the entropies and mutual information between two segmentations S and S', respectively.

When the final segmentation is close to the ground truth segmentation, the PRI and CV are larger while the *BDE* and VI are smaller. All these algorithms are evaluated on BSDS500, and the average values of *PRI*, *BDE*, *CV*, and VI of segmentation results are given in Table 5.5. c is set from 2 to 6 for

Table 5.5 Average performance of 10 algorithms on BSDS500 that includes 500 images (The best values are highlighted) [1] / with permission of IEEE.

Algorithm	PRI	BDE	CV	VI
FCM	0.72	14.06	0.39	3.15
FCM_S1	0.69	13.45	0.44	2.75
FCM_S2	0.73	13.89	0.40	3.04
EnFCM	0.69	13.47	0.44	2.74
FGFCM	0.73	13.76	0.40	3.06
FLICM	0.72	13.94	0.39	3.11
NWFCM	0.72	13.95	0.39	3.14
KWFLICM	0.72	13.88	0.39	3.10
NDFCM	0.73	13.71	0.40	3.05
FRFCM	**0.76**	**13.03**	**0.46**	**2.59**

each image in BSDS500. We choose a best c corresponding to the highest PRI. The average values of *PRI*, *VI*, *CV*, and *BDE* obtained by different algorithms are presented in Table 5.5. We can see that our FRFCM clearly outperforms other algorithms on *PRI*, *BDE*, *CV,* and *VI* values.

From experiments in Sections 5.5.1–5.5.3, the presented FRFCM is able to provide good segmentation results for different types of images. Moreover, it has a better performance than other algorithms.

5.5.4 Running Time

Based on the analysis above, the computational complexity of different algorithms are given in Table 5.6, where N is the number of pixels in an image, c is the number of clustering prototypes, b is the number of iterations, w is the size of the filtering window, and τ is the number of gray levels in the image. Generally, $\tau \ll N$.

According to Table 5.6, EnFCM, FGFCM, and FMFFCM have low computational complexity due to $\tau \ll N$ (the gray level of the tested image is $\tau = 256$, and the number of pixels in the tested image is

Table 5.6 Computational complexity of 10 algorithms [1] / with permission of IEEE.

Algorithm	Computational complexity
FCM	$O(N{\times}c{\times}b)$
FCM_S1	$O(N{\times}w^2+N{\times}c{\times}b)$
FCM_S2	$O(N{\times}w^2+N{\times}c{\times}b)$
EnFCM	$O(N{\times}w^2+\tau{\times}c{\times}b)$
FGFCM	$O(N{\times}w^2+\tau{\times}c{\times}b)$
FLICM	$O(N{\times}c{\times}b{\times}w^2)$
NWFCM	$O(N{\times}(w+1)^2+N{\times}c{\times}b)$
NDFCM	$O(N{\times}w^2+N{\times}c{\times}b)$
FRFCM	$O(N{\times}w^2+\tau{\times}c{\times}b)$

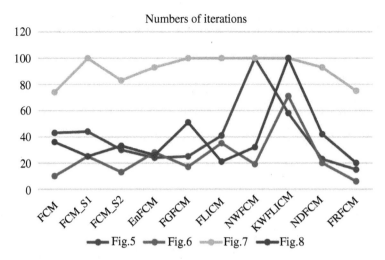

Figure 5.11 Comparison of numbers of iterations of 10 algorithms on tested images [1] / with permission of IEEE.

Table 5.7 Comparison of execution times (in seconds) of 10 algorithms on tested images [1] / with permission of IEEE. The best values are highlighted.

Image	FCM	FCM_S1	FCM_S2	EnFCM	FGFCM	FLICM	NWFCM	KWFLICM	NDFCM	FRFCM
Figure 5.5	0.62	0.28	0.23	1.99	12.34	26.04	36.66	50.42	1.85	**0.06**
Figure 5.6	0.22	0.28	0.16	0.25	0.27	38.63	13.8	87.49	2.15	**0.03**
Figure 5.7	1.66	1.19	1.00	2.25	8.31	110.49	59.06	123.57	3.69	**0.12**
Figure 5.8	0.44	0.20	0.21	0.39	5.12	10.07	9.88	54.15	1.48	**0.05**

$N = 256{\times}256$). Moreover, to estimate the applicability of different algorithms, we compared their running times. All experiments are performed on a workstation with an Intel core (TM) i7-6700, 3.4 GHz CPU, and 16G memory using MATLAB. Figure 5.11 shows the number of iterations, and Table 5.7 shows execution times (in seconds) of different algorithms on tested images.

From Table 5.7, it is clear that KWFLICM and FLICM have a very high computational complexity compared to other algorithms. NWFCM is also slow because the computation of neighborhood weights based on patch distance is complex. FCM_S1 and FCM_S2 are fast because mean-filtered images and median-filtered images are computed in advance. EnFCM is fast due to the introduction of gray level, which is far less than the number of pixels in an image. FGFCM is not fast because the computation of filtered image is complex. FRFCM only employs MR and membership filtering, where MR is performed in advance, and median filtering is implemented only once after clustering. Moreover, the idea of histogram is also used in FRFCM. Therefore, the objective function of FRFCM converges very fast, and FRFCM requires a very small computational time. In addition, we presented the comparison of the number of iterations in Figure 5.11. We can see that FRFCM requires the least number of iterations.

In Section 5.5.3, the algorithms mentioned above are extended to color images, and Figures 5.9 and 5.10 show segmentation results. In contrast with gray image segmentation, they require much more time to segment color image due to the increase of the dimension of data. Table 5.8 shows the comparison of computational complexity of different algorithms for Figure 5.9.

Table 5.8 Comparison of numbers of iterations and execution times (in seconds) of 10 algorithms on color images (Figure 5.9) [1] / with permission of IEEE.

	FCM	FCM_S1	FCM_S2	EnFCM	FGFCM	FLICM	NWFCM	KWFLICM	NDFCM	FRFCM
Numbers of iterations	36	100	70	99	46	68	50	76	37	44
Running time	1.52	4.71	3.33	4.466	4.91	366.22	160.78	451.90	12.32	2.73

From Table 5.8, we can see that the computational cost of FLICM, KWFLICM, and NWFCM is extremely large for color image segmentation. Although FCM_S1, FCM_S2, EnFCM, FGFCM, NDFCM, and FMFFCM have similar computational complexity for color images ($q = N$), FRFCM is significantly faster than these algorithms, as shown in Table 5.8, and obtains better segmentation results, as shown in Figures 5.9 and 5.10.

5.5.5 Summary

In this chapter, a significantly fast and robust FRFCM algorithm [1] for image segmentation has been presented to improve the segmentation quality and reduce the influence of image noise. By introducing the MR operation, the local spatial information of images has been utilized to improve the segmentation effect. Because the MR is able to suppress noise while preserving the contour of objects, a trade-off has easily been achieved between noise suppression and detail preservation. Moreover, the MR is able to provide good reconstructed results for images corrupted by different types of noises. Furthermore, FRFCM employs the membership filtering to exploit the local spatial constraint. We demonstrated that the membership filtering is able to provide similar results compared with a local spatial constraint, but a local spatial constraint requires much more time than the membership filtering in each iteration. Experimental results show that the presented FRFCM is able to provide better segmentation results without tuning parameters for different grayscale or color images.

However, similar to other improved FCM algorithms, the number of clusters is also set experimentally in FRFCM. In the future, we will explore new FCM algorithms that automatically set the number of clusters. In addition, the selection of a mask image or marker image is also an open problem; some better results can be achieved by changing the mask or marker image.

References

1 Lei, T., Jia, X., Zhang, Y. et al. (2018). Significantly fast and robust fuzzy c-means clustering algorithm based on morphological reconstruction and membership filtering. *IEEE Trans. Fuzzy Syst.* **26** (5): 3027–3041.

2 Pont-Tuset, J., Arbelaez, P., Barron, J.T. et al. (2016). Multiscale combinatorial grouping for image segmentation and object proposal generation. *IEEE Trans. Pattern Anal. Mach. Intell.* **39** (1): 128–140.

3 Cao, H., Deng, H.W., and Wang, Y.P. (2011). Segmentation of M-FISH images for improved classification of chromosomes with an adaptive fuzzy C-means clustering algorithm. *IEEE Trans. Fuzzy Syst.* **20** (1): 1–8.

4 Javed, A., Kim, Y.C., Khoo, M.C. et al. (2015). Dynamic 3-D MR visualization and detection of upper airway obstruction during sleep using region-growing segmentation. *IEEE Trans. Biomed. Eng.* **63** (2): 431–437.

5 Grau, V., Mewes, A.U.J., Alcaniz, M. et al. (2004). Improved watershed transform for medical image segmentation using prior information. *IEEE Trans. Med. Imaging* **23** (4): 447–458.

6 Gong, M., Li, H., Zhang, X. et al. (2015). Nonparametric statistical active contour based on inclusion degree of fuzzy sets. *IEEE Trans. Fuzzy Syst.* **24** (5): 1176–1192.

7 ComaniciuD, M. (2002). Arobust approach toward feature space analysis. *IEEE Trans. Pattern Anal. Mach. Intell.* **24** (5): 313–329.

8 Mahapatra, D. (2017). Semi-supervised learning and graph cuts for consensus based medical image segmentation. *Pattern Recogn.* **63**: 700–709.

9 Li, Z., and Chen, J. (2015). Superpixel segmentation using linear spectral clustering. In *Proceedings of the IEEE Conference on Computer Vision and Pattern Recognition* (pp. 1356–1363).

10 Chatzis, S.P. and Varvarigou, T.A. (2008). A fuzzy clustering approach toward hidden Markov random field models for enhanced spatially constrained image segmentation. *IEEE Trans. Fuzzy Syst.* **16** (5): 1351–1361.

11 Pathak, D., Krahenbuhl, P., and Darrell, T. (2015). Constrained convolutional neural networks for weakly supervised segmentation. In *Proceedings of the IEEE International Conference on Computer Vision* (pp. 1796–1804).

12 Bezdek, J.C., Ehrlich, R., and Full, W. (1984). FCM: the fuzzy c-means clustering algorithm. *Comput. Geosci.* **10** (2–3): 191–203.

13 Ahmed, M.N., Yamany, S.M., Mohamed, N. et al. (2002). A modified fuzzy c-means algorithm for bias field estimation and segmentation of MRI data. *IEEE Trans. Med. Imaging* **21** (3): 193–199.

14 Chen, S. and Zhang, D. (2004). Robust image segmentation using FCM with spatial constraints based on new kernel-induced distance measure. *IEEE Trans. Syst. Man Cybern. B Cybern.* **34** (4): 1907–1916.

15 Szilagyi, L., Benyo, Z., Szilágyi, S. M., and Adam, H. S. (2003, September). MR brain image segmentation using an enhanced fuzzy c-means algorithm. In *Proceedings of the 25th Annual International Conference of the IEEE Engineering in Medicine and Biology Society* (IEEE Cat. No. 03CH37439) (Vol. 1, pp. 724–726). IEEE.

16 Cai, W., Chen, S., and Zhang, D. (2007). Fast and robust fuzzy c-means clustering algorithms incorporating local information for image segmentation. *Pattern Recogn.* **40** (3): 825–838.

17 Krinidis, S. and Chatzis, V. (2010). A robust fuzzy local information C-means clustering algorithm. *IEEE Trans. Image Process.* **19** (5): 1328–1337.

18 Gong, M., Zhou, Z., and Ma, J. (2011). Change detection in synthetic aperture radar images based on image fusion and fuzzy clustering. *IEEE Trans. Image Process.* **21** (4): 2141–2151.

19 Gong, M., Liang, Y., Shi, J. et al. (2012). Fuzzy c-means clustering with local information and kernel metric for image segmentation. *IEEE Trans. Image Process.* **22** (2): 573–584.

20 Nguyen, M.P. and Chun, S.Y. (2017). Bounded self-weights estimation method for non-local means image denoising using minimax estimators. *IEEE Trans. Image Process.* **26** (4): 1637–1649.

21 Saranathan, A.M. and Parente, M. (2015). Uniformity-based superpixel segmentation of hyperspectral images. *IEEE Trans. Geosci. Remote Sens.* **54** (3): 1419–1430.

22 Zaixin, Z., Lizhi, C., and Guangquan, C. (2014). Neighbourhood weighted fuzzy c-means clustering algorithm for image segmentation. *IET Image Process.* **8** (3): 150–161.

23 Guo, F.F., Wang, X.X., and Shen, J. (2016). Adaptive fuzzy c-means algorithm based on local noise detecting for image segmentation. *IET Image Process.* **10** (4): 272–279.

24 Vincent, L. (1993). Morphological grayscale reconstruction in image analysis: applications and efficient algorithms. *IEEE Trans. Image Process.* **2** (2): 176–201.

25 Najman, L. and Schmitt, M. (1996). Geodesic saliency of watershed contours and hierarchical segmentation. *IEEE Trans. Pattern Anal. Mach. Intell.* **18** (12): 1163–1173.

26 Gong, M., Su, L., Jia, M., and Chen, W. (2013). Fuzzy clustering with a modified MRF energy function for change detection in synthetic aperture radar images. *IEEE Trans. Fuzzy Syst.* **22** (1): 98–109.

27 Mendonca, A.M. and Campilho, A. (2006). Segmentation of retinal blood vessels by combining the detection of centerlines and morphological reconstruction. *IEEE Trans. Med. Imaging* **25** (9): 1200–1213.

28 Chen, J.J., Su, C.R., Grimson, W.E.L. et al. (2011). Object segmentation of database images by dual multiscale morphological reconstructions and retrieval applications. *IEEE Trans. Image Process.* **21** (2): 828–843.

29 Soille, P. (2013). *Morphological Image Analysis: Principles and Applications.* Springer Science & Business Media.

30 Yang, X., Gao, X., Tao, D. et al. (2015). Shape-constrained sparse and low-rank decomposition for auroral substorm detection. *IEEE Trans. Neural Netw. Learn. Syst.* **27** (1): 32–46.

31 Lei, T., Zhang, Y., Wang, Y. et al. (2017). A conditionally invariant mathematical morphological framework for color images. *Inform. Sci.* **387**: 34–52.

32 Arbelaez, P., Maire, M., Fowlkes, C., and Malik, J. (2010). Contour detection and hierarchical image segmentation. *IEEE Trans. Pattern Anal. Mach. Intell.* **33** (5): 898–916.

33 Unnikrishnan, R., Pantofaru, C., and Hebert, M. (2007). Toward objective evaluation of image segmentation algorithms. *IEEE Trans. Pattern Anal. Mach. Intell.* **29** (6): 929–944.

34 Wang, X., Tang, Y., Masnou, S., and Chen, L. (2015). A global/local affinity graph for image segmentation. *IEEE Trans. Image Process.* **24** (4): 1399–1411.

6

Fast Image Segmentation Using Watershed Transform

We introduce the image segmentation algorithm based on fuzzy clustering in Chapter 5. In this chapter, we will focus on the watershed transform (WT) and power watershed (PW), but they have a serious over-segmentation phenomenon. Morphological reconstruction (MR) is typically used to filter the seeds (regional minimum) in a gradient image to reduce over-segmentation. However, MR might mistakenly filter meaningful seeds that are required for generating accurate segmentation results, and it is also sensitive to the scale parameters because a single-scale structuring element is employed. In this chapter, a novel adaptive morphological reconstruction (AMR) operation is presented that has three advantages. This is based on the recently published article by Lei et al. [1]. Firstly, AMR can adaptively filter useless seeds while preserving meaningful ones. Secondly, AMR is insensitive to the scale of structuring elements because multiscale structuring elements are employed. Finally, AMR has two attractive properties: monotonic increasing-ness and convergence, which help seeded segmentation algorithms to achieve a good hierarchical segmentation. Experiments clearly demonstrate that AMR is useful for improving algorithms of seeded image segmentation and seed-based spectral segmentation. Compared to several state-of-the-art algorithms, the presented algorithms can provide better segmentation results but require less computing time.

6.1 Introduction

MR [2] is a powerful operation in mathematical morphology. It has been widely used in image filtering, image segmentation, feature extraction, and so forth. In these applications, one kind of the most important applications is seeded segmentation algorithms, such as WT [3] and PW [4]. However, there are two drawbacks [5] when MR is used in seeded segmentation algorithms.

First, it is difficult to reduce over-segmentation while obtaining a high segmentation accuracy for seeded segmentation algorithms (we use MR-WT to denote MR-based watershed transform and use MR-PW to denote MR-based PW). Although MR is able to filter noise in gradient images, some important contour details are smoothed out as well.

Second, MR is sensitive to the scale of structuring elements. In practical applications, if the scale is too small, the reconstructed gradient image suffers from a serious over-segmentation. Conversely, if the scale is too large, the reconstructed gradient image suffers from an under-segmentation.

Generally, MR is used in the watershed transform to improve the segmentation effect by employing a structuring element to filter regional minima [6]. However, it is very difficult to filter useless

Image Segmentation: Principles, Techniques, and Applications, First Edition. Tao Lei and Asoke K. Nandi.
© 2023 John Wiley & Sons Ltd. Published 2023 by John Wiley & Sons Ltd.

regional minima while preserving meaningful ones by simply considering one single-scale structuring element. Although H-min imposition [7] is a simple and efficient method for over-segmentation reduction, it relies on a threshold choice and is likely to miss some important boundaries. Region merging [8] is also a popular method for this, but it requires iterating and renewing edge weight, leading to a high computing burden. In addition, some researchers employ reasonable contour detection methods, e.g. globalized probability of boundary (gPb) [9], which combines the multiscale information from brightness, color, and texture, to achieve better image segmentation. However, the gPb is computationally expensive because it combines too many feature cues for contour detection. To speed up the algorithm of contour detection, Dollar and Zitnick [10] took advantage of the structure present in regional image patches and random decision forests and proposed a fast and structured edge (SE) detection approach using structured forests. This algorithm obtains real-time performance and state-of-the-art edge detection but requires a huge memory for training data. To reduce the memory requirement, Hallman and Fowlkes [11] proposed a simple and efficient model for learning boundary detection, namely oriented edge forests (OEF). Although these improved contour detectors are superior to traditional detectors, e.g. Sobel or Canny, and they are helpful for improving subsequent image segmentation, they still generate a large number of seeds leading to serious over-segmentation.

In practice, contour detection methods are usually combined with other approaches to improve the image segmentation effect. For example, Fu et al. [12] proposed a robust image segmentation approach using contour-guided color palettes by integrating contour and color cues, where SE, mean-shift algorithm [13], region merging, and spectral clustering [14] are combined to achieve better segmentation results. However, the approach is complex because it combines several different algorithms that require many parameters.

In this chapter, we present an AMR operation [1] that is able to generate a better seed image than MR to improve seeded segmentation algorithms. Firstly, AMR employs multiscale structuring elements to reconstruct a gradient image. Secondly, a pointwise maximum operation on these reconstructed gradient images is performed to obtain the final adaptive reconstruction result. Because AMR employs small structuring elements to reconstruct pixels of large gradient magnitudes while employing large structuring elements to reconstruct pixels of small gradient magnitudes in a gradient image, AMR is able to obtain better seed images to improve the seeded segmentation algorithms.

The main advantages are summarized as follows.

- Multiscale structuring elements are employed by AMR, and different scaled structuring elements are adaptively adopted by pixels of different gradient magnitudes without computing the local features of a gradient image.
- AMR has a convergence property and a monotonic increasing property, and the two properties help seeded segmentation algorithms to achieve a hierarchical segmentation.
- AMR has a low computational complexity, and it can help seed-based spectral segmentation to achieve better image segmentation results than state-of-the-art algorithms.

The rest of the chapter is organized as follows. In the next section, the research background related with AMR is introduced and analyzed. On this basis, AMR is presented, and its two properties, monotonic increasing-ness and convergence, are carefully analyzed in Section 6.3. To demonstrate the superiority of AMR, AMR is used for seeded image segmentation and seed-based spectral segmentation. Experiments are presented in Section 6.4, followed by discussion and summary in Section 6.5.

6.2 Related Work

6.2.1 Morphological Opening and Closing Reconstructions

MR is an image transformation that requires two input images, a marker image and a mask image. Let two grayscale images f and g denote the marker image that is the starting point for the transformation and the mask image that constrains the transformation, respectively [15]. If $f \leq g$, which means f is pointwise less than or equal to g, the morphological dilation reconstruction (R^{δ_m}) of g from f is denoted by

$$R_g^{\delta_m}(f) = \delta_{mg}^{(n)}(f), \tag{6.1}$$

where $\delta_{mg}^{(1)}(f) = \delta_m(f) \wedge g$, $\delta_{mg}^{(k)}(f) = \delta_m\left(\delta_{mg}^{(k-1)}(f)\right) \wedge g$ for $2 \leq k \leq n$, $k, n \in N^+$ satisfies $\delta_{mg}^{(n)}(f) = \delta_{mg}^{(n-1)}(f)$. The symbol δ_m represents the elementary morphological dilation operation, and $\wedge$ stands for the pointwise minimum at each pixel of two images, as shown in Figure 6.1a.

Similarly, if $f \geq g$, the morphological erosion reconstruction (R^{ε_m}) of g from f, which is the dual operation of R^{δ_m}, is defined as

$$R_g^{\varepsilon_m}(f) = \varepsilon_{mg}^{(n)}(f), \tag{6.2}$$

where $\varepsilon_{mg}^{(1)}(f) = \varepsilon_m(f) \vee g$, $\varepsilon_{mg}^{(k)}(f) = \varepsilon_m\left(\varepsilon_{mg}^{(k-1)}(f)\right) \vee g$ for $2 \leq k \leq n$, $k, n \in N^+$ satisfies $\varepsilon_{mg}^{(n)}(f) = \varepsilon_{mg}^{(n-1)}(f)$. The symbol ε_m represents the elementary morphological erosion operation, and $\vee$ stands for the pointwise maximum at each pixel of two images, as shown in Figure 6.1b.

To further illustrate the principle of MR for image transformation, we present an example for the binary MR, as shown in Figure 6.2, where the red regions denote seeds, i.e. the marker image f.

According to Figure 6.2 and (6.1)–(6.2), a suitable marker image is important for MR. We have known that $f \leq g$ for R^{δ_m} while $f \geq g$ for R^{ε_m}. Thus, there are lots of choices for f. Different marker images correspond to different reconstruction results, as shown in Figure 6.3. To obtain an efficient f in practice, the marker image is usually obtained by performing a transformation on the corresponding mask image [16]. For example, the erosion or dilation result of a mask image is often

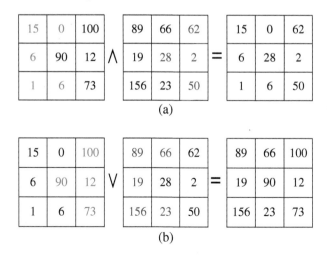

Figure 6.1 An example for pointwise extremum operation [1] / with permission of IEEE. (a) Pointwise minimum. (b) Pointwise maximum.

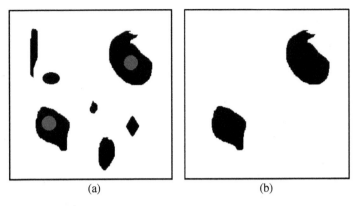

(a) (b)

Figure 6.2 Binary MR from markers [1] / with permission of IEEE. (a) A mask image. (b) Reconstructed result.

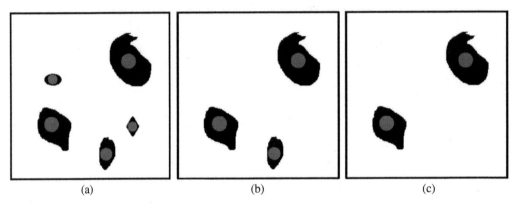

(a) (b) (c)

Figure 6.3 Binary MR from different markers [1] / with permission of IEEE. (a) Mask 1. (b) Mask 2. (c) Mask 3.

considered as a marker image [17], i.e. $f = \varepsilon_{mb_i}(g)$ or $f = \delta_{mb_i}(g)$, where b_i is a disk-shaped structuring element, and the radius of b_i is i, $i \in N^+$. Therefore, MR is sensitive to the parameter i because the marker image is decided by the scale of the structuring element.

As compositional morphological opening and closing operations show better performance than elementary morphological erosion and dilation operations for image filtering, feature extraction, and so forth, we present the definition of compositional morphological opening and closing reconstructions (R^O and R^C) of g from f as follows

$$\begin{cases} R_g^O(f) = R_g^{\delta_m}\left(R_g^{\varepsilon_m}(f)\right) \\ R_g^C(f) = R_g^{\varepsilon_m}\left(R_g^{\delta_m}(f)\right) \end{cases}. \tag{6.3}$$

6.2.2 Multiscale and Adaptive Mathematical Morphology

For image filtering and enhancement using morphological operators, a large-scale structuring element can suppress noise but may also blur the image details, whereas a small-scale structuring element can preserve image details but may fail to suppress noise. Some researchers have proposed multiscale and adaptive morphological operators to improve the performance of traditional morphological operators. However, most multiscale morphological operators [18], such as morphological gradient operators and morphological filtering operators, average all scales of morphological operation results as final output:

$$y = \frac{1}{\lambda} \sum_{j=1}^{\lambda} g_j, \tag{6.4}$$

where y is the final output result, and j is the radius of the structuring element, $1 \leq j \leq \lambda, j, \lambda \in N^+$. Although the average result is superior to the result based on single-scale morphological operators, it causes contour offset and mistakes. Some researchers have improved multiscale morphological operators by introducing a weighted coefficient to (6.4), and they have defined adaptive multiscale morphological operators as follows [19]

$$y = \frac{1}{\lambda} \sum_{j=1}^{\lambda} \omega_{cj} g_j, \tag{6.5}$$

where ω_{cj} is the weighted coefficient on the jth scale result. However, because the computing of weighted coefficient is complex, the adaptive multiscale morphological operators have low computational efficiency. Moreover, the weighted average result is similar to average result because it is difficult to obtain the optimal weighted coefficient, even though the former is slightly better than the latter.

Although lots of adaptive multiscale morphological operators [20] have been proposed, it can be seen from (6.4)–(6.5) that both the multiscale and adaptive morphological operators employ a linear combination of different scale results to improve single-scale morphological gradient or filtering operators. Because the linear combination is unsuitable for multiscale MR operation, in this chapter we try to employ a nonlinear combination (i.e. the pointwise maximum operation denoted by $\vee$) to design AMR operators. These operators are different from conventional multiscale and adaptive morphological operators employing linear combination in (6.4)–(6.5). We use the nonlinear operation $\vee$ instead of a linear combination since the former is more suitable than the latter for the removal of useless seeds in seeded image segmentation.

6.2.3 Seeded Segmentation

Seeded segmentation algorithms, such as graph cuts [21], random walker [22], watersheds [3], and PW [4], have been widely used in complex image segmentation tasks due to their good performance. It is not required to give seed images for both graph cuts and random walker because they usually consider each pixel as a seed. However, a seed image is necessary for WT and PW by computing the regional minima of a gradient image.

Since both WT and PW obtain seeds from a gradient image that often includes a huge number of seeds generated by noise and unimportant texture details, they usually suffer from over-segmentation. A larger number of approaches for addressing over-segmentation has been proposed, and these approaches can be categorized into two groups.

- Feature extraction or feature learning is used to obtain a better gradient image that enhances important contours while smoothing noise and texture details [9].
- MR is used to gradient reconstruction to reduce the number of regional minima [23].

BSDS (http://www.eecs.berkeley.edu/Research/Projects/CS/vision/bsds) is a very popular image dataset, and it is often used for the evaluation of image segmentation algorithms. For each image in BSDS, there are 4–9 ground truths segmentations that are delineated by different human subjects.

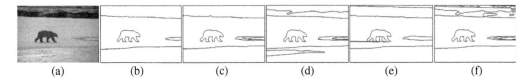

Figure 6.4 The original image and ground truths (GT) from BSDS500 [1] / with permission of IEEE. (a) "100007." (b) GT 1. (c) GT 2. (d) GT 3. (e) GT 4. (f) GT 5.

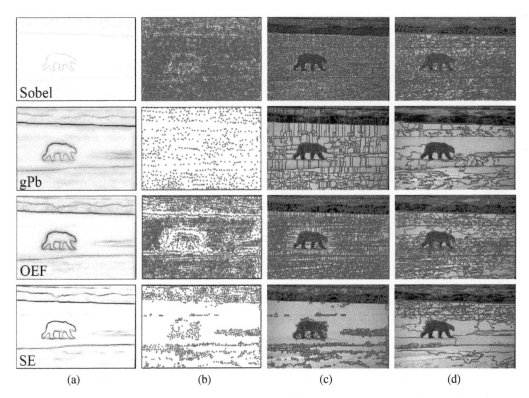

Figure 6.5 Over-segmentation reduction by improving the gradient image of "100007" [1]. (a) Different gradient images. (b) Seed images (regional minima). (c) WT. (d) PW (p = 2) [4].

For the first group of approaches, gPb, OEF, and SE are popular for reducing over-segmentation, as shown in Figures 6.4–6.5. In Figure 6.5, although gPb, OEF, and SE provide better gradient images that can reduce over-segmentation for WT and PW, the segmentation results are still poor compared to ground truths shown in Figure 6.4.

The second group of approaches depends on MR and WT and is denoted by MR-WT. Najman and Schmitt [23] employed MR to remove regional minima to reduce over-segmentation. Furthermore, a dynamic threshold is used to change the gradient magnitude that is smaller than the threshold, and then a hierarchical segmentation result is obtained. Wang [18] proposed a multiscale morphological gradient algorithm (MMG) for image segmentation using watersheds. The proposed MMG employs multiple structuring elements to obtain a better gradient image and uses MR to remove regional minima to improve watershed segmentation.

Figure 6.6 illustrates the principle of seeded segmentation framework based on MR-WT. We can see that when the number of the regional minima of the gradient image decreases rapidly with the increase of i, the boundary is also destroyed simultaneously. It is clear that the larger structuring

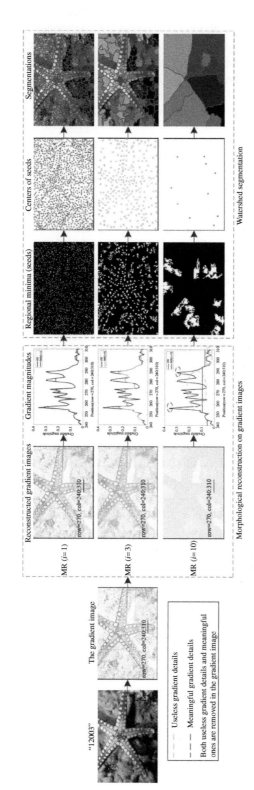

Figure 6.6 A Seeded segmentation framework based on MR-WT [1]. $(R_g^C(f))$ is employed to reconstruct gradient image, and original gradient [OG] denotes a row of original gradient image) [1] / with permission of IEEE.

element corresponds to fewer seeds. One major reason is that MR employs a single-scale structuring element, which equally treats all pixels of different gradient magnitudes in the gradient image. For example, in dilation reconstruction, the marker image $f = \varepsilon_{mb_i}(g)$ converges to the minimum grayscale value of pixels in the mask image as the value of i increases. Obviously, both large and small structuring elements lead to poor reconstruction results while a moderate structuring element achieves a rough balance via sacrificing contour precision. Therefore, it is difficult to obtain a good seed image by employing a single-scale structuring element. Although many researchers employ multiscale structuring elements to generate a better gradient image, there are few studies on multiscale MR for gradient images. Moreover, the fusion of different scale results is also a problem.

6.2.4 Spectral Segmentation

It is well known that spectral clustering [14] is greatly successful due to the fact that it does not make strong assumptions on data distribution, and it is implemented efficiently even for large data sets, as long as we make sure that the affinity matrix is sparse. However, since the size of the affinity matrix is $(M \times N)^2$ for an image of size $M \times N$, and it is not sparse because of Gaussian similarity measure, spectral clustering is often inefficient for image segmentation due to eigenvalue decomposition of the huge affinity matrix. To address the issue, a great number of algorithms have been proposed to construct a smaller affinity matrix and thus to improve the computational efficiency of spectral clustering [24]. Most of these algorithms employ pre-segmentation (superpixel) methods such as the simple linear iterative clustering (SLIC) [25], meanshift [13], linear spectral clustering (LSC) [26], and superpixel hierarchy [27], to reduce the number of pixels of the original image and, in turn, reduce the size of the affinity matrix. As an example, Zhang et al. [28] proposed a fast image segmentation approach that is a reexamination of spectral clustering on image segmentation. The approach provides better image segmentation results yet requires a long running time.

The popular superpixel approaches have some drawbacks for spectral segmentation. Firstly, the mean-shift algorithm involves three parameters and is sensitive to these parameters. Secondly, SLIC only generates superpixels that include regular regions, and these regions have a similar shape and size. Finally, LSC is superior to SLIC because LSC successfully connects a local feature with a global optimization objective function so that LSC can generate more reasonable segmentation results. However, similar to SLIC, LSC also provides superpixels that include regular regions with a similar shape and size.

As seed-based spectral segmentation algorithms are sensitive to pre-segmentation results, an excellent pre-segmentation algorithm can improve segmentation results generated by seed-based spectral segmentation algorithms.

6.3 Adaptive Morphological Reconstruction (AMR)

6.3.1 The Presented AMR

To overcome the drawback of MR on regional minima filtering, this chapter presents an AMR [1] that is able to filter useless regional minima and maintain meaningful ones generated by salient objects. Figure 6.7 shows the motivation of AMR in which multiscale structuring elements are employed to reconstruct gradient images, i.e. small structuring elements are chosen by pixels of large gradient magnitude while large structuring elements are chosen by pixels of small gradient agnitudee.

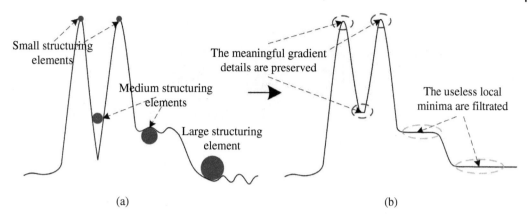

Small structuring elements

Medium structuring elements

Large structuring element

The meaningful gradient details are preserved

The useless local minima are filtrated

(a)

(b)

Figure 6.7   The motivation of AMR [1] / with permission of IEEE. (a) Gradient. (b) Reconstructed gradient.

Definition 6.1   Let $b_s \subseteq \cdots b_i \subseteq b_{i+1} \cdots \subseteq b_t$ be a series of nested structuring elements, where i is the scale parameter of a structuring element, $1 \leq s \leq i \leq t$, s, i, $t \in N^+$. For a gradient images g such that $f = \varepsilon_{b_i}(g)$ and $f \leq g$, the AMR denoted by ψ of g from f is defined as

$$\psi(g, s, t) = \bigvee_{s \leq i \leq t} \left\{ R_g^C(f)_{b_i} \right\}. \tag{6.6}$$

Note that the pointwise maximum operation is only suitable for R^C but not suitable for R^O. Because $\lim_{t \to \infty} R_g^O(f)_{b_t} = max(g)$ and $\lim_{t \to \infty} \bigvee_{s \leq i \leq t} \left\{ R_g^O(f)_{b_i} \right\} = max(g)$, $\psi(g, s, t)$ is unable to obtain a significantly convergent gradient image if $\psi(g, s, t) = \bigvee_{s \leq i \leq t} \left\{ R_g^O(f)_{b_i} \right\}$.

We apply AMR to the gradient image shown in Figure 6.6. The reconstruction and segmentation results are shown in Figure 6.8, where the adopted structuring elements are a disk and $s = 1$. More detailed comparisons are shown in Figure 6.9. By comparing Figure 6.8 with Figure 6.6, it is obvious that AMR obtains better seed images than MR due to the fact that the nonlinear operation $\vee$ is able to efficiently remove useless seeds.

To further show the influence of s on AMR, Figure 6.10 shows segmentation results provided by AMR through changing the value of s. We can see that there are some small segmented areas when the value of s is small. These small areas are merged by increasing the value of s. However, although a large s leads to the merging of small areas, the precision of object contours will be decreased, as shown in Figure 6.6. Therefore, we usually set $1 \leq s \leq 3$ for a moderate-sized image.

6.3.2   The Monotonic Increasing-ness Property of AMR

AMR is an approach that aims at finding meaningful regional minima by merging or filtering useless regional minima. AMR includes two parameters, s and t. When we increase the value of t, gradient images reconstructed by AMR keep the increasing order as shown in Theorem 6.1.

Theorem 6.1   Let ψ be an AMR operator. If ψ is increasing with respect to the scale of structuring elements, i.e. for a gradient image g such that $f = \varepsilon_{mb_i}(g)$ and $f \leq g$, $1 \leq p$, $q \leq t$, p, q, $t \in N^+$, then we have

$$p \leq q \Rightarrow \psi(g, s, p) \leq \psi(g, s, q). \tag{6.7}$$

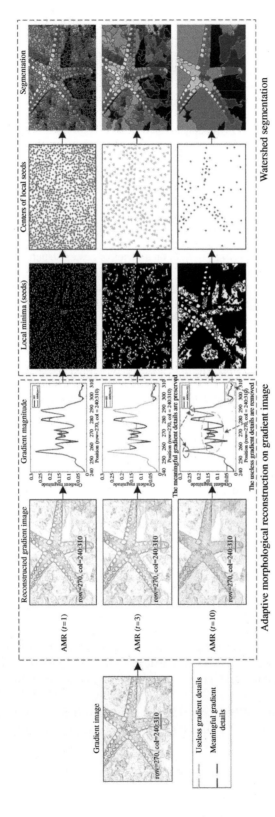

Figure 6.8 Seeded segmentation framework based on AMR-WT [1] / with permission of IEEE.

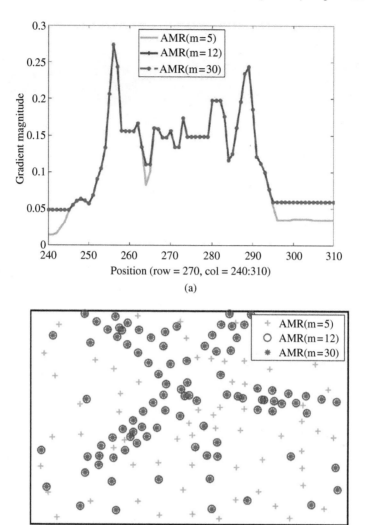

Figure 6.9 Comparison of gradient reconstruction and seed filtering with variant value of *t* [1] / with permission of IEEE. Because AMR has an important property of convergence, the seed image is unchanged when the value of *t* is large enough. The seed image is unchanged when $t \geq 12$ for the image "12003". (a) The variation of gradient magnitudes. (b) The variation of seed images.

Theorem 6.1 shows that the gradient image processed by AMR is monotonic increasing in *t*. Figure 6.9 demonstrates Theorem 6.1. We can see that if *t* is enlarged, the more unimportant seeds are removed, and important seeds are preserved. Actually, the result is equivalent to region merging. However, the method is simpler than region merging. According to the result, it can be seen that AMR can help seeded segmentation algorithms to achieve a hierarchical segmentation [29]. Hierarchical segmentation is a multilevel segmentation scheme, and it usually outputs a coarse-to-fine hierarchy of segments ordered by the level of details. Multiscale combinatorial grouping (MCG), proposed by Pont-Tuset et al. [30] is an excellent hierarchical segmentation approach that employs a fast normalized cut algorithm and an efficient algorithm for combinatorial merging of hierarchical regions. Based on the hierarchical segmentation results provided by MCG, some improved approaches are also proposed. These improved approaches achieve better segmentation effect but have lower computational efficiency than MCG.

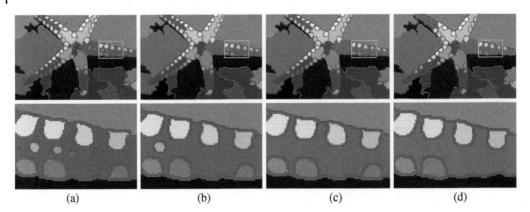

Figure 6.10 Segmentation results using AMR-WT by changing the value of s [1] / with permission of IEEE. (a) $s = 1, t = 10$. (b) $s = 2, t = 10$. (c) $s = 3, t = 10$. (d) $s = 5, t = 10$.

Before analyzing the relationship between AMR-WT and hierarchical segmentation, we first review some basic concepts of hierarchical segmentation. Let Ω be a finite set. A hierarchy H on Ω is a set of parts of Ω such that:

- $\Omega \in H$;
- For every $\omega \in \{\Omega\}$, $\{\omega\} \in H$; and
- For each pair $(h, h') \in H^2$, $h \cap h' = \varnothing \Rightarrow h \subseteq h' \text{or} h' \subseteq h$.

Note that H is a chain of nested partitions. Let H_0 be the initial partition of Ω, which corresponds to the finest partition of Ω, and H_n be the coarsest partition of Ω, which segments the image as a single region. A partition H_z, $0 \leq z \leq n$, on Ω has the property that

$$H_z = H_0, \quad if \ z \leq 0, \tag{6.8}$$

$$\exists n \in N^+, \ H_z = \{\Omega\}, \forall z \geq n, \tag{6.9}$$

$$p \leq q \Rightarrow H_p \subseteq H_q, 1 \leq p, q < n, \tag{6.10}$$

where $H_p \subseteq H_q$ denotes the partition. H_p is finer than the partition H_q. Derived from Theorem 6.1 and Figure 6.9, we obtain

$$\psi(g, s, p) \leq \psi(g, s, q) \Rightarrow S^a(\psi(g, s, p)) \subseteq S^a(\psi(g, s, q)), \tag{6.11}$$

where S^a denotes seeded segmentation algorithms such as WT or PW. Suppose that $H_0 = S^a(g)$, $H_t = S^a(\psi(g, s, t))$, and $s = 1$. Then

$$H_0 \subseteq H_1 \subseteq \cdots \subseteq H_t. \tag{6.12}$$

According to (6.12), the principle of the hierarchical segmentation based on AMR is shown in Figure 6.11, in which the data points represent regions obtained by the hierarchical segmentation at different levels.

6.3.3 The Convergence Property of AMR

By comparing Figure 6.6 with Figure 6.8, it can be observed that AMR provides significant gradient images, and AMR-WT generates convergent segmentation results via enlarging the scale of structuring elements. An important convergence property of AMR is described in the following.

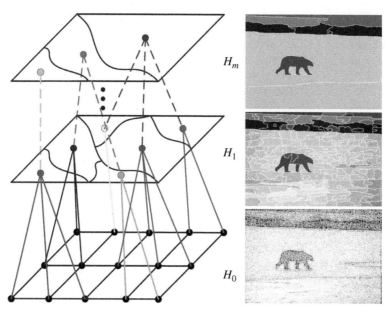

Figure 6.11 The principle of hierarchical segmentation, $H_0 \subseteq H_1 \subseteq \cdots \subseteq H_t$ [1] / with permission of IEEE.

Theorem 6.2 Let ψ be an AMR operator that is convergent when increasing the scale parameter m, i.e. for any gradient images f and g such that $b_s \subseteq \cdots b_i \subseteq b_{i+1} \cdots \subseteq b_t$, $min\,(\psi(g,s,t)) \geq max\left(R_g^C(f)_{b_{t+1}}\right)$. Then

$$\psi(g,s,t) = \psi(g,s,t+j),$$

i.e.

$$\bigvee_{s\,\leq\,i\,\leq\,t}\left\{R_g^C(f)_{b_i}\right\} = \bigvee_{s\,\leq\,i\,\leq\,t+j}\left\{R_g^C(f)_{b_i}\right\}, 1 \leq s \leq t, j \in N^+. \tag{6.13}$$

As can be seen in Figure 6.9, the gradient result and the corresponding seed image will remain unchanged when $t \geq 12$. This empirically illustrates that the gradient image reconstructed by AMR is convergent when increasing the value of t. Besides, it is observed that the large gradient magnitudes converge to themselves while the small gradient magnitudes converge to ones larger than themselves for the gradient images reconstructed by AMR. However, for the gradient images reconstructed by MR, the large gradient magnitudes converge to ones smaller than themselves while the small gradient magnitudes converge to ones larger than themselves when the structuring element is small. With the increase of the value of t, the value of gradient magnitude finally converges to the minimum of the original gradient image, i.e. $\lim_{t\to\infty} R_g^C(f)_{b_t} = min\,(g)$. Consequently, MR removes all regional minima while AMR only filtrates useless regional minima and preserves significant ones when $t \to \infty$.

Furthermore, we analyze how to determine the parameter m for AMR. The computational efficiency of AMR is influenced by the parameter t A small t means a low computational complexity. According to Theorem 6.2, the reconstructed gradient image and the corresponding segmentation remain unchanged when $min\,(\psi(g,s,t)) \geq max\left(R_g^C(f)_{b_{t+1}}\right)$, but the obtained t is usually large. As the chapter aims at employing AMR to generate a good seed image for seeded segmentation framework, we replace the convergence condition $min\left(\psi(g,s,t)\right) \geq max\left(R_g^C(f)_{b_{t+1}}\right)$ checking the difference between $\psi(g, s, t)$ and $\psi(g, s, t-1)$. This chapter presents an objective function used for justifying the convergence of AMR

$$J(g,s) = max\,|\psi(g,s,t) - \psi(g,s,t-1)|, \tag{6.14}$$

where $t \geq 2, t \in N^+$. It is clear that the segmentation result will remain unchanged when $J \leq \eta', \eta'$ is a minimal threshold error, and it is a constant used for J, but t is a variant for $\psi(g, s, t)$. Consequently, only a parameter s needs to be tuned for obtaining different reconstruction results.

6.3.4 The Algorithm of AMR

AMR only involves the parameter s and η', as described in the detailed steps of AMR in Algorithm 6.1. To speed up the convergence of Algorithm 6.1, the three parameters s, t, and η' are used for AMR because the iteration can be stopped according to t or η'. The computational complexity of AMR depends on the values of t or η'. A large value of t corresponds to a small value of η'. The larger is the value of t, the longer is the execution time of AMR. Since we have known that AMR has a fast convergent property as shown in Figure 6.8, a small t is enough for moderate-sized images in practical applications. A small t indicates that AMR has a low computational complexity.

Note that the parameter t is unnecessary theoretically since we use two convergent conditions t and η' to speed up the convergence of Algorithm 6.1. We applied Algorithm 6.1 to images with

---

Algorithm 6.1 Adaptive Morphological Reconstruction (AMR) [1] / with permission of IEEE.

Input: g (a gradient image).
Output: ψ (an adaptive morphological reconstruction result).

1: Initialize: set values for s and m (the scale of the minimal and maximal structuring element) and η', both t and η' are the convergent condition used for AMR.
2: for $i = s, s+1, \cdots, m$ do
3: Compute $R_g^C(f)_{b_i}$ where $f = \varepsilon_{mb_i}(g)$, b_i is a structuring element.
4: Update $\psi(g, s, i)$, and $J(g, s)$,
5: **if** $i = s$, **then**
6: $\psi(g,s,i) = \left\{R_g^C(f)_{b_1}\right\}$
7: $J(g, s) = max\,|\psi(g, s, i)|$
8: **else**
9: $\psi(g,s,i) = \vee\left\{\psi(g,s,i-1), R_g^C(f)_{b_i}\right\}$
10: $J(g, s) = max\,|\psi(g, s, i) - \psi(g, s, i-1)|$
11: **end if**
12: **if** $J \leq \eta'$ **then**
13: break
14: end if
15: end for

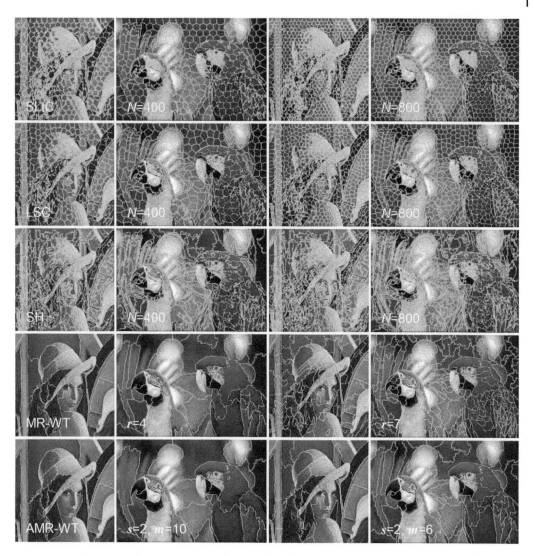

Figure 6.12 Segmentation results on images with complex texture [1] / with permission of IEEE. Here, *N* denotes the number of superpixel areas used for SLIC, LSC, and SH; *N* is 400 for the left two images and *N* is 800 for the right two images. Also, r_s denotes the radius of structuring element used for MR-WT; values of r_s are 7 and 4, respectively. For AMR-WT, *s* = 2 and *t* = 10 are used for the left two images, while *s* = 2 and *t* = 6 are used for the right two images.

complex texture content to demonstrate that the presented AMR is effective for reducing over-segmentation, as shown in Figure 6.12. AMR-WT not only overcomes the problem of over-segmentation but also obtains better contours than MR-WT and state-of-the-art superpixel methods. Furthermore, we test Algorithm 6.1 on images with text to show the monotonic increasing and convergent properties of AMR. Figure 6.13 shows the comparison results. We can see that the segmentation results are nested, which demonstrates the monotonic increasing property of AMR. Moreover, the segmentation results are unchanged when $t \geq 11$, which demonstrates the convergent property of AMR.

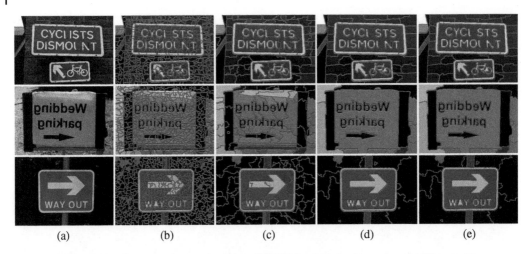

(a) (b) (c) (d) (e)

Figure 6.13 Segmentation results using AMR-WT by changing the value of t. The results shows that AMR is monotonic increasing by increasing the value of t. Moreover, AMR is convergent because the segmentation result is unchanged when $t \geq 11$ [1] / with permission of IEEE. (a) Images. (b) $s = 1, t = 1$. (c) $s = 1, t = 3$. (d) $s = 1, t = 11$. (e) $s = 1, t = 50$.

6.4 AMR for Seeded Image Segmentation

To demonstrate the effectiveness and efficiency of the presented AMR [1], we apply AMR to seeded image segmentation and spectral segmentation. We conduct experiments on the BSDS500 dataset. The experiments are performed on a workstation with an Intel core (TM) i7-6700MQ, 3.4 GHz CPU, and 16 GB memory.

We compare the presented algorithms with state-of-the-art algorithms including a multiscale morphological gradient for watersheds (MMG-WT) [18], multiscale ncut (MNCut) [31], oriented-watershed transform-ultrametric contour map (gPb-owt-ucm) [9], the algorithm recovering occlusion boundaries from an image proposed by Hoiem (gPb-Hoiem) [32], spectral segmentation algorithms proposed by Kim et al. (FNCut, cPb-owt-ucm) [24], higher-order correlation clustering (HO-CC) [33], global/regional affinity graph (GL-graph) [34], single-scale combinatorial grouping (SCG) [30], and MCG [30]. The open source codes and model parameters suggested by the corresponding authors are used. Because the author did not present specific parameter values for MMGR-WT, we set $r_s = 5$ and $0.1 \leq h \leq 0.3$, where h is a threshold, and it is used to generate a marker image, and r_s is the radius of the structuring element used for MR. For the presented approaches, we set $1 \leq s \leq 3$, $t = 50$, and $\eta' = 10^{-4}$.

We report the experimental results using three evaluation metrics to quantitatively measure the performance of segmentation algorithms: probabilistic rand index (PRI), segmentation covering (CV), and variation of information (VI). The PRI and CV are similarity measures, and they are large while the VI is small when the final segmentation is close to ground truth segmentation.

6.4.1 Seeded Image Segmentation

AMR is useful for improving seeded image segmentation because it employs multiscale structuring elements to obtain a convergent seed image without pre-setting many parameters. To show the capability of AMR, it is applied to different gradient images to filter seeds. Figure 6.14 shows reconstructed gradient images by AMR and the corresponding segmentation results by WT/PW. These results are clearly better than the ones shown in Figure 6.5. The problem of over-segmentation for

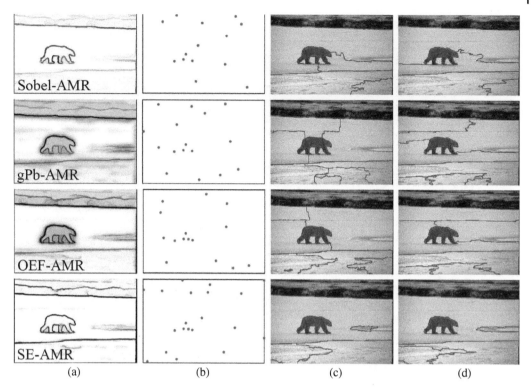

Figure 6.14 Comparison of segmentation results using AMR-WT/PW (s = 2) [1] / with permission of IEEE. (a) Gradient images. (b) Seeds (regional minimum). (c) WT. (d) PW (q = 2) [4].

Table 6.1 Comparison of the number of seeds generated by gradient images [1] / with permission of IEEE.

Images	Sobel	gPb [9]	OEF [11]	SE [10]
Original gradient images	9175	746	5348	1347
Gradient images reconstructed by AMR	15	16	15	19

seeded segmentation algorithms is therefore addressed. Furthermore, comparing Figures 6.6–6.14, although both MR and AMR are able to filter seeds, AMR is able to maintain meaningful seeds that correspond to important contours.

Furthermore, Table 6.1 shows the number of seeds generated by gradient images. We can see that the reconstructed gradient images generate fewer seeds than original gradient images, which demonstrates AMR is efficient for the filtering of useless seeds. Moreover, AMR is robust for different gradient images obtained by Sobel, gPb, OEF, and SE because the final segmentation results are similar.

In Figure 6.14, we set $s = 2$ because the segmentation result includes too many small regions when $s = 1$. Clearly, s controls the number of small regions in segmentation results. Generally, the value of s depends on the resolution of the images to be segmented, e.g. $1 \leq s \leq 3$ for BSDS500.

To demonstrate that the presented AMR is robust for different images, we implement AMR-WT/PW on the BSDS500. Figure 6.15 shows the comparison of segmentation results using different algorithms, i.e. Sobel-AMR-WT/PW, gPb-AMR-WT/PW, OEF-AMR-WT/PW, and SE-AMR-WT/PW.

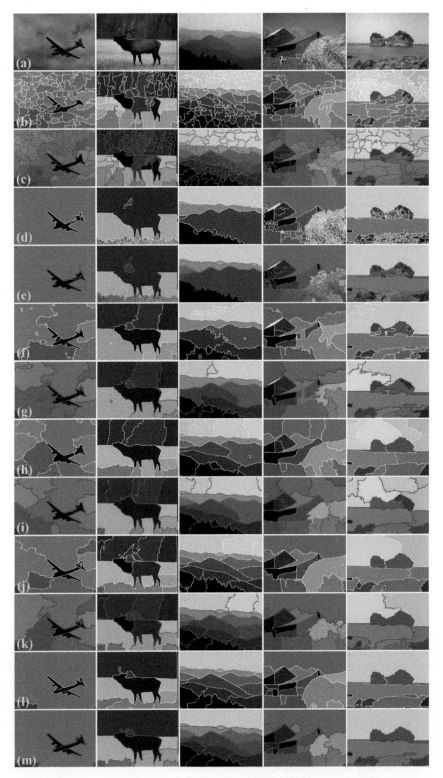

Figure 6.15 Comparison of segmentation results using different algorithms ($s = 2$) [1] / with permission of IEEE. (a) Images. (b) Sobel-MR-WT ($r_s = 5$). (c) Sobel-MR-PW ($r_s = 5$). (d) MMG-MR-WT ($r_s = 5$ and $h = 0.2$). (e) MMG-MR-PW ($r_s = 5$ and $h = 0.2$). (f) Sobel-AMR-WT. (g) Sobel-AMR-PW. (h) gPb-AMR-WT. (i) gPb-AMR-PW. (j) OEF-AMR-WT. (k) OEF-AMR-PW. (l) SE-AMR-WT. (m) SE-AMR-PW.

The segmentation results demonstrate the effectiveness of AMR for the filtering of useless seeds, Moreover, AMR is effective for both WT and PW.

To compare the performance of different algorithms on the BSDS500, Table 6.2 shows experimental results of three evaluation metrics: PRI, CV, and VI. We can see that AMR is more efficient for improving segmentation results obtained by WT or PW compared to MR. MR is sensitive to r_s while AMR is insensitive to s. Although MMG-MR-WT/PW is effective for over-segmentation reduction by introducing the parameter h, segmentation results are sensitive to both r_s and h. The gPb-AMR-WT/PW, OEF-AMR-WT/PW, and SE-AMR-WT/PW obtain better performance than Soble AMR-WT/PW since the former provides better gradient images. The SE-AMR-WT/PW obtains the best performance. In addition, AMR-WT obtains higher PRI and CV and lower VI than AMR-PW in the same situation.

Because AMR converges quickly, it has a high computation efficiency for gradient reconstruction. Table 6.3 shows the comparison of running time of AMR-WT on different gradient images obtained by Sobel, gPb, OEF, and SE, respectively. We only present the running time of AMR-WT here because AMR-PW has a similar running time as AMR-WT. It can be seen from Table 6.3 that AMR-WT has a short running time to achieve image segmentation on the BSDS500. The SE-AMR-WT requires the shortest running time because the corresponding gradient image converges quicker under AMR. Tables 6.2 and 6.3 show AMR is effective and efficient for improving seeded segmentation algorithms such as WT and PW.

Additional evidence of the superiority of AMR can be found in Figure 6.16, which shows experimental results on images with rich texture and faded boundaries. According to Figures 6.15 and 6.16, we can see that the presented AMR is effective for different kinds of images.

Table 6.2 Quantitative results (PRI, CV, VOI) on the BSDS500. Larger is better for PRI and CV while smaller is better for VI. The best values are in bold [1] / with permission of IEEE.

Methods	PRI↑	CV↑	VI↓
Sobel-MR-WT/PW ($r_s = 5$)	0.71/0.71	0.16/0.15	4.02/4.12
Sobel-MR-WT/PW ($r_s = 8$)	0.73/0.73	0.28/0.28	3.08/3.05
Sobel-MR-WT/PW ($r_s = 12$)	0.69/0.68	0.38/0.39	2.67/2.33
MMG-MR-WT/PW [18] ($h = 0.1$)	0.76/0.76	0.27/0.27	4.47/4.45
MMG-MR-WT/PW [18] ($h = 0.2$)	0.74/0.74	0.38/0.38	3.50/3.45
MMG-MR-WT/PW [18] ($h = 0.3$)	0.62/0.62	0.42/0.42	2.95/2.92
Sobel-AMR-WT/PW ($s = 1$)	0.76/0.75	0.39/0.34	2.54/2.79
Sobel-AMR-WT/PW ($s = 2$)	0.76/0.75	0.39/0.36	2.51/2.70
Sobel-AMR-WT/PW ($s = 3$)	0.76/0.76	0.39/0.36	2.52/2.66
gPb-AMR-WT/PW ($s = 1$)	0.75/0.75	0.35/0.32	2.55/2.77
gPb-AMR-WT/PW ($s = 2$)	0.75/0.74	0.35/0.33	2.55/2.77
gPb-AMR-WT/PW ($s = 3$)	0.75/0.74	0.36/0.33	2.55/2.76
OEF-AMR-WT/PW ($s = 1$)	0.77/0.75	0.39/0.34	2.45/2.72
OEF-AMR-WT/PW ($s = 2$)	0.77/0.75	0.39/0.34	2.43/2.72
OEF-AMR-WT/PW ($s = 3$)	0.77/0.76	0.39/0.35	2.41/2.70
SE-AMR-WT/PW ($s = 1$)	0.80/0.79	0.45/0.41	2.25/2.52
SE-AMR-WT/PW ($s = 2$)	0.80/0.79	0.45/0.41	2.33/2.51
SE-AMR-WT/PW ($s = 3$)	0.80/0.79	0.45/0.41	2.21/2.50

Table 6.3 Comparison of average running time of AMR-WT on the BSDS500 (in seconds). Lower is better the best values are in bold ($\eta' = 10^{-4}$) [1] / with permission of IEEE.

s	Sobel	gPb	OEF	SE
$s = 1$	0.801	1.278	0.861	0.565
$s = 2$	0.774	1.241	0.825	0.534
$s = 3$	0.689	1.169	0.766	0.480

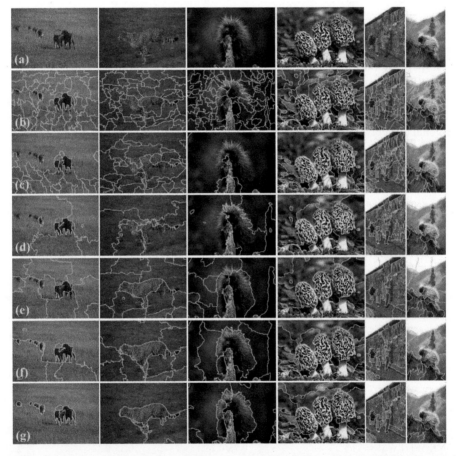

Figure 6.16 Comparison of segmentation results using different algorithms ($s = 2$) [1][1] / with permission of IEEE. (a) Images with rich texture or faded boundaries. (b) Sobel-MR-WT ($r_s = 5$). (c) MMG-MR-WT ($r_s = 5$ and $h = 0.2$). (d) Sobel-AMR-WT. (e) gPb-AMR-WT. (f) OEF-AMR-WT. (g) SE-AMR-WT.

6.4.2 Seed-Based Spectral Segmentation

In this section, we directly construct the affinity matrix on a pre-segmentation image provided by AMR-WT to reduce the size of the affinity matrix and then compute the subsequent steps of spectral segmentation (we name it AMR-SC). Note that we employ AMR-WT instead of AMR-PW because the former is able to provide better pre-segmentation result than the latter, as shown in Table 6.2. As the pre-segmentation image only consists of dozens of regions, we consider color feature in $L * ab$

color space and Gaussian function as the criterion to measure the similarity of two regions. Throughout the chapter, we let $\sigma = 1$. It is clear that the affinity matrix produced by AMR is a small matrix. Therefore, the clusters can be detected easily and fast with the k-means algorithm.

In this chapter, the pre-segmentation depends on AMR. According to Table 6.2, we set $s = 2$ and 3, and we set the number of clusters for k-means according to [24, 34]. The presented AMR- SC is evaluated on the BSDS500 and compared with algorithms such as gPb-owt-ucm, FNCut, GL-graph, SCG and MCG. Figures 6.17 and 6.18 show that the presented AMR-SC generates better segmentation results than the other algorithms. The result demonstrates that AMR is useful for improving

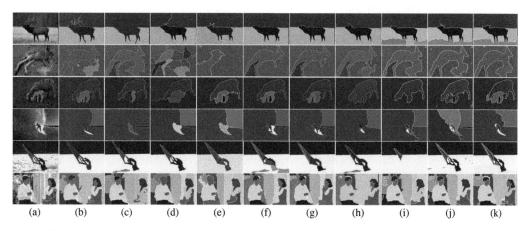

(a)   (b)   (c)   (d)   (e)   (f)   (g)   (h)   (i)   (j)   (k)

Figure 6.17 Comparison of segmentation results on the BSDS500 using different algorithms (s = 2) [1] / with permission of IEEE. (a) Images. (b) Ground truths. (c) gPb-owt-ucm. (d) FNCut. (e) GL-graph. (f) SCG. (g) MCG. (h) Sobel-AMR-SC. (i) gPb-AMR-SC. (j) OEF-AMR-SC. (k) SE-AMR-SC.

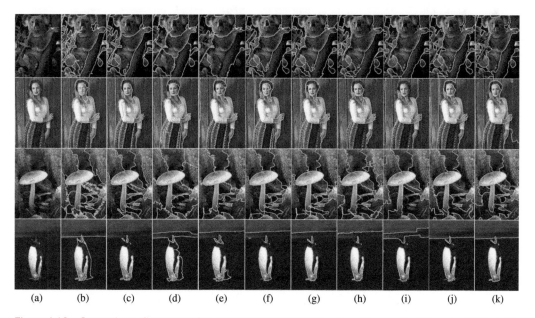

(a)   (b)   (c)   (d)   (e)   (f)   (g)   (h)   (i)   (j)   (k)

Figure 6.18 Comparison of segmentation results on the BSDS500 using different algorithms (s = 2) [1] / with permission of IEEE. Compared to Figures 6.15 and 6.17, we use an overlay of the segmented result with respect to the original image to show the accuracy of the boundary. (a) Images. (b) Ground truths. (c) gPb-owt-ucm. (d) FNCut. (e) GL-graph. (f) SCG. (g) MCG. (h) Sobel-AMR-SC. (i) gPb-AMR-SC. (j) OEF-AMR-SC. (k) SE-AMR-SC.

spectral segmentation because of two reasons. The first is that the regional spatial information of an image provided by pre-segmentation is integrated into spectral segmentation, and the second is that the affinity graph is reduced efficiently by removing useless seeds.

Furthermore, we employ the three measures: PRI, CV, and VI to compare the presented AMR-SC with nine state-of-the-art image segmentation algorithms. Table 6.4 show the region benchmarks on SDS500. In Table 6.4, the presented AMR-SC clearly dominates other algorithms on PRI and CV and is on par with SCG on VI mainly due to accurate pre-segmentation provided by AMR-WT. The OEF-AMR-SC and SE-AMR-SC provide higher performance than gPb-AMR-SC and Sobel-AMR-SC because OEF and SE obtain better gradient images than gPb and Sobel. In addition, AMR-SC is insensitive to the parameter s.

We tested the running time complexity on the BSDS500 data set. The running time comparison is shown in Table 6.4. On average, generating a pre-segmentation result with SE-AMR-WT takes 0.54 seconds (SE generates a gradient image requiring 0.06 seconds. AMR-WT takes 0.48 seconds, for $s = 3$ and $\eta' = 10^{-4}$), and constructing an affinity graph and spectral clustering takes 0.059 seconds. Consequently, SE-AMR-SC takes about 0.60 second to segment an image from BSDS500. In contrast, the gPb-owt-ucm takes almost 106.38 seconds, and FNCut takes about 10.58 seconds. As the GL-graph has four steps, i.e., over-segmentation, feature extraction, bipartite graph construction, and graph partition using spectral clustering, it is more complex than SE-AMR-SC and takes almost 7.41 seconds. MCG takes about 18.60 seconds per image to compute the multiscale hierarchy, but SCG takes only 2.21 seconds per image. It is clear that our SE-AMR-SC is the fastest because AMR-SC only depends on the gradient information, and the generated affinity matrix is small.

Table 6.4 Quantitative results (PRI, CV, and VI) on the BSDS500. Larger is better for PRI and CV while smaller is better for VI and running time. The best values are in bold [1] / with permission of IEEE.

Methods	PRI↑	CV↑	VI↓	Time↓
MNCut [31]	0.78	0.45	2.23	37.25
gPb-owt-ucm [9]	0.83	0.59	1.69	106.38
gPb-Hoiem [32]	0.81	0.56	1.78	109.77
FNCut [24]	0.81	0.53	1.86	10.58
cPb-owt-ucm [24]	0.83	0.59	1.65	107.13
HO-CC [33]	0.83	0.60	1.79	35.18
GL-graph [34]	0.84	0.59	1.80	7.41
SCG [30]	0.83	0.60	1.63	2.21
MCG [30]	0.83	0.61	1.57	18.60
Sobel-AMR-SC ($s = 2$)	0.82	0.61	1.77	0.86
Sobel-AMR-SC ($s = 3$)	0.82	0.61	1.77	0.81
gPb-AMR-SC ($s = 2$)	0.82	0.61	1.73	102.94
gPb-AMR-SC ($s = 3$)	0.82	0.61	1.73	102.92
OEF-AMR-SC ($s = 2$)	0.85	0.63	1.62	6.16
OEF-AMR-SC ($s = 3$)	0.84	0.63	1.64	6.07
SE-AMR-SC ($s = 2$)	0.85	0.63	1.62	0.62
SE-AMR-SC ($s = 3$)	0.85	0.63	1.62	0.60

6.5 Discussion and Summary

6.5.1 Discussion

AMR has two parameters, η' and s. η' relates to the convergent condition. Generally, a large value of η' means a few iterations (a small t, where t is the number of iterations) while a small value of η' corresponds to many iterations (a large t). Table 6.5 shows the influence of η' on t for test images. We can see that t increases with the decrease of η', but m is unchanged when $\eta' \leq 10^{-4}$.

Furthermore, to show the influence of η' on AMR, we implement AMR on the BSDS500 data set by setting different values of η', and Tables 6.6, 6.7 show the results. It is clear that the number of iterations for AMR-WT is smaller and running time is shorter if the value of η' is larger. However, the number of iterations and running time are unchanged when $\eta' \leq 10^{-4}$. Therefore, in practical applications, users can select different values of η' according to their requirements.

Furthermore, we implemented SE-AMR-WT on BSDS500 with different values of η'. The performance indices of segmentations are shown in Table 6.8. By comparing Tables 6.5–6.8, we can see that the average number of iterations, running time, and segmentation accuracy are unchanged for AMR-WT when $\eta' \leq 10^{-4}$. Therefore, the presented AMR is insensitive to η'.

Table 6.5 The number of iterations of SE-ARM-WT under different values of η, ($s = 3$). The number of iterations m is unchanged when $\eta' \leq 10^{-4}$, and the invariant values of m are in bold [1] / with permission of IEEE.

Images	$\eta' = 10^{-2}$	$\eta' = 10^{-3}$	$\eta' = 10^{-4}$	$\eta' = 10^{-5}$
"2092"	9	16	16	16
"8023"	6	12	19	19
"8049"	13	16	19	19
"12 074"	9	17	17	17
"12 084"	9	13	13	13
"15 004"	11	21	21	21

Table 6.6 The average number of iterations of SE-AMR-WT under different values of η', $s = 3$. The average number of iterations is unchanged when $\eta' \leq 10^{-4}$, and the invariant values of m are in bold [1] / with permission of IEEE.

	$\eta' = 10^{-1}$	$\eta' = 10^{-2}$	$\eta' = 10^{-3}$	$\eta' = 10^{-4}$	$\eta' = 10^{-5}$
m	2.0	9.3	17.0	18.9	18.9

Table 6.7 The average running time of SE-AMR-WT on the BSDS500 (in seconds), $s = 3$. The average running time is unchanged when $\eta' \leq 10^{-4}$, and the invariant values are in bold [1] / with permission of IEEE.

	$\eta' = 10^{-1}$	$\eta' = 10^{-2}$	$\eta' = 10^{-3}$	$\eta' = 10^{-4}$	$\eta' = 10^{-5}$
Time	0.089	0.228	0.430	0.480	0.480

Table 6.8 Quantitative results (PRI, CV and VI) of SE-AMR-WT on the BSDS500 under different values of η', $s = 3$. Larger value is better for PRI and CV while smaller value is better for VI [1] / with permission of IEEE.

η'	PRI↑	CV↑	VI↓
10^{-1}	0.77	0.30	3.21
10^{-2}	0.79	0.39	2.52
10^{-3}	0.80	0.45	2.23
10^{-4}	0.80	0.45	2.21
10^{-5}	0.80	0.45	2.21

Table 6.9 Quantitative results (PRI, CV, and VI) of SE-AMR-WT on the BSDS500 under different values of s, $\eta' = 10^{-4}$. Larger value is better for PRI and CV while smaller value is better for VI [1] / with permission of IEEE.

s	PRI↑	CV↑	VI↓
1	0.80	0.45	2.25
2	0.80	0.45	2.23
3	0.80	0.46	2.21
4	0.80	0.46	2.22
5	0.80	0.46	2.25
6	0.79	0.46	2.30

The value of s controls the initial gradient value of images. A large s will cause the contour offset while a small value of s will cause too many unexpected small regions. Therefore, we choose $s = 2$ and $s = 3$ for BSDS500 in Table 6.4. To further show the influence of s on AMR, Table 6.9 presents the performance indices of segmentations on BSDS500 by setting different values of s. It can be seen from Table 6.9 that SE-AMR-WT is insensitive to s if $1 \leq s \leq 6$.

6.5.2 Summary

In this chapter, we have studied the advantages and disadvantages of MR on seeded segmentation algorithms. The chapter presented an efficient AMR algorithm that can preferably improve seeded segmentation algorithms. The presented algorithm, AMR [1], has two significant properties, i.e. it is monotonically increasing and convergent. The property of monotonic increase helps AMR to achieve a hierarchical segmentation. The property of convergence is able to alleviate the drawback of MR by filtering out useless regional minima in a gradient image and guarantees a convergent result. Moreover, we have explored the applications of AMR and have found that AMR is not only able to improve seeded image segmentation results but also can obtain better spectral segmentation results than state-of-the-art algorithms. Furthermore, the presented AMR-SC is computationally efficient because a small affinity matrix is used for spectral clustering. Experimental results clearly demonstrate that the presented AMR-WT generates satisfactory and convergent segmentation results without hard-tuning parameters, and AMR-SC outperforms most of the state-of-the-art algorithms for image segmentation, performing the best in two metrics: PRI and CV.

The segmentation results generated by AMR-WT or AMR-SC can be directly used in object recognition and scene labeling. However, neither AMR-WT nor AMR-SC can obtain semantic segmentation results compared to the popular convolutional neural network (CNN), e.g. fully convolutional network (FCN) [35]. To further improve the contour quality of segmentation results, traditional algorithms such as conditional random field, image superpixel, and spatial pyramid pooling are used to improve the performance of CNN on image segmentation. AMR can be also used in CNN to improve semantic segmentation results. For our future work, we plan to investigate how to combine AMR and FCN effectively and efficiently.

References

1 Lei, T., Jia, X., Liu, T. et al. (2019). Adaptive morphological reconstruction for seeded image segmentation. *IEEE Trans. Image Process.* **28** (11): 5510–5523.

2 Vincent, L. (1993). Morphological grayscale reconstruction in image analysis: applications and efficient algorithms. *IEEE Trans. Image Process.* **2** (2): 176–201.

3 Vincent, L. and Soille, P. (1991). Watersheds in digital spaces: an efficient algorithm based on immersion simulations. *IEEE Trans. Pattern Anal. Machine Intell.* **13** (06): 583–598.

4 Couprie, C., Grady, L., Najman, L., and Talbot, H. (2010). Power watershed: a unifying graph-based optimization framework. *IEEE Trans. Pattern Anal. Mach. Intell.* **33** (7): 1384–1399.

5 Lei, T., Jia, X., Zhang, Y. et al. (2018). Significantly fast and robust fuzzy c-means clustering algorithm based on morphological reconstruction and membership filtering. *IEEE Trans. Fuzzy Syst.* **26** (5): 3027–3041.

6 Lei, T., Zhang, Y., Wang, Y. et al. (2017). A conditionally invariant mathematical morphological framework for color images. *Inform. Sci.* **387**: 34–52.

7 Cheng, J. and Rajapakse, J.C. (2008). Segmentation of clustered nuclei with shape markers and marking function. *IEEE Trans. Biomed. Eng.* **56** (3): 741–748.

8 Peng, B., Zhang, L., and Zhang, D. (2011). Automatic image segmentation by dynamic region merging. *IEEE Trans. Image Process.* **20** (12): 3592–3605.

9 Arbelaez, P., Maire, M., Fowlkes, C., and Malik, J. (2010). Contour detection and hierarchical image segmentation. *IEEE Trans. Pattern Anal. Mach. Intell.* **33** (5): 898–916.

10 Dollár, P. and Zitnick, C.L. (2014). Fast edge detection using structured forests. *IEEE Trans. Pattern Anal. Mach. Intell.* **37** (8): 1558–1570.

11 Hallman, S. and Fowlkes, C.C. (2015). Oriented edge forests for boundary detection. In: *Proceedings of the IEEE Conference on Computer Vision and Pattern Recognition*, 1732–1740.

12 Fu, X., Wang, C.Y., Chen, C. et al. (2015). Robust image segmentation using contour-guided color palettes. In: *Proceedings of the IEEE International Conference on Computer Vision*, 1618–1625.

13 Comaniciu, D. and Meer, P. (2002). Mean shift: a robust approach toward feature space analysis. *IEEE Trans. Pattern Anal. Mach. Intell.* **24** (5): 603–619.

14 Ng, A.Y., Jordan, M.I., and Weiss, Y. (2002). On spectral clustering: analysis and an algorithm. In: *Advances in Neural Information Processing Systems*, 849–856.

15 Serra, J. (1988). *Image Analysis and Mathematical Morphology*, vol. **2**. New York, NY, USA: Academic.

16 Liao, W., Dalla Mura, M., Chanussot, J. et al. (2015). Morphological attribute profiles with partial reconstruction. *IEEE Trans. Geosci. Remote Sens.* **54** (3): 1738–1756.

17 Soille, P. (2013). *Morphological Image Analysis: Principles and Applications*. Springer Science & Business Media.

18 Wang, D. (1997). A multiscale gradient algorithm for image segmentation using watershelds. *Pattern Recogn.* **30** (12): 2043–2052.

19 Li, Y., Xu, M., Liang, X., and Huang, W. (2017). Application of bandwidth EMD and adaptive multiscale morphology analysis for incipient fault diagnosis of rolling bearings. *IEEE Trans. Industrial Electronics* **64** (8): 6506–6517.

20 Shih, H.C. and Liu, E.R. (2016). Automatic reference color selection for adaptive mathematical morphology and application in image segmentation. *IEEE Trans. Image Process.* **25** (10): 4665–4676.

21 Boykov, Y. and Funka-Lea, G. (2006). Graph cuts and efficient ND image segmentation. *Int. J. Computer Vision* **70** (2): 109–131.

22 Grady, L. (2006). Random walks for image segmentation. *IEEE Trans. Pattern Anal. Mach. Intell.* **28** (11): 1768–1783.

23 Najman, L. and Schmitt, M. (1996). Geodesic saliency of watershed contours and hierarchical segmentation. *IEEE Trans. Pattern Anal. Mach. Intell.* **18** (12): 1163–1173.

24 Kim, T.H., Lee, K.M., and Lee, S.U. (2012). Learning full pairwise affinities for spectral segmentation. *IEEE Trans. Pattern Anal. Mach. Intell.* **35** (7): 1690–1703.

25 Achanta, R., Shaji, A., Smith, K. et al. (2012). SLIC superpixels compared to state-of-the-art superpixel methods. *IEEE Trans. Pattern Anal. Mach. Intell.* **34** (11): 2274–2282.

26 Chen, J., Li, Z., and Huang, B. (2017). Linear spectral clustering superpixel. *IEEE Trans. Image Process.* **26** (7): 3317–3330.

27 Wei, X., Yang, Q., Gong, Y. et al. (2018). Superpixel hierarchy. *IEEE Trans. Image Process.* **27** (10): 4838–4849.

28 Zhang, Z., Xing, F., Wang, H. et al. (2018). Revisiting graph construction for fast image segmentation. *Pattern Recogn.* **78**: 344–357.

29 Xu, Y., Carlinet, E., Géraud, T., and Najman, L. (2016). Hierarchical segmentation using tree-based shape spaces. *IEEE Trans. Pattern Anal. Mach. Intell.* **39** (3): 457–469.

30 Pont-Tuset, J., Arbelaez, P., Barron, J.T. et al. (2016). Multiscale combinatorial grouping for image segmentation and object proposal generation. *IEEE Trans. Pattern Anal. Mach. Intell.* **39** (1): 128–140.

31 Cour, T., Benezit, F., and Shi, J. (2005). Spectral segmentation with multiscale graph decomposition. In: *2005 IEEE Computer Society Conference on Computer Vision and Pattern Recognition* (CVPR'05), vol. **2**, 1124–1131. IEEE.

32 Hoiem, D., Efros, A.A., and Hebert, M. (2011). Recovering occlusion boundaries from an image. *Int. J. Computer Vision* **91** (3): 328–346.

33 Kim, S., Yoo, C.D., Nowozin, S., and Kohli, P. (2014). Image segmentation usinghigher-order correlation clustering. *IEEE Trans. Pattern Anal. Mach. Intell.* **36** (9): 1761–1774.

34 Wang, X., Tang, Y., Masnou, S., and Chen, L. (2015). A global/local affinity graph for image segmentation. *IEEE Trans. Image Process.* **24** (4): 1399–1411.

35 Long, J., Shelhamer, E., and Darrell, T. (2015). Fully convolutional networks for semantic segmentation. In: *Proceedings of the IEEE Conference on Computer Vision and Pattern Recognition*, 3431–3440.

7

Superpixel-Based Fast Image Segmentation

In Chapter 5, we introduced a fast and robust fuzzy *c*-means clustering algorithm based on morphological reconstruction and membership filtering (FRFCM). The presented FRFCM not only integrates the spatial structure information of images but also greatly reduces the time complexity and improves the practicality. However, FRFCM only considers a regular neighboring window, which breaks up the real local spatial structure of images and thus leads to poor image segmentation effect. Inspired by the adaptive morphological watershed algorithm in Chapter 6, this chapter fuses superpixel blocks with similar texture, color, and brightness characteristics into the fuzzy clustering algorithm to improve the final segmentation result.

In this chapter, we present a superpixel-based fast FCM clustering algorithm that is significantly faster and more robust than state-of-the-art clustering algorithms for color image segmentation. This is based on the recently published article by Lei et al. [1]. In contrast to traditional neighboring window with fixed size and shape, the superpixel image provides better adaptive and irregular local spatial neighborhoods that are helpful for improving color image segmentation. Second, based on the obtained superpixel image, the original color image is simplified efficiently and its histogram is computed easily by counting the number of pixels in each region of the superpixel image. Finally, we implement FCM with a histogram parameter on the superpixel image to obtain the final segmentation result. Experiments performed on synthetic images and real images demonstrate that the presented algorithm provides better segmentation results and takes less time than state-of-the-art clustering algorithms for color image segmentation.

7.1 Introduction

Image segmentation algorithms can be roughly grouped into two categories—unsupervised and supervised image segmentation. Unsupervised approaches, such as graph cut [2], active contour model [3], watershed transform (WT) [4], hidden Markov random field (HMRF) [5], and fuzzy entropy [6], are useful and popular due to their simplicity without depending on training samples and labels. In contrast to unsupervised image segmentation approaches, although some supervised approaches such as convolutional neural networks (CNNs) [7] and fully convolution networks (FCNs) [8] are able to achieve image segmentation by using feature learning, but they require a lot of training samples and labeled images. In addition, the segmentation result has a coarse contour since CNNs and FCNs essentially achieve image classification. In this chapter, we mainly discuss unsupervised image segmentation.

Image Segmentation: Principles, Techniques, and Applications, First Edition. Tao Lei and Asoke K. Nandi.
© 2023 John Wiley & Sons Ltd. Published 2023 by John Wiley & Sons Ltd.

In unsupervised algorithms, clustering represents one kind of important and popular algorithms for grayscale and color image segmentation because it is suitable and useful for both low- and high-dimensional data. Generally, clustering algorithms can be roughly categorized into three groups—minimizing an objective function [9], decomposing a density function [10], and graph theory [11]. In this chapter, we will focus on image segmentation based on clustering by minimizing an objective function. It is well known that k-means and FCM are clustering algorithms that minimize an objective function. Because k-means is a hard or crisp clustering algorithm, it is sensitive to initial clustering centers or membership. In contrast, FCM is a soft algorithm that improves the shortcomings of k-means at the cost of increasing iterations. However, both k-means and FCM are sensitive to noise because the local spatial information of pixels is missed for image segmentation.

To address this shortcoming, a great number of improved clustering algorithms that incorporate local spatial information into their objective function have been proposed in recent years. These algorithms can be grouped into two groups. The first group employs neighborhood information of a center pixel using a window of fixed size to improve the image segmentation effect, e.g. FCM algorithm with spatial constraints (FCM_S) [12], FCM_S1, FCM_S2 [13], fast generalized FCM algorithm (FGFCM) [14], fuzzy local information c-means clustering algorithm (FLICM) [15], neighborhood weighted FCM clustering algorithm (NWFCM) [16], FCM algorithm based on noise detection (NDFCM) [17], Memon's algorithm [18], and the FLICM based on kernel metric and weighted fuzzy factor (KWFLICM) [19]. The advantage of these algorithms is that the neighborhood information can be computed in advance, except for FCM_S and FLICM, to reduce the computational complexity. However, a neighborhood window of fixed size and shape is unable to satisfy the requirement of robust image segmentation. The second group employs adaptive neighborhood information instead of the window of fixed size and shape, e.g. Liu's algorithm [20], Bai's algorithm [21], and adaptive FLICM [22]. As adaptive neighborhood information is consistent with real image structuring information, the second group of algorithms obtains a better robustness for noisy images and a better segmentation effect than the first group.

Though improved FCM algorithms consider the neighborhood information of an image, the neighborhood information of the corresponding membership, which is helpful for improving classification effect, is ignored. HMRF is a popular algorithm for addressing the issue. In [23], the current membership called posterior probability depends on clustering centers and the prior probability of neighborhood. Because HMRF considers the previous state of current membership, it obtains better result than FCM for image segmentation [23]. Based on the idea, Zhang et al. [24] incorporated the local spatial information of membership into the objective function of FCM, which obtains better results for image segmentation than the algorithm proposed in [23]. Furthermore, Liu et al. [20] improved the FCM algorithm by integrating the distance between different regions obtained by mean-shift and the distance of pixels into its objective function. Although these HMRF-based clustering algorithms [20, 23, 24] effectively improve the effect of image segmentation, they have a high computational complexity caused by the computation of neighborhood information provided by original image and previous state's membership in every iteration.

It is clear that the algorithms mentioned above improve the image segmentation effect at the cost of increasing the computational complexity. Therefore, the question arises: How one can maintain local spatial information while reducing the computational complexity efficiently? Lei et al. [25] presented a fast and robust FCM algorithm (FRFCM) to address the problem by employing morphological reconstruction [26] and membership filtering. Because the repeated distance computation between pixels within neighborhood window and clustering centers is removed, the algorithm is very fast and provides a better segmentation result than state-of-the-art algorithms. Nevertheless, the FRFCM requires much execution time for color image segmentation because it is difficult to compute the histogram of color images. To address this issue, this chapter presents a superpixel-based fast FCM (SFFCM) for color image segmentation. The presented algorithm is able to achieve color image segmentation with a very low computational cost yet achieve a high segmentation precision.

This chapter can be summarized as following:

1) The chapter presents a multiscale morphological gradient reconstruction (MMGR) operation to generate superpixel image with accurate boundaries, which is helpful for integrating adaptive neighboring information and reducing the number of different pixels in a color image.
2) Based on a superpixel image obtained by MMGR, the chapter presents a simple color histogram computational method that can be used to achieve a fast FCM algorithm for color image segmentation.

The rest of this chapter is organized as follows. In Section 7.2, we describe related work. Section 7.3 presents FCM with superpixel and analyzes its superiority. The discussion and summary are presented in Section 7.4.

7.2 Related Work

FCM often misses spatial information, leading to a poor result for image segmentation. Although a great number of improved algorithms address the problem by incorporating local spatial information into the objective function, this, in turn, increases the computational complexity of algorithms. Fortunately, superpixel is able to address the problem. Superpixel is an image preprocessing tool that over-segments an image into a number of small regions. A superpixel region is usually defined as perceptually uniform and homogenous regions in the image. Superpixel is able to improve the effectiveness and efficiency of image segmentation due to two advantages. On the one hand, superpixel is able to achieve a pre-segmentation based on the local spatial information of images. The pre-segmentation provides better local spatial information than traditional neighboring windows employed by FCM_S, FLICM, FGFCM, KWFLICM, NWFCM, NDFCM, and FRFCM. On the other hand, superpixel is able to reduce the number of different pixels in an image by replacing all pixels in a region with the mean value of the superpixel region. In this chapter, we will employ superpixel technology to obtain adaptive local spatial information and then compute the histogram of superpixel image to achieve fast color image segmentation.

7.2.1 Fuzzy Clustering with Adaptive Local Information

In early improved FCM algorithms, local spatial information is often insufficient in a neighboring window of fixed size and shape. If the window is too small, the local spatial information will be limited for improving the segmentation effect. But if the window is too large, the computational complexity of the corresponding algorithm will be very high. Recently, some improved FCM algorithms [22] incorporate adaptive local spatial information into their objective function to obtain better robustness and higher performance for image segmentation. Adaptive local spatial information means that the pixels within a neighboring region have variable weighting factors depending on local characteristics of an image. For example, in Liu's algorithm [20], the adaptive neighborhood of a pixel is decided by its neighboring window and the corresponding region obtained by a prior mean-shift algorithm [10].

In [20], the objective function denoted by J_m is defined as

$$J_m = \sum_{i=1}^{N}\sum_{k=1}^{c} u_{ki}D_{ki} + f^{\varpi}\sum_{i=1}^{N}\sum_{k=1}^{c} u_{ki} \log \frac{u_{ki}}{\pi_{ki}}, \tag{7.1}$$

where u_{ki} is the membership between the i-th pixel and the k-th clustering center, $1 \leq i \leq N, 1 \leq k \leq c$, N is the number of data items, c is the number of clusters, $N, c \in N^{+}, f^{\varpi}$ is the degree of fuzziness of u_{ki}, π_{ki} is defined in (7.3) below, and the distance function D_{ki} is the combination of the pixel dissimilarity and region dissimilarity

$$D_{ki} = \frac{(d_{ki} + d_{kR_i})}{2}. \tag{7.2}$$

In (7.2), d_{ki} is the dissimilarity distance between the i-th pixel and the k-th clustering center, d_{kR_i} is the region dissimilarity between the region R_i obtained by mean-shift and the k-th clustering center, $d_{kR_i} = \frac{1}{sum(R_i)}\sum_{j \in R_i} d_{kj}$, R_i is the region that contains the i-th pixel, and the sum(R_i) denotes the number of pixels in the region R_i. Furthermore,

$$\pi_{ki} = \frac{\sum_{j \in N_i} \omega_j \zeta_j u_{kj}}{\sum_{k=1}^{c}\sum_{j \in N_i} \omega_j \zeta_j u_{kj}}. \tag{7.3}$$

In (7.3), ω_j is a weighting parameter of the neighborhood pixels, N_i is the neighborhood of the i-th pixel, and $\in N_i$, and ζ is the region-level iterative strength

$$\zeta_j = \frac{1}{Z}\left(E_{R_i,R_j} + 1\right)^{-1}, \tag{7.4}$$

where $E_{R_i,R_j} = mean(R_i) - mean(R_j)$ is the Euclidean distance between the mean values of region R_i and R_j. $Z = \sum_{j \in N_i}\left(E_{R_i,R_j} + 1\right)^{-1}$ is a normalized constant.

Clearly, a prior over-segmentation obtained by mean-shift is necessary for Liu's algorithm [20]. However, mean-shift is sensitive to parameters. Moreover, the fuzzy membership depends on both the pixel's neighboring window and the region containing the pixel.

Based on the analysis above, although Liu's algorithm is able to improve the image segmentation effect by incorporating adaptive local spatial information into the objective function, it has a high computational complexity due to the repeated computation of adaptive neighboring information in every iteration. Although we also employ adaptive neighboring information obtained by a superpixel algorithm to improve the segmentation effect, significantly different from Liu's algorithm is the fact that the presented superpixel algorithm has a lower computational complexity.

7.2.2 FCM Based on Histogram of Gray Images

The traditional FCM algorithm has to compute the distance between each pixel and clustering centers, which leads to a high computational complexity when the resolution of an image is high. The enhanced FCM (EnFCM) proposed by Szilágyi et al. [27] solves the problem by performing clustering on gray levels instead of pixels. The idea is efficient for the reduction of the computational complexity because the repeated distance computation is removed by integrating a histogram to its objective function. The objective function of EnFCM is defined as

$$J_m = \sum_{l=1}^{\tau}\sum_{k=1}^{c} \gamma_l u_{kl}^{m} \| \xi_l - v_k \|^2, \tag{7.5}$$

where u_{kl} represents the fuzzy membership of gray value l with respect to the k-th clustering center v_k, m is the weighting exponent, ξ is a grayscale image, ξ_l is the gray level, $1 \leq l \leq \tau$, τ denotes the number of the gray levels of ξ (it is generally far smaller than N), γ_l is the number of pixels whose gray level equals to ξ_l, and

$$\sum_{l=1}^{\tau} \gamma_l = N. \tag{7.6}$$

Clearly, the introduction of a histogram is able to reduce the computational complexity of FCM. Because the level of the histogram is far less than the number of pixels in an image, it is faster to implement FCM on gray levels than pixels for grayscale image segmentation. However, it is difficult to extend this idea of EnFCM to FCM for color image segmentation [28] because the number of different colors is usually close to the number of pixels in a color image. This is also the reason that FRFCM [25] usually requires a longer execution time to segment a color image than the corresponding grayscale image.

To address the issue, in this chapter, we will compute the histogram of a color image according to the corresponding superpixel image since the number of regions in the superpixel image is far smaller than the number of pixels in the original color image. We will use the mean value of all pixels within an area instead of these pixels to reduce the number of different colors in the original color image. It is easy to compute the histogram of the superpixel image because it contains only a small number of different colors. And then, the fast FCM algorithm will be achieved for color image segmentation, which will be presented in detail in Section 7.3.

7.3 Superpixel Integration to FCM

Since a superpixel image is able to provide better local spatial information than a neighboring window of fixed size and shape, superpixel technologies such as mean-shift [10], simple linear iterative clustering (SLIC) [29], and WT [30], are usually considered as pre-segmentation algorithms for improving segmentation results generated by clustering algorithms [31]. Compared to SLIC, mean-shift and WT produce irregular superpixel areas that are better than hexagonal regions obtained by SLIC. In practical applications, mean-shift is more popular than WT since the latter is sensitive to noise leading to a serious over-segmentation.

Even though mean-shift is able to provide better superpixel results, it is sensitive to parameter values, e.g. the spatial bandwidth, denoted by h_s; the range bandwidth, denoted by h_r; and the minimum size of final output regions, denoted by h_k. Moreover, the computational complexity of mean-shift is higher than WT. Therefore, we need to develop a fast superpixel algorithm that can provide better pre-segmentation result and require less time than mean-shift. Because WT only depends on region minima of gradient images to obtain pre-segmentation, it has a very low computational complexity. In this work, we employ a novel WT based on MMGR (MMGR-WT) to produce superpixel images. The MMGR-WT is able to provide more appropriate pre-segmentation results using shorter execution time than mean-shift. Moreover, it is insensitive to parameters.

Based on the superpixel image obtained by MMGR-WT, we compute the histogram of superpixel images to achieve fast FCM algorithm. The computation of the histogram of superpixel images is easy because the number of different colors from superpixel images is far smaller than that from the original color image. Finally, the histogram is considered as a parameter of the objective function to achieve fast color image segmentation. The framework of the algorithm presented in the chapter is shown in Figure 7.1.

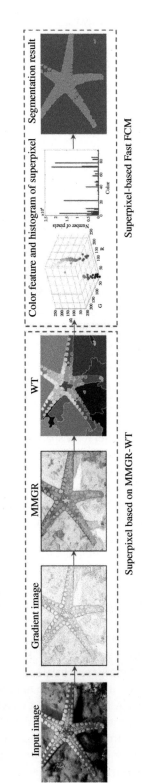

Figure 7.1 Framework of the presented algorithm [1] / with permission of IEEE.

7.3.1 Superpixel Based on Local Feature

WT is a fast algorithm used for image segmentation via computing local minima of a gradient image and searching the watershed line between adjacent local minima. The algorithm easily causes an over-segmentation because it is sensitive to noise. To address the problem, many algorithms have been proposed by modifying the gradient image of the original image. Among these algorithms, morphological gradient reconstruction (MGR) [32] is a simple and efficient algorithm for overcoming over-segmentation because it is able to preserve the contour details of objects while removing noise and useless gradient details. The basic definition of morphological reconstruction was presented in Chapter 6.

According to Chapter 5, it is not difficult to find that both morphological opening reconstruction R^O and morphological closing reconstruction R^C are able to remove region minima in a gradient image to reduce over-segmentation. For instance, we use R^C to reduce over-segmentation, as shown in Figure 7.2.

In Figure 7.2, the SE is defined as a disk, where r_{SE} is the radius of the SE. Figure 7.2 shows that the number of segmentation regions decreases quickly by increasing the value of r_{SE}. However, a small SE easily leads to over-segmentation, while a large SE easily leads to under-segmentation. Therefore, it is difficult to obtain a superpixel image with both fewer regions and accurate contour by using MGR. To balance the number of regions in superpixel image and contour precision, a suitable SE is required, but it is difficult to choose a suitable SE for different images.

To solve the problem, we try to use different SEs to reconstruct a gradient image and then fuse these reconstructed gradient images to remove the dependency of segmentation result on SEs. Thus, this chapter presents an MMGR operation denoted by R^{MC} that is defined as follows

$$R_g^{MC}(f, s, t) = \vee \left\{ R_g^C(f)_{B_s}, R_g^C(f)_{B_{s+1}}, \cdots, R_g^C(f)_{B_t} \right\}, \tag{7.7}$$

where s and t represent minimal and maximal r_{SE}, respectively, $s \leq r_{SE} \leq t$, $r \in N^+$, and $f \leq g$.

We can see that R^{MC} employs multiscale SEs to reconstruct a gradient image to obtain multiple reconstructed images. By computing the pointwise maximum of these reconstructed gradient images, an excellent gradient image that removes most of useless local minima while preserving important edge details is obtained.

The presented MMGR includes two parameters, s and t, where s controls the size of the minimal region and t controls the size of the maximal region. If s is too small, there will be many small regions in segmentation results, but if s is too large, the boundary precision will be low. An example is shown in Figure 7.3 It can be seen that the superpixel result has a high contour precision but includes some small regions when s = 1, it has a high contour precision and excludes small regions when s = 2 or s = 3, and it has a clearly low contour precision when s = 8. Consequently, we choose $1 \leq s \leq 3$ here. Because t controls the size of the maximal region, the superpixel image is better when the value of t is larger, as shown in Figure 7.4 However, the superpixel image is unchanged when the value of t is larger than a threshold; for example, the threshold is 11 in Figure 7.4. Clearly, the

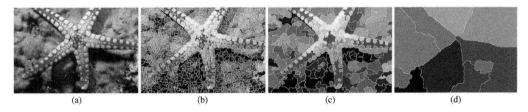

(a)   (b)   (c)   (d)

Figure 7.2 Watershed segmentation based on MGR with different SEs [1] / with permission of IEEE. (a) Original image "12003" (image size: 481 × 321). (b) r_{SE} = 1. (c) r_{SE} = 3. (d) r_{SE} = 10.

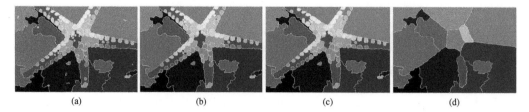

(a) (b) (c) (d)

Figure 7.3 Segmentation results using MMGR-WT with different s, where t = 10 [1] / with permission of IEEE. (a) s = 1. (b) s = 3. (c) s = 5. (d) s = 8.

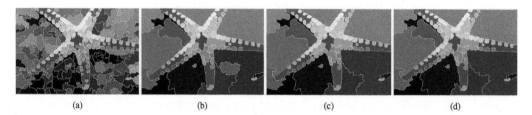

(a) (b) (c) (d)

Figure 7.4 Watershed segmentation based on MMGR-WT with different sized SEs [1] / with permission of IEEE. (a) s = 2, t = 3. (b) s = 2, t = 7. (c) s = 2, t = 11. (d) s = 2, t = 20.

Table 7.1 Comparison of the number of superpixel regions for WT based on MGR and MMGR respectively [1] / with permission of IEEE.

	MGR			MMGR		
Parameters	r_{SE} = 1	r_{SE} = 3	r_{SE} = 10	s = 2, t = 7	s = 2, t = 11	s = 2, t = 20
Number	1210	263	10	264	95	95

superpixel image is convergent via increasing the value of t. Moreover, the convergent result is perfect because it includes fewer regions and yet provides accurate contour. Therefore, the MMGR is insensitive to the change of t when t is larger than a threshold. Table 7.1 shows the comparison of the number of superpixel regions for WT-MGR and WT-MMGR, respectively.

In Table 7.1, t is a variable. It is difficult to set different values of t for each image. In practical applications, t is adaptive, and it is not required for MMGR as long as we set a minimal error threshold denoted by η' instead of t, i.e.

$$max\left\{ R_g^{MC}(f,s,t) - R_g^{MC}(f,s,t+1) \right\} \leq \eta'. \tag{7.8}$$

In (7.8), t can be replaced by η' because t is supposed to have different values for each image in a data set, but a fixed value of η' can be used for all images in the data set. Note that if η' is too large, t will be small, but the error will be large. On the contrary, if η' is too small, the error will be small, but t will be large, leading to a high computational burden for MMGR. Therefore, it is important to choose an appropriate η' for a data set. We perform MMGR on 10 images from the Berkeley segmentation dataset and benchmark (BSDS), we can obtain different values of t according to a fixed value of η', as shown in Table 7.2.

Table 7.2 shows that the values of t will be larger when decreasing η'. However, t will be unchanged when η' is smaller or equal to 10^{-4}. Therefore, we set $\eta' = 10^{-4}$ in this chapter.

Table 7.2 Values of t for 10 images from BSDS for different values of η' [1] / with permission of IEEE.

Image	$\eta' = 10^{-2}$	$\eta' = 10^{-3}$	$\eta' = 10^{-4}$	$\eta' = 10^{-5}$
"2092"	12	17	26	26
"3096"	10	10	10	10
"8023"	10	10	14	14
"8049"	14	19	22	22
"8143"	7	10	10	10
"12003"	12	18	18	18
"12074"	10	18	24	24
"12084"	14	15	15	15
"14037"	10	14	17	17
"15004"	14	18	18	18

To demonstrate the effectiveness of the MMGR, Figure 7.5 shows superpixel images obtained by SLIC, mean-shift, and MMGRWT, respectively, where s_k is the number of desired superpixels, s_m is the weighting factor between color and spatial differences, and s_s is the threshold used for region merging. These parameters are selected depending on [29] and [31]. It can be seen from Figure 7.5 that the superpixel images generated by SLIC include lots of areas with similar shape and size, but the superpixel images generated by the mean-shift and MMGR-WT include lots of areas with different shapes sizes. It is clear that the latter two algorithms provide better visual effect for the segmentation of real images.

Although SLIC and mean-shift are able to generate superpixel images according to task requirements by changing parameters, they have a longer execution time than the presented MMGR as shown in Table 7.3, where SLIC corresponds to Figure 7.5b, mean-shift1 corresponds to Figure 7.5c, mean-shift2 corresponds to Figure 7.5d, and MMGR-WT corresponds to Figure 7.5e. Because our purpose is to present a fast FCM algorithm for color image segmentation, MMGR is more appropriate than SLIC and mean-shift for our task requirement.

7.3.2 Superpixel-Based Fast FCM

In Section 7.3.1, we presented MMGR-WT to obtain better local spatial information used for fuzzy clustering. Because MMGR-WT depends on the local feature of an image, while FCM depends on the global feature, the combination of MMGRWT and FCM is able to improve image segmentation. This section presents an SFFCM algorithm by incorporating adaptive local spatial information into the objective function of FCM.

EnFCM is popular and efficient for achieving fast image segmentation because a gray image only includes 256 Gy levels, which is usually far smaller than the number of pixels in an image, but the number of different colors in a color image is far larger than 256. The quantization technology is typically used to reduce the number of colors in an image. The basic idea of quantization technology is that a clustering algorithm is performed on each channel of a color image to obtain an image with fewer color levels than before. However, the traditional color quantization only reduces the number of different colors, but the color distribution of the quantized image is still similar to that of the original image because the local spatial information is ignored. Because a superpixel image carries

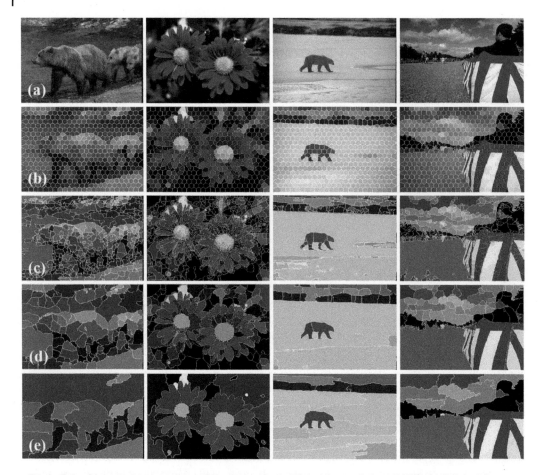

Figure 7.5 Superpixel images using different methods [1] / with permission of IEEE. (a) Original images. (b) Superpixel images obtained by SLIC (s_k = 500, s_m = 50, s_s = 1). (c) Superpixel images obtained by mean-shift1 with h_s = 7, h_r = 7, h_k = 30. (d) Superpixel images obtained by mean-shift2 with h_s = 15, h_r = 15, h_k = 50. (e) Superpixel images obtained by MMGR-WT (s = 2).

Table 7.3 Comparison of execution time (in seconds) of different methods used to superpixel images [1] / with permission of IEEE.

Algorithms	"100075"	"124084"	"100007"	"145086"	Average
SLIC	3.86	4.07	3.89	3.88	3.93
mean-shift1	7.02	7.22	0.94	7.20	7.10
mean-shift2	2.66	7.22	2.67	2.86	2.79
MMGR-WT	**0.32**	**0.32**	**0.31**	**0.36**	**0.33**

the spatial information of the image and reduces the number of different colors, the superpixel image is superior to images quantized by clustering algorithms. We applied the clustering algorithm proposed in [28] and the presented MMGR-WT to quantize a color image and then computed the histogram of the quantized image as shown in Figure 7.6, where the number of different colors is

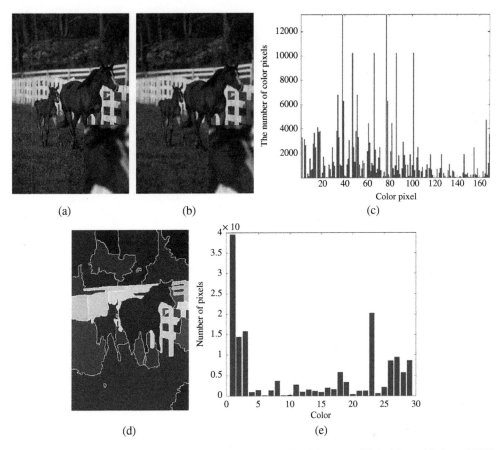

Figure 7.6 Quantization of color image and the corresponding histogram [1] / with permission of IEEE. (a) Original image. (b) Color quantization using algorithm proposed in [28] (*c* = 10). (c) Histogram of Figure 7.6b. (d) Superpixel image using MMGR-WT (*s* = 2). (e) Histogram of Figure 7.6c.

57 214 in the original image. Furthermore, Figure 7.7, the color distribution of Figure 7.6, shows that the presented MMGR-WT is more appropriate than the clustering algorithm proposed in [28] for subsequent image segmentation.

It is clear that the histograms of Figure 7.6b, d are simpler with only a small number of different colors appearing in the quantized images. According to Figure 7.6c, e, we can extend EnFCM to color image segmentation easily. Compared to Figure 7.6c, e has even fewer color levels. In addition, it is clear that the color distributions of Figure 7.7c is different from Figure 7.7a, b, and the former is helpful for subsequent pixel classification.

Based on the superpixel image obtained by MMGR-WT, this chapter presents the objective function of fast SFFCM for color image segmentation as follow

$$J_m = \sum_{l=1}^{\tau} \sum_{k=1}^{c} S_l u_{kl}^m \left\| \left(\frac{1}{S_l} \sum_{j \in R_l} x_j \right) - v_k \right\|^2, \tag{7.9}$$

where l is the color level, $1 \leq l \leq \tau$, τ the number of regions of the superpixel image, l, $\tau \in N^+$, S_l is the number of pixels in the l-th region R_l, and x_j is the color pixel within the l-th region of the superpixel image obtained by MMGR-WT. The new objective function only introduces histogram

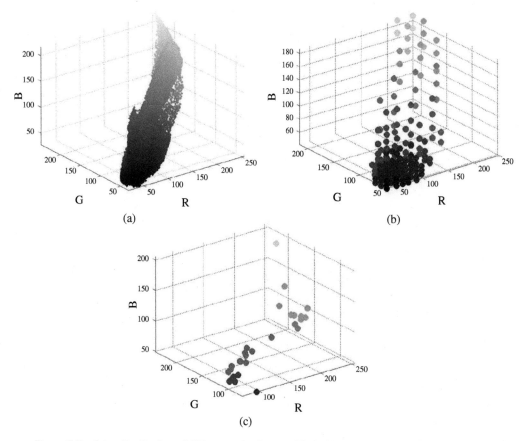

Figure 7.7 Color distribution of different color images [1] / with permission of IEEE. (a) The color distribution of Figure 7.6a. (b) The color distribution of Figure 7.6b. (c) Color distribution of Figure 7.6d.

information compared with the old one in FCM. Because each color pixel in the original image is replaced by the mean value of color pixels within the corresponding region of the superpixel image, the number of color level is equivalent to the number of regions in the superpixel image. Thus, the computational complexity is efficiently reduced due to $l \ll N$.

Utilizing the Lagrange multiplier technique, the aforementioned optimization problem can be converted to an unconstrained optimization problem that minimizes the following objective function

$$\tilde{J}_m = \sum_{l=1}^{\tau} \sum_{k=1}^{c} S_l u_{kl}^m \left\| \left(\frac{1}{S_l} \sum_{j \in R_l} x_j \right) - v_k \right\|^2 - \lambda \left(\sum_{k=1}^{c} u_{kl} - 1 \right), \tag{7.10}$$

where λ is a Lagrange multiplier. We compute the partial derivative of $\tilde{J}_m$ with respect to u_{kl} and v_k, respectively:

$$\frac{\partial \tilde{J}_m}{\partial u_{kl}} = \sum_{l=1}^{\tau} \sum_{k=1}^{c} \frac{\partial S_l u_{kl}^m \left\| \left(\frac{1}{S_l} \sum_{j \in R_l} x_j \right) - v_k \right\|^2}{\partial u_{kl}} - \lambda$$

$$= \sum_{l=1}^{\tau} \sum_{k=1}^{c} m S_l u_{kl}^{m-1} \left\| \left(\frac{1}{S_l} \sum_{j \in R_l} x_j \right) - v_k \right\|^2 - \lambda$$

$$= 0, \tag{7.11}$$

$$\frac{\partial \tilde{J}_m}{\partial v_k} = \sum_{l=1}^{\tau} \sum_{k=1}^{c} \partial S_l \frac{u_{kl}^m \left\| \left(\frac{1}{S_l} \sum_{j \in R_l} x_j \right) - v_k \right\|^2}{\partial v_k}$$

$$= \sum_{l=1}^{\tau} \sum_{k=1}^{c} S_l u_{kl}^m \frac{\partial \left\| \left(\frac{1}{S_l} \sum_{j \in R_l} x_j \right) - v_k \right\|^2}{\partial v_k}$$

$$= \sum_{l=1}^{\tau} S_l u_{kl}^m \frac{\partial \left\| \left(\frac{1}{S_l} \sum_{j \in R_l} x_j \right) - v_k \right\|^2}{\partial v_k}$$

$$= -2 \sum_{l=1}^{\tau} S_l u_{kl}^m \left\| \left(\frac{1}{S_l} \sum_{j \in R_l} x_j \right) - v_k \right\|$$

$$= 0. \tag{7.12}$$

Combing (7.11) and (7.12) together, the corresponding solutions for u_{kl} and v_k are obtained as

$$v_k = \frac{\sum_{l=1}^{\tau} u_{kl}^m \sum_{j \in R_l} x_j}{\sum_{i=1}^{\tau} S_l u_{kl}^m}, \tag{7.13}$$

$$u_{kl} = \frac{\left\| \left(\frac{1}{S_l} \sum_{j \in R_l} x_j \right) \right\|^{-2/(m-1)}}{\sum_{j=1}^{c} \left\| \left(\frac{1}{S_l} \sum_{j \in R_l} x_j \right) - v_j \right\|^{-2/(m-1)}}. \tag{7.14}$$

Based on (7.7)–(7.14), the presented SFFCM algorithm can be summarized as follows:

Step 1: Set values for c, m, s, η', η_c, where η_c is the convergence condition used for SFFCM.
Step 2: Compute a superpixel image using (7.7)–(7.8), and then compute its histogram.

1) Compute the gradient image using Sobel operators.
2) Implement MMGR using (7.7)–(7.8) and η'.
3) Implement WT to obtain the superpixel image.

Step 3: Initialize randomly the membership partition matrix $U^{(o)}$ according to the superpixel image.
Step 4: Set the loop counter $b = 0$.
Step 5: Update the clustering centers using (7.13).
Step 6: Update the membership partition matrix $U^{(t)}$ using (7.14).
Step 7: If $max|U^{(b)} - U^{(b+1)}| < \eta_c$ then stop; otherwise, set $b = b + 1$ and go to Step 5.

We applied the presented SFFCM to Figure 7.6a following the previous steps. Then, the segmentation result is shown in Figure 7.8. We can see that the presented SFFCM is able to obtain better segmentation result than the traditional algorithm. Based on the analysis mentioned above, we conclude that the presented SFFCM has the following advantages.

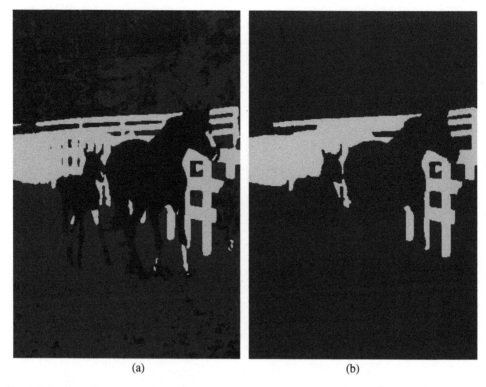

(a)   (b)

Figure 7.8   Segmentation results on Figure 7.6a [1] / with permission of IEEE. (a) Segmentation result using FCM for quantized image. (b) Segmentation result using the presented SFFCM.

1) SFFCM is very fast for color image segmentation because the number of different colors is reduced efficiently due to superpixel and color histogram.
2) SFFCM is insensitive to the change of parameters because the superpixel image obtained by MMGR-WT is convergent.
3) SFFCM obtains an excellent result for color image segmentation because both adaptive local spatial information and global color feature are incorporated into the objective function.

7.4   Discussion and Summary

We conduct experiments on two synthetic color images of size 256×256 and real color images from the BSDS [33] and the Microsoft Research Cambridge (MSRC) data set [34]. The first synthetic image includes four different colors, while the second includes five different colors. The experiments are conducted on a DELL desktop with Intel core CPU, i7-6700, 3.4 GHz, and 16 GB RAM.

7.4.1   Comparison with Other Algorithms

To assess the effectiveness and efficiency of the presented SFFCM, we compare it with nine algorithms based on clustering used for color image segmentation: FCM [9], FGFCM [14], HMRF-FCM [23], FLICM [15], NWFCM [16], KWFLICM [19], NDFCM [17], Liu's algorithm [20], and FRFCM [25]. Since these algorithms employ different local spatial neighborhoods to improve segmentation results, they have different advantages and disadvantages.

7.4.2 Parameter Setting

Since the algorithms we use for comparison and the presented SFFCM algorithm are clustering algorithms based on objective function optimization, three indispensable parameters—the weighting exponent, the convergence condition, and the maximal number of iteration—must be set before iterations. In our experiments, the three parameters are 2, 10^{-5}, and 50, respectively. In addition, the value of the minimal error threshold used for MMGR is 10^{-4}. The algorithms we selected for comparison employ a window of size 3×3, requiring a neighboring window of fixed size for fair comparison. Moreover, computational complexity is also an important reason for the choice of the window of size 3×3. In addition, a neighborhood window is unnecessary for FCM. According to the criterion of parameters setting mentioned for these algorithms, the spatial scale factor and the gray-level scale factor in FGFCM and NDFCM are $\lambda_s = 3$ and $\lambda_g = 5$, respectively. The third parameter of the NDFCM, a new scale factor, is $\lambda_a = 3$. The NWFCM only refers to the gray-level scale factor, $\lambda_g = 5$. Because Liu's algorithm requires a pre-segmentation obtained by mean-shift, the three parameters $h_s = 10$, $h_r = 10$, and $h_k = 100$ follow the original paper. Except the three indispensable parameters mentioned above and the number of the cluster prototypes, HMRF-FCM, FLICM, and KWFLICM do not require any other parameter. In FRFCM, the SE used for multivariate morphological reconstruction is a square of size 3×3, and the filtering window used for membership filtering is also a square of size 3×3. As the presented SFFCM needs a minimal SE for MMGR, we set $s = 2$ for MMGR.

7.4.3 Results on Synthetic Image

First, we test these comparative algorithms and the presented SFFCM on two synthetic color images to show their robustness to noise. In this experiment, three kinds of different noise, Gaussian, salt-and-pepper, and uniform noise, are added to these synthetic images. All algorithms mentioned above are implemented, and segmentation results are shown in Figures 7.9 and 7.10.

FCM, HMRF-FCM, FLICM, and NWFCM provide poor results as shown in Figure 7.9c, e–g and 7.10c, e–g, which show that they are sensitive to both Gaussian and salt-and-pepper noise. HMRF-FCM, FLICM, and NWFCM cannot improve the FCM algorithm for color images. FGFCM, NDFCM, and FRFCM obtain good segmentation results as shown in Figure 7.9d, i, k for the image corrupted by Gaussian noise but poor segmentation results as shown in Figure 7.10d, i, k for the image corrupted by salt-and-pepper noise. It is clear that the three algorithms are insensitive to Gaussian noise, but they are sensitive to salt-and-pepper noise of high density. KWFLICM, Liu's algorithm, and the presented SFFCM provide better results as shown in Figures 7.9h, j, l and 7.10h, j, l, which demonstrates that they are robust against both Gaussian noise and salt-and-pepper noise as adaptive neighboring information is employed by the three algorithms.

To assess the performance of different algorithms on noisy image segmentation, two performance indices, the quantitative score (S), which is the degree of equality between pixel sets A_k and the ground truth (GT) C_k, and the optimal segmentation accuracy (SA), which is the sum of the correctly classified pixels divided by the sum of the total number of the pixels [19], are adopted. S and SA are defined as

$$S = \sum_{k=1}^{c} \frac{A_k \cap C_k}{A_k \cup C_k}, \tag{7.15}$$

$$SA = \sum_{k=1}^{c} \frac{A_k \cap C_k}{\sum_{j=1}^{c} C_j}, \tag{7.16}$$

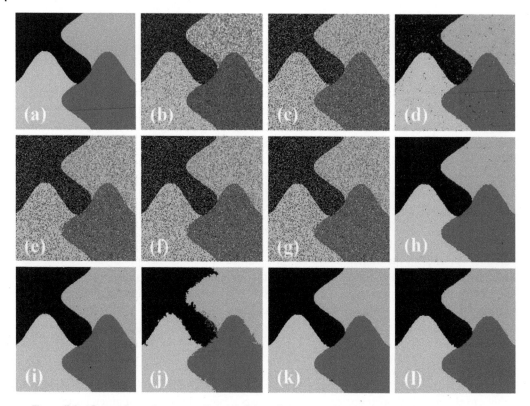

Figure 7.9 Comparison of segmentation results on the symmetric color image [1] / with permission of IEEE. (a) First synthetic image. (b) Noisy image (Gaussian noise, the noise level is 10%). (c) FCM. (d) FGFCM. (e) HMRF-FCM. (f) FLICM. (g) NWFCM. (h) KWFLICM. (i) NDFCM. (j) Liu's algorithm. (k) FRFCM. (l) SFFCM.

where A_k is the set of pixels belonging to the k-th class found by the algorithm, while C_k is the set of pixels belonging to the class in the GT. We implemented each of these algorithms on two synthetic images and computed the mean value and the root mean square error (RMSE) of S and SA, as shown in Tables 7.4 and 7.5.

In Tables 7.4 and 7.5, FCM, HMRF-FCM, FLICM, and NWFCM obtain similar S values as well as SA values, which further demonstrates that HMRF-FCM, FLICM, and NWFCM are inefficient for color image segmentation. FCM misses the local spatial information leading to poor segmentation results. HMRF-FCM, FLICM, and NWFCM only employ a small neighboring window to incorporate local spatial information into their objective function, which is helpful for segmenting images corrupted by low-density noise but not useful for segmenting images corrupted by high-density noise. FGFCM and NDFCM obtain higher values of S and SA than FCM, HMRF-FCM, FLICM, and NWFCM because the tested images are synthetic and the added noise is known. Because FGFCM and NDFCM employ a filter to suppress noise before iterations in clustering, they obtain larger S and SA than FCM, HMRF-FCM, FLICM, and NWFCM for synthetic images corrupted by known noise. FRFCM obtains high S and SA when noisy density is low but small S and SA when noisy density is high because FRFCM employ multivariate morphological reconstruction to simplify the image and use the membership filtering to improve segmentation results.

As KWFLICM, Liu's algorithm, and the presented SFFCM employ adaptive local spatial information to improve segmentation results, they obtain larger S and SA than those algorithms we

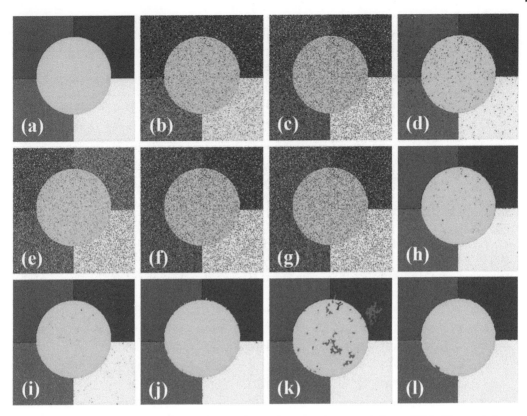

Figure 7.10 Comparison of segmentation results on the second symmetric image [1] / with permission of IEEE. (a) Second synthetic image. (b) Noisy image (salt-and-pepper, the noise level is 40%). (c) FCM. (d) FGFCM. (e) HMRF-FCM. (f) FLICM. (g) NWFCM. (h) KWFLICM. (i) NDFCM. (j) Liu's algorithm. (k) FRFCM. (l) SFFCM.

compare them with, which employ local spatial information in a window of fixed size. Liu's algorithm obtains higher values of S and SA because of the combination of mean-shift, FCM, and HMRF. In some cases, the presented SFFCM provides smaller S and SA than Liu's algorithm but higher values than the algorithms we use for comparison because contour details are smoothed in segmentation results obtained by the SFFCM. However, SFFCM provides the best mean value of S for two synthetic images and the best RMSE of SA for the second synthetic image, which shows that SFFCM is able to obtain good segmentation results for images corrupted by different noises.

7.4.4 Results on Real Images

To demonstrate that the presented SFFCM is useful for real image segmentation, we further conducted experiments on BSDS and MSRC. BSDS is a popular benchmark that has been widely used by researchers for the task of image segmentation [33]. The early BSDS is named BSDS300 and is composed of 300 images, and the current BSDS is an extended version that is composed of 500 images, called BSDS500. For each image in BSDS, there are four to nine GT segmentations. These GT segmentations are delineated by different human subjects. For instance, there are five GT segmentations delineated by five subjects on image "12003" and "113009," as shown in Figure 7.11. The MSRC data set contains 23 object classes and comprises 591 natural images. For each image in MSRC, there is only one GT segmentation that is pixel-wise labeled.

Table 7.4 Comparison scores (*S%*) of the 10 algorithms on the first synthetic image corrupted by noise of different levels (*c* = 4). (The best values are in bold [1] / with permission of IEEE.)

Noise	FCM	FGFCM	HMRF-FCM	FLICM	NWFCM	KWFLICM	NDFCM	Liu's algorithm	FRFCM	SFFCM
Gaussian 5%	95.53	99.85	95.60	95.60	95.61	99.78	99.87	99.41	99.82	99.20
Gaussian 10%	85.69	99.75	85.70	85.69	85.70	99.68	99.61	98.46	99.81	99.17
Gaussian 15%	66.35	98.20	66.24	66.26	66.25	99.24	99.50	97.54	99.62	99.15
Gaussian 20%	55.22	94.05	55.29	55.31	55.29	98.45	99.53	96.76	99.48	98.98
Salt-and-pepper 10%	80.21	95.56	80.21	80.21	80.21	99.65	87.91	99.67	99.68	99.18
Salt-and-pepper 20%	65.54	87.55	65.54	65.54	65.54	98.66	77.28	99.23	98.58	98.91
Salt-and-pepper 30%	57.51	77.21	54.07	54.07	54.07	95.21	69.39	98.55	97.11	98.68
Salt-and-pepper 40%	42.19	65.94	44.51	42.19	44.51	87.69	66.60	96.76	78.64	92.83
Uniform 10%	84.41	99.49	84.44	84.41	84.45	99.81	82.31	99.77	99.81	99.22
Uniform 20%	77.08	97.95	77.02	77.08	77.02	99.57	77.59	99.28	99.68	99.19
Uniform 30%	59.83	94.31	60.00	60.08	60.00	98.91	89.23	98.72	99.39	98.99
Uniform 40%	50.05	87.46	49.94	50.05	49.94	96.50	99.46	97.23	98.44	98.83
Mean value	67.30	97.44	67.71	67.54	67.72	97.76	86.36	98.45	97.51	98.53
RMSE	16.49	10.48	15.98	16.28	15.98	3.48	13.15	7.11	6.00	7.80

Figures 7.12–7.14 show segmentation results of images from BSDS and MSRC. The parameters setting is the same as that in Section 7.4.2. Since the size of images in BSDS and MSRC is different from the size of synthetic images, the SE used for multivariate morphological reconstruction in FRFCM is a disk of size 5 × 5 for BSDS and MSRC. The value of *s* is set to 3 for real images in this section. In addition, the CIE-Lab color space is used for all algorithms for fair comparison.

As can be seen from Figures 7.12 and 7.13, segmentation results obtained by FCM, FGFCM, HMRF-FCM, FLICM, NWFCM, KWFLICM, and NDFCM include a great number of small regions because only a small local neighboring window is employed (a large neighboring window will cause a very high computational complexity). FRFCM obtains better results than the algorithms mentioned above due to the introduction of multivariate morphological reconstruction and membership filtering. However, Liu's algorithm and the presented SFFCM obtain better results than FRFCM due to the use of adaptive local spatial information provided by pre-segmentation. Although Liu's algorithm provides a better segmentation result than the presented SFFCM on the left image, the latter provides better results than the former on four other images in Figure 7.12. In practical applications, since it is difficult to present an algorithm to achieve the best segmentation result for every image in a data set, researchers typically use the average result on all images in the data set, e.g. BSDS and MSRC, to estimate the algorithm performance.

In Figure 7.14, all algorithms are efficient for images in which the foreground is clearly different from the background as shown in the first row. A large number of small regions appear in segmented images except images obtained by SFFCM as shown in the second to fifth rows. Liu's algorithm,

Table 7.5 *SA (SA%)* of 10 algorithms on the second synthetic image corrupted by noise of different levels (*c* = 5). (The best values are in bold [1] / with permission of IEEE.)

Noise	FCM	FGFCM	HMRF-FCM	FLICM	NWFCM	KWFLICM	NDFCM	Liu's algorithm	FRFCM	SFFCM
Gaussian 3%	93.10	99.43	93.09	93.10	93.10	99.47	99.87	99.96	99.64	99.59
Gaussian 5%	87.03	98.44	86.85	87.01	86.86	99.04	99.70	99.96	99.55	99.52
Gaussian 10%	74.71	95.74	74.24	74.64	74.33	97.52	99.52	98.58	97.09	99.38
Gaussian 15%	66.02	92.30	65.70	66.06	65.70	95.69	98.84	97.09	94.87	99.41
Salt-and-pepper 10%	86.42	97.33	86.46	86.42	73.86	99.44	97.26	99.84	99.57	99.55
Salt-and-pepper 20%	74.40	92.41	74.54	74.40	74.54	98.55	89.52	99.68	97.08	99.48
Salt-and-pepper 30%	63.52	84.59	58.15	63.52	58.15	95.59	86.49	99.03	92.60	99.00
Salt-and-pepper 40%	49.86	75.76	48.89	49.86	48.89	76.97	62.54	97.56	85.48	98.99
Uniform 10%	90.41	99.26	90.38	90.42	90.38	99.47	94.13	99.91	99.62	99.59
Uniform 20%	87.09	97.77	80.97	87.09	80.97	99.16	90.17	99.81	98.95	99.53
Uniform 30%	72.34	94.96	62.67	72.34	72.02	98.13	99.59	99.64	96.71	99.41
Uniform 40%	63.42	89.73	62.80	63.42	54.73	82.16	99.82	82.75	93.71	99.20
Mean value	75.19	93.14	73.73	75.19	72.79	95.10	96.62	97.81	96.24	**99.39**
RMSE	13.10	7.02	14.16	13.01	14.00	7.46	10.67	4.84	4.17	**0.21**

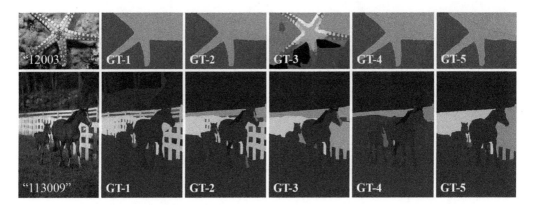

Figure 7.11 GT segmentations of images "12003" and "113009" from BSDS [1] / with permission of IEEE.

FRFCM, and SFFCM obtain better results than other algorithms as shown in the sixth row. All algorithms fail to segment images except SFFCM, as shown in the last three rows.

To evaluate segmentation results obtained by different algorithms, five performance measures [33], namely, the probabilistic rand index (PRI), the covering (CV), the variation of information (VI), the global consistency error (GCE), and the boundary displacement error (BDE), are computed

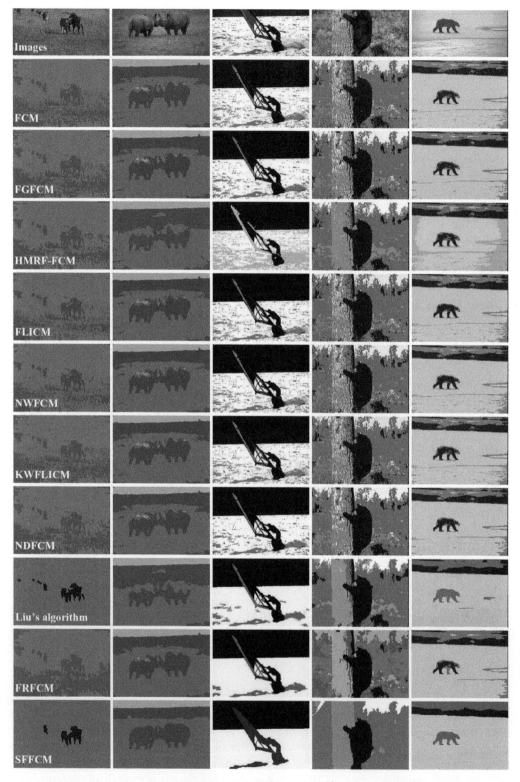

Figure 7.12 Comparison of segmentation results on color images from BSDS using different models [1] / with permission of IEEE.

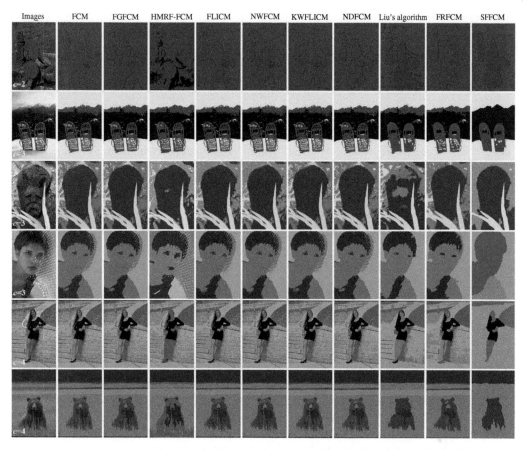

Figure 7.13 Comparison of segmentation results on color images from BSDS using different models [1] / with permission of IEEE.

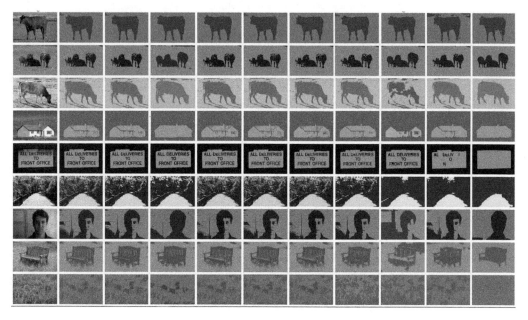

Figure 7.14 Comparison of segmentation results on color images from MSRC using different models (*c* = 2) [1] / with permission of IEEE.

in this experiment. The PRI is a similarity measure that counts the fraction of pairs of pixels whose labels are consistent between the computed segmentation and the corresponding GT segmentation. The CV is an overlap measure that can be also used to evaluate the segmentation effect. The VI is a similarity measure that is always used to measure the distance between two segmentations in terms of their average conditional entropy. The GCE computes the degree to which two segmentations are mutually consistent. The BDE is an error measure that is used to measure the average displacement error of boundary pixels between two segmentations. If the segmentation result is more similar to the GT, PRI and CV will be larger but VI, GCE, and BDE will be smaller. In the BSDS, each image corresponds to multiple GT segmentations, which leads to a result that a segmentation result corresponds to multiple groups of performance index. Therefore, the average value of multiple groups of performance index is usually considered as the final performance index of the segmentation result.

All these algorithms are evaluated on the BSDS and MSRC data sets. The value of c is set from 2 to 6 for each image in the BSDS, while its value is set from 2 to 4 for each image in the MSRC. We choose the best value of c corresponding to the highest PRI. Because the BSDS and MSRC include lots of images, the average values of PRI, CV, VI, GCE, and BDE corresponding to segmentation results of all images in the BSDS or MSRC are presented in Tables 7.6–7.8. In Tables 7.6, 7.7, we can see that FCM, FGFCM, FLICM, NWFCM, and KWFLICM have similar values of the PRI, CV, VI, GCE, and BDE. NDFCM has a similar performance to HMRF-FCM. FRFCM clearly outperforms other algorithms on PRI and BDE due to the introduction of multivariate morphological reconstruction. Liu's algorithm obtains better performance than FRFCM because it computes the distance between pixels and clustering centers according to the combination of superpixel image and the original image. Similarly, Table 7.8 shows that FLICM, NWFCM, and KWFLICM have similar values of the PRI, CV, VI, GCE, and BDE. The performance of FGFCM is similar to NDFCM. Different from Tables 7.6, 7.7, HMRF-FCM obtains better performance than Liu's algorithm and FRFCM as shown in Table 7.8. Clearly, the presented SFFCM is the best one because it obtains the best values of PRI, CV, VI, and GCE and is within 0.04 of the

Table 7.6 Average performance of 10 algorithms on the BSDS300 that includes 300 images. (The best values are in bold [1] / with permission of IEEE.)

Algorithms	PRI↑	CV↑	VI↓	GCE↓	BDE↓
FCM	0.74	0.43	2.87	0.41	13.78
FGFCM	0.74	0.43	2.80	0.40	13.63
FLICM	0.74	0.43	2.82	0.40	13.69
NWFCM	0.74	0.43	2.78	0.41	13.78
KWFLICM	0.74	0.43	2.82	0.40	13.70
NDFCM	0.74	0.44	2.87	0.39	13.52
HMRF_FCM	0.74	0.43	2.77	0.40	13.71
Liu's algorithm	0.77	0.48	2.53	0.35	**12.57**
FRFCM	0.75	0.46	2.62	0.36	12.87
SFFCM	**0.78**	**0.55**	**2.02**	**0.26**	12.90

Table 7.7   Average performance of 10 algorithms on the BSDS500 that includes 500 images. (The best values are in bold [1] / with permission of IEEE.)

Algorithm	PRI↑	CV↑	VI↓	GCE↓	BDE↓
FCM	0.74	0.43	2.88	0.40	13.48
FGFCM	0.75	0.44	2.81	0.39	13.28
FLICM	0.74	0.43	2.83	0.40	13.38
NWFCM	0.74	0.43	2.88	0.40	13.47
KWFLICM	0.74	0.44	2.83	0.40	13.40
NDFCM	0.75	0.44	2.78	0.39	13.13
HMRF_FCM	0.75	0.43	2.78	0.40	13.22
Liu's algorithm	0.76	0.47	2.58	0.36	**12.31**
FRFCM	0.76	0.45	2.67	0.37	12.35
SFFCM	**0.78**	**0.54**	**2.06**	**0.26**	12.80

Table 7.8   Average performance of 10 algorithms on the MSRC that includes 591 images. (The best values are in bold [1][1] / with permission of IEEE.)

Algorithm	PRI↑	CV↑	VI↓	GCE↓	BDE↓
FCM	0.70	0.55	7.93	0.32	12.67
FGFCM	0.70	0.56	7.85	0.31	12.39
FLICM	0.72	0.59	7.73	0.28	12.29
NWFCM	0.69	0.55	7.90	0.32	12.61
KWFLICM	0.69	0.55	7.93	0.32	12.67
NDFCM	0.69	0.55	7.90	0.32	12.54
HMRF_FCM	0.70	0.56	7.84	0.31	12.38
Liu's algorithm	0.71	0.54	7.77	0.34	12.43
FRFCM	0.71	0.58	7.79	0.30	**12.23**
SFFCM	**0.73**	**0.62**	**7.58**	**0.25**	12.49

best value obtained for BDE, as shown in Tables 7.6–7.8 and the best segmentation results as shown in Figure 7.12–7.14.

To demonstrate that the presented SFFCM is insensitive to parameters, we further discussed the relationship between the weighting exponent m and the SFFCM. We have known that the FCM algorithm is insensitive to m when used for image segmentation. The presented SFFCM has the same objective function as FCM. The difference between them is that the presented SFFCM employs a color histogram created by MMGRWT to speed up the FCM algorithm. Therefore, theoretically, the performance of the presented SFFCM is also insensitive to the value of m. Tables 7.9 and 7.10 show the performance of SFFCM for different values of m. Figure 7.15 shows the plot of Tables 7.9 and 7.10. It is clear that the performance of SFFCM is changed slightly via changing the value of m.

Table 7.9 Average performance of SFFCM on BSDS300. (The best values are in bold [1] / with permission of IEEE.)

Exponent	PRI↑	CV↑	VI↓	GCE↓	BDE↓
$m = 2$	0.78	0.55	2.02	0.26	12.90
$m = 5$	0.78	0.55	2.00	0.26	12.91
$m = 10$	0.78	0.55	2.01	0.25	12.87
$m = 30$	0.78	0.55	2.02	0.26	13.03
$m = 100$	0.78	0.55	2.01	0.25	12.89

Table 7.10 Average performance of SFFCM on BSDS500. (The best values are in bold [1] / with permission of IEEE.)

Exponent	PRI↑	CV↑	VI↓	GCE↓	BDE↓
$m = 2$	0.78	0.54	2.06	0.26	12.80
$m = 5$	0.78	0.54	2.04	0.26	12.72
$m = 10$	0.78	0.54	2.06	0.26	12.72
$m = 30$	0.78	0.54	2.06	0.26	13.80
$m = 100$	0.78	0.54	2.05	0.26	12.82

7.4.5 Execution Time

Execution time is an important index used to measure the performance of an algorithm. Table 7.11 shows execution time of different algorithms on two synthetic images and real images used in Sections 7.4.3–7.4.4. We computed the average execution time of algorithms on images from BSDS and MSRC, respectively.

It can be seen in Table 7.11 that FCM is faster than other algorithms, except SFFCM, because no additional computation is implemented. FGFCM and NDFCM are faster than FLICM, NWFCM, and KWFLICM because the neighboring information is computed in advance. FLICM, NWFCM, and KWFLICM repeatedly compute the neighboring information in each iteration, leading to a high computational complexity. Both HMRF-FCM and Liu's algorithm require a long execution time because a prior probability used for the HMRF model must be computed in each iteration. FRFCM is fast because multivariate morphological reconstruction and membership filtering are implemented only once. The presented SFFCM is very fast, even faster than FCM for some images because the number of different colors in the superpixel image obtained by MMGR-WT is decreased efficiently and the color histogram is integrated into SFFCM.

7.4.6 Conclusions

In this chapter, an SFFCM algorithm for color image segmentation has been presented. Two main contributions are presented. The first one is that we showed the MMGR operation to obtain a good superpixel image. The second one is that we incorporated a color histogram into the objective function to achieve fast image segmentation. The presented SFFCM is tested on synthetic and real images. The experimental results demonstrate that the presented SFFCM is superior to state-of-

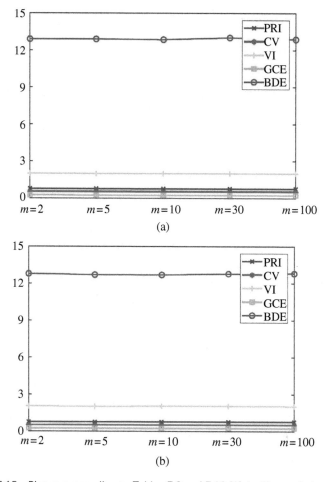

Figure 7.15 Plot corresponding to Tables 7.9 and 7.10 [1] / with permission of IEEE. (a) BSDS300. (b) BSDS500.

Table 7.11 Comparison of execution time (in seconds) of 10 algorithms. (The best values are in bold [1] / with permission of IEEE.)

	FCM	FGFCM	HMRF-FCM	FLICM	NWFCM	KWFLICM	NDFCM	Liu's algorithm	FRFCM	SFFCM
The first image	0.48	7.65	43.31	57.88	48.45	64.46	4.62	23.49	0.69	**0.19**
The second image	0.87	2.06	47.43	127.05	80.39	133.9	4.82	23.69	0.91	**0.21**
BSDS500	7.15	4.67	73.43	195.53	136.20	206.70	17.26	79.88	7.87	**0.74**
MSRC	**0.22**	7.59	19.35	37.76	30.07	39.83	4.60	17.75	0.53	0.29

the-art clustering algorithms because it provides the best segmentation results and requires the shortest running time.

Our algorithm is very fast for color image segmentation but similar to other k-means clustering algorithms, and it has limitations in practical applications since the number of clusters must be set in advance.

References

1 Lei, T., Jia, X., Zhang, Y. et al. (2019). Superpixel-based fast fuzzy c-means clustering for color image segmentation. *IEEE Trans. Fuzzy Syst.* **27** (9): 1753–1766.

2 Ma, J., Li, S., Qin, H., and Hao, A. (2016). Unsupervised multi-class co-segmentation via joint-cut over $ L_ \{1\} $-manifold hyper-graph of discriminative image regions. *IEEE Trans. Image Process.* **26** (3): 1216–1230.

3 Gong, M., Li, H., Zhang, X. et al. (2015). Nonparametric statistical active contour based on inclusion degree of fuzzy sets. *IEEE Trans. Fuzzy Syst.* **24** (5): 1176–1192.

4 Bai, M. and Urtasun, R. (2017). Deep watershed transform for instance segmentation. In: *Proceedings of the IEEE Conference on Computer Vision and Pattern Recognition*, 5221–5229.

5 Pereyra, M. and McLaughlin, S. (2017). Fast unsupervised bayesian image segmentation with adaptive spatial regularisation. *IEEE Trans. Image Process.* **26** (6): 2577–2587.

6 Yin, S., Qian, Y., and Gong, M. (2017). Unsupervised hierarchical image segmentation through fuzzy entropy maximization. *Pattern Recogn.* **68**: 245–259.

7 Krizhevsky, A., Sutskever, I., and Hinton, G.E. (2012). Imagenet classification with deep convolutional neural networks. *Adv. Neural Inform. Process. Syst.* **25**: 1097–1105.

8 Long, J., Shelhamer, E., and Darrell, T. (2015). Fully convolutional networks for semantic segmentation. In: *Proceedings of the IEEE Conference on Computer Vision and Pattern Recognition*, 3431–3440.

9 Pal, N.R. and Bezdek, J.C. (1995). On cluster validity for the fuzzy c-means model. *IEEE Trans. Fuzzy Syst.* **3** (3): 370–379.

10 Comaniciu, D. and Meer, P. (2002). Mean shift: a robust approach toward feature space analysis. *IEEE Trans. Pattern Anal. Mach. Intell.* **24** (5): 603–619.

11 Ng, A.Y., Jordan, M.I., and Weiss, Y. (2002). On spectral clustering: Analysis and an algorithm. In: *Advances in Neural Information Processing Systems*, 849–856.

12 Ahmed, M.N., Yamany, S.M., Mohamed, N. et al. (2002). A modified fuzzy c-means algorithm for bias field estimation and segmentation of MRI data. *IEEE Trans. Med. Imaging* **21** (3): 193–199.

13 Chen, S. and Zhang, D. (2004). Robust image segmentation using FCM with spatial constraints based on new kernel-induced distance measure. *IEEE Trans. Syst. Man Cybern. B Cybern.* **34** (4): 1907–1916.

14 Cai, W., Chen, S., and Zhang, D. (2007). Fast and robust fuzzy c-means clustering algorithms incorporating local information for image segmentation. *Pattern Recogn.* **40** (3): 825–838.

15 Krinidis, S. and Chatzis, V. (2010). A robust fuzzy local information C-means clustering algorithm. *IEEE Trans. Image Process.* **19** (5): 1328–1337.

16 Zaixin, Z., Lizhi, C., and Guangquan, C. (2014). Neighbourhood weighted fuzzy c-means clustering algorithm for image segmentation. *IET Image Process.* **8** (3): 150–161.

17 Guo, F.F., Wang, X.X., and Shen, J. (2016). Adaptive fuzzy c-means algorithm based on local noise detecting for image segmentation. *IET Image Process.* **10** (4): 272–279.

18 Memon, K.H. and Lee, D.H. (2018). Generalised kernel weighted fuzzy C-means clustering algorithm with local information. *Fuzzy Set. Syst.* **340**: 91–108.

19 Gong, M., Liang, Y., Shi, J. et al. (2012). Fuzzy c-means clustering with local information and kernel metric for image segmentation. *IEEE Trans. Image Process.* **22** (2): 573–584.

20 Liu, G., Zhang, Y., and Wang, A. (2015). Incorporating adaptive local information into fuzzy clustering for image segmentation. *IEEE Trans. Image Process.* **24** (11): 3990–4000.

21 Bai, X., Chen, Z., Zhang, Y. et al. (2015). Infrared ship target segmentation based on spatial information improved FCM. *IEEE Trans. Cybern.* **46** (12): 3259–3271.

22 Zhang, H., Wang, Q., Shi, W., and Hao, M. (2017). A novel adaptive fuzzy local information c-means clustering algorithm for remotely sensed imagery classification. *IEEE Trans. Geosci. Remote Sens.* **55** (9): 5057–5068.

23 Chatzis, S.P. and Varvarigou, T.A. (2008). A fuzzy clustering approach toward hidden Markov random field models for enhanced spatially constrained image segmentation. *IEEE Trans. Fuzzy Syst.* **16** (5): 1351–1361.

24 Zhang, H., Wu, Q.M.J., Zheng, Y. et al. (2014). Effective fuzzy clustering algorithm with Bayesian model and mean template for image segmentation. *IET Image Process.* **8** (10): 571–581.

25 Lei, T., Jia, X., Zhang, Y. et al. (2018). Significantly fast and robust fuzzy c-means clustering algorithm based on morphological reconstruction and membership filtering. *IEEE Trans. Fuzzy Syst.* **26** (5): 3027–3041.

26 Lei, T., Zhang, Y., Wang, Y. et al. (2017). A conditionally invariant mathematical morphological framework for color images. *Inform. Sci.* **387**: 34–52.

27 Szilagyi, L., Benyo, Z., Szilágyi, S.M., and Adam, H.S. (2003). MR brain image segmentation using an enhanced fuzzy c-means algorithm. In: *Proceedings of the 25th Annual International Conference of the IEEE Engineering in Medicine and Biology society* (IEEE Cat. No. 03CH37439), vol. **1**, 724–726. IEEE.

28 Özdemir, D. and Akarun, L. (2002). A fuzzy algorithm for color quantization of images. *Pattern Recogn.* **35** (8): 1785–1791.

29 Achanta, R., Shaji, A., Smith, K. et al. (2012). SLIC superpixels compared to state-of-the-art superpixel methods. *IEEE Trans. Pattern Anal. Mach. Intell.* **34** (11): 2274–2282.

30 Hu, Z., Zou, Q., and Li, Q. (2015). Watershed superpixel. In: *2015 IEEE International Conference on Image Processing (ICIP)*, 349–353. IEEE.

31 Kim, T.H., Lee, K.M., and Lee, S.U. (2012). Learning full pairwise affinities for spectral segmentation. *IEEE Trans. Pattern Anal. Mach. Intell.* **35** (7): 1690–1703.

32 Vincent, L. (1993). Morphological grayscale reconstruction in image analysis: applications and efficient algorithms. *IEEE Trans. Image Process.* **2** (2): 176–201.

33 Arbelaez, P., Maire, M., Fowlkes, C., and Malik, J. (2010). Contour detection and hierarchical image segmentation. *IEEE Trans. Pattern Anal. Mach. Intell.* **33** (5): 898–916.

34 Shotton, J., Winn, J., Rother, C., and Criminisi, A. (2006). Textonboost: Joint appearance, shape and context modeling for multi-class object recognition and segmentation. In: *European Conference on Computer Vision*, 1–15. Berlin, Heidelberg: Springer.

Part III

Applications

8

Image Segmentation for Traffic Scene Analysis

Traffic scene images usually include drivable areas, lane lines, pedestrians, vehicles, traffic lights, traffic signs, and other elements. The perception of traffic scenes is of great importance in autonomous driving technology. At present, researchers use data collected by sensors to accurately detect and identify traffic environment information, which provides a basis for decision-making and control of the autonomous driving system. Autonomous driving technology [1], taking traffic scene understanding as the core module, is committed to providing intelligent control for vehicles. It has great research value and has attracted extensive attention from academia and industry. However, due to the particularity of some types of roads (such as tunnels and viaducts), there are some special traffic restrictions. For example, drivers cannot change lanes in the tunnel, which may make the tunnel more likely to be congested. In order to further improve the intelligence of autonomous driving, vehicles should be allowed to automatically perceive the scene and make corresponding decisions.

This chapter mainly focuses on traffic scene image semantic segmentation as shown in Figure 8.1. It firstly introduces the development of traffic scene image semantic segmentation based on machine learning, especially deep learning, and then describes in detail the latest traffic scene image segmentation methods using deep learning. Finally, the advantages and disadvantages of the current traffic scene image segmentation methods are summarized, and the opportunities and challenges in this field are discussed, which provides a clear research direction for researchers in related fields.

8.1 Introduction

The four basic modules of the autonomous driving system include scene understanding, positioning navigation, path planning, and control execution, in which scene understanding module is the most important. Like human eyes, traffic scene understanding is a prerequisite for autonomous driving. It mainly involves multiple subtasks, such as target detection, scene classification, depth estimation, tracking, event classification, and behavior analysis [2]. This chapter focuses more on scene classification tasks, that is, traffic scene image semantic segmentation tasks based on deep learning method.

Due to the limitation of computing power, early image segmentation methods usually adopted unsupervised learning methods. These methods mainly extract the low-level features of images through manual design. The quality of the manually extracted features largely determines the quality of the entire system. The classical image segmentation methods such as clustering [3], watershed transform [4], and graph cut [5] usually cannot provide image semantic annotation.

Image Segmentation: Principles, Techniques, and Applications, First Edition. Tao Lei and Asoke K. Nandi.
© 2023 John Wiley & Sons Ltd. Published 2023 by John Wiley & Sons Ltd.

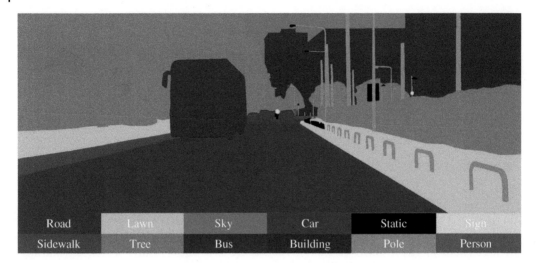

| Road | Lawn | Sky | Car | Static | Sign |
| Sidewalk | Tree | Bus | Building | Pole | Person |

Figure 8.1 Traffic scene segmentation.

In recent years, with the development of neural network techniques and the improvement of hardware computing power, machine learning methods based on neural networks have developed from shallow machine learning to deep learning. Deep learning is a type of representation learning that learns high-level abstract representations of data and automatically extracts features from training data. Its learning ability increases exponentially with the increase of the network depth. A hidden layer in deep learning is equivalent to a linear combination of input features. Convolutional neural network is one of the most popular deep learning methods for traffic scene semantic segmentation. A convolutional layer in convolutional neural networks extract the information from input images, which is called image features. These features are reflected by each pixel from images through combination or independent ways, such as texture features and color features. Finally, a classifier is used for pixel classification.

Deep learning methods based on convolutional neural networks can learn high-level abstract representations of data and automatically extract features. Due to these advantages, deep learning methods based on convolutional neural networks have been widely used in autonomous driving and achieved great success. The following sections of this chapter not only discuss the early semantic segmentation models for traffic scenes but also analyze the newly presented convolutional neural networks based on attention and multi-scale feature fusion for scene segmentation and further summarize the characteristics of different methods.

8.2 Related Work

8.2.1 Convolutional Neural Networks for Image Classification

The traffic scene semantic segmentation method based on deep learning originated from image classification tasks. In other words, we can use image classification networks to solve image semantic segmentation tasks. The earliest deep convolutional neural network can be traced back to LeNet [6] proposed by LeCun et al. This network has a simple architecture including convolutional layers, pooling layers, fully connected layers, and activation functions. The network is used for handwriting digital recognition, and it is a pioneering work in the development history of neural networks.

Table 8.1 Summary of classic convolutional neural network (CNN) models.

Methods	Year	Background	Characteristic
LeNet [6]	1998	Proposed for character recognition	5-layer CNN
AlexNet [7]	2012	The championship in ILSVRC 2012	Use ReLU and dropout functions
VGGNet [9]	2014	The second place in ILSVRC 2014	Network layers is deepened
ResNet [11]	2015	The championship in ILSVRC 2015	Residual connection
MobileNet [12]	2017	Proposed by Google	Depth-wise separable convolutions
ShuffleNet [13]	2018	Proposed by Megvii	Channel shuffle for grouped convolutions
GhostNet [14]	2020	Proposed by Huawei	Generate feature maps from cheap operation

To improve the performance of convolutional neural networks and extend its application fields, Krizhevsky et al. proposed AlexNet [7], which was deeper than LeNet. As the network can learn more complex features from a large image dataset [8], it provides high accuracy for image classification tasks. After that, Simonyan et al. proposed a deeper network named VGG-Net [9], which explored the relationship between the depth of a convolutional neural network and its performance. Although the previous networks are successful for image classification tasks, it is difficult to build deeper CNNs due to the network training difficulty [10]. To address the problem, He et al. proposed ResNet [11], which presented a residual learning framework to solve the network training difficulty due to gradient explosion and gradient vanishing when building a very deep network.

It is well known that a deep neural network usually has a large mountain of parameters. To reduce the number of parameters of networks to deploy these networks on mobile devices, many researchers focus on lightweight CNNs such as MobileNet [12], ShuffleNet [13], and GhostNet [14]. These networks not only achieve model compression but also improve the accuracy of image classification to some extent. Moreover, they are extended to other tasks such as object detection and semantic segmentation, where they also show clear advantages.

Table 8.1 reviews some popular CNNs in terms of years, backgrounds, and contributions. These networks are all suitable as basic feature extraction models for traffic scene classification.

8.2.2 Traffic Scene Semantic Segmentation Using Convolutional Neural Networks

Different from image classification networks mentioned in the section 8.2.1, since the task of image semantic segmentation aims to achieve pixel-level classification, current mainstream traffic scene semantic segmentation methods mainly depend on encoder-decoder networks. As the beginning of the encoder-decoder network structure, Long et al. proposed a fully convolutional neural network (FCN) [15] to realize end-to-end pixel-level semantic segmentation. Its key is to build fully convolutional layers to automatically extract image features used for image segmentation. FCN makes two important contributions: (i) It uses convolutional layers instead of fully connected layers at the end of CNN so that the entire network is used as a feature encoding process, and the convolutional layer outputs a segmentation map with a reduced size after feature encoding. (ii) The segmented image is restored to its original size using transpose convolution, and a skip connection is used to fuse high- and low-level feature information to achieve better segmentation results. However, this method lacks sensitivity to image details, resulting in rough and blurry image segmentation.

To solve the above problems, image segmentation methods based on encoder-decoder networks have begun to focus on the improvement of decoders. For example, SegNet [16] performs upsampling for low-resolution input feature maps in its decoder. Specifically, the decoder uses the pooling indices computed in the corresponding encoder's max-pooling step to perform nonlinear upsampling. An illustration of the SegNet architecture is shown in Figure 8.2. On this basis, many traffic scene-oriented semantic segmentation models adopt a complete encoder-decoder network structure, such as U-Net [17] and RefineNet [18]. However, these networks suffer from complex architectures and a large number of parameters. Therefore, lightweight encoder-decoder segmentation networks have attracted more attention from researchers. Paszke et al. proposed ENet [19] in 2016 to improve computational efficiency. Experiments show that the ENet achieves faster inference speed than SegNet.

Fully convolutional encoder-decoder networks classify images at the pixel level but ignore contextual information. Therefore, similarity between pixels may lead to recognition confusion. Zhao et al. proposed PSPNet [20] to obtain global contextual information using a pyramid pooling strategy. In order to reduce the loss of contextual information between different subregions, as shown in Figure 8.3, they use a pyramid pooling module containing information of different scales and plug it on the feature map of the last layer of the network. At the same time, the proposed global prior representation works well on scene parsing tasks, and PSPNet thus provides a superior framework for pixel-level prediction. However, the methods based on atrous convolution still obtain information from a few surrounding points and cannot form dense context information. Moreover, these pooling-based methods aggregate context information in a nonadaptive manner, which may lead to poor segmentation results because the same context information is used for different pixels.

The attention-based models have been widely used in traffic scene semantic segmentation tasks. Among them, squeeze-and-excitation networks (SENet) [21] enhance the network's feature representation ability by modeling the relationships among feature map channels. The core idea of the SENet is to use two full connection layers to learn the feature map weight so that the important feature maps obtain large weights, and the unimportant feature map obtain small weights. Chen et al. [22] used attention masks to fuse feature maps or predictions from different branches, and the network was able to distinguish where the feature information is more important in the image. In order to obtain dense pixel-level contextual information, PSANet [23] learns to summarize the contextual information of each location by predicting attention maps, Wang et al. [24] proposed a nonlocal module that uses the self-attention mechanism to build the relationship between a pixel and

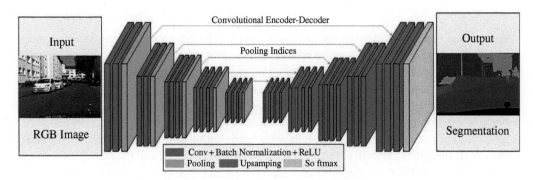

Figure 8.2 The architecture of SegNet.

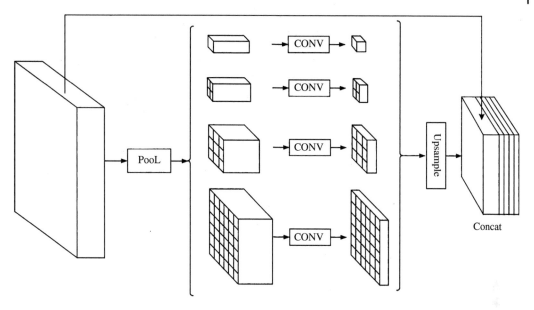

Figure 8.3 The architecture of PSP module.

all pixels in the image and can produce a more powerful pixel-level representation ability. However, the current attention-based semantic segmentation model still has limitations. On the one hand, the single modeling on feature map's spatial or channel relationship cannot achieve the unity of their forms. So the simultaneous acquisition of spatial and channel relationships can only be approximately obtained through a serial sequence of the spatial attention and channel attention. On the other hand, using the full connected layer and self-attention to obtain the feature map's channel or spatial relationship usually requires much computational and storage costs, which is very unfavorable to the industrial deployment of the model.

8.3 Multi-Scale Feature Fusion Network for Scene Segmentation

For the basic convolutional neural network structure, too many pooling layers may lead to the loss of image detail information due to the reduction of image resolution. To address the above problem, researchers have done a lot of work in solving the above problem by improving the standard convolutional operation [25–29]. Chen et al. proposed the use of dilated convolution to construct atrous spatial pyramid pooling (ASPP) to extract multi-scale contextual information [28]. The ASPP module can capture multi-scale contextual information through feature fusion using dilated convolutions with different dilation rates. However, the ASPP module is concerned with more parameters, which is not conducive to model deployment. So Chen et al. applied depth-wise separable convolution [30] to the ASPP module to reduce the number of parameters and further designed the encoder-decoder network named DeepLabV3+ [29]. The encoder-decoder structure can better recover the edge of the target by gradually reconstructing spatial information, so as to obtain better semantic segmentation accuracy. The following contents will summarize the advantages and disadvantages of DeepLabV3+ from three aspects: multi-scale feature fusion strategy, encoder-decoder structure design, and experiment results.

8.3.1 Multi-Scale Feature Fusion Using Dilated Convolution

Earlier we mentioned that in deep convolutional neural networks, continuous downsampling and repeated pooling will significantly reduce the resolution of feature maps, and the global contextual information will be seriously lost. Therefore, the researchers used atrous convolution, as shown in Figure 8.4, to widen the receptive field [28] on the premise of avoiding a large loss of image features so that the output of each convolution layer contains much more information. The difference between atrous convolution and vanilla convolution is that null values are inserted between the parameters of a convolution kernel to expand the size of the convolution kernel. The number of null values inserted in each row and column is determined by the hyperparameter dilation rate.

Along with the increases of dilation rate of dilated convolution, the effective weights of the filter will gradually decrease. When the dilation rate approaches the size of feature maps, the filter has only one effective weight, that is, the center point, which degrades to a 1×1 convolution kernel and can no longer obtain global contextual information. To address this problem, researchers have proposed ASPP [28], which combines a 1×1 convolution operation and a global average pooling operation. As shown in Figure 8.5, the ASPP module is divided into two parts as follows.

1) One 1×1 convolution and three 3×3 dilated convolutions with different dilation rates of 6, 12, and 18, respectively.
2) Global average pooling. Specifically, the input feature map is averaged over all pixels, and the result is passed through to 1×1 convolution, and then upsampled by using bilinear interpolation to restore it to the desired spatial size for concatenate with other branch outputs. Image-level features can be effectively extracted through global average pooling, which can overcome the problem of feature loss caused by a large value of dilation rate..

8.3.2 Encoder-Decoder Architecture

Dilated convolution can obtain receptive fields of different scales without adding additional computation and parameters. On this basis, the researchers used the depth-wise separable convolution proposed in the Xception model. The vanilla convolution operation can be regarded as the joint

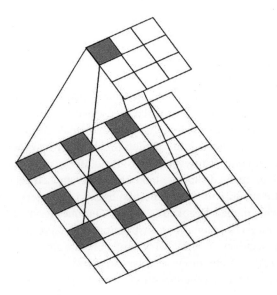

Figure 8.4 The dilated convolution.

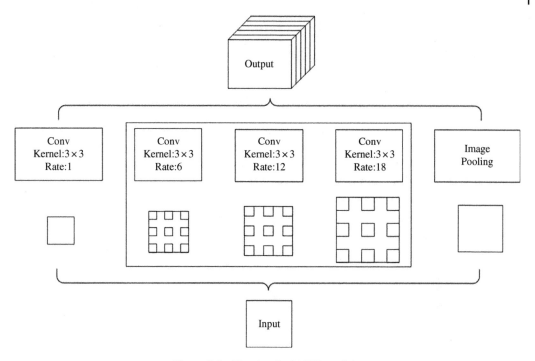

Figure 8.5 The detail of ASPP module.

mapping of the two dimensions of space and channel, in contrast, the view of the Xception[30] is that, the spatial correlation and channel correlation in convolution can be decoupled. Separable spatial and channel convolution may obtain the same or even better effect and can greatly improve the convolution efficiency.

Specifically, the depth-wise separable convolution can be divided into two parts, namely, the depth-wise convolution and point-wise convolution. The depth-wise convolution performs channel-by-channel correlation mapping with input feature maps, and the obtained results are aggregated mapping with the point-wise convolution between channels to obtain output feature maps [29]. Because of its low cost of computing and memory, researchers apply depth-wise separable convolution to the ASPP module mentioned above to further compress the network model, reduce the number of parameters, and ensure the effective fusion of multi-scale features, so as to achieve the effect of coexistence of efficiency and accuracy. On this basis, the researchers further designed an encoder-decoder network. The encoder-decoder model has played a key role in semantic segmentation. It can accept images of any size and output segmentation results of the same size as the output. In DeepLabV3+, the encoder module encodes the multi-scale context information by applying dilated convolution at multiple different scales, and the decoder module obtains the segmentation result of target boundary refinement. This encoder-decoder design ensures sufficient feature extraction and complete restoration of edge details and reduces the computational complexity. The specific encoder-decoder structure of DeepLabv3+ is shown in Figure 8.6.

8.3.3 Experiments

In order to evaluate the convolutional neural network of multi-scale feature fusion proposed in this chapter, a comprehensive experiment is conducted on the Cityscapes data set [31]. The experiment shows that the proposed DeepLabV3+ achieved advanced performance on the Cityscapes data set.

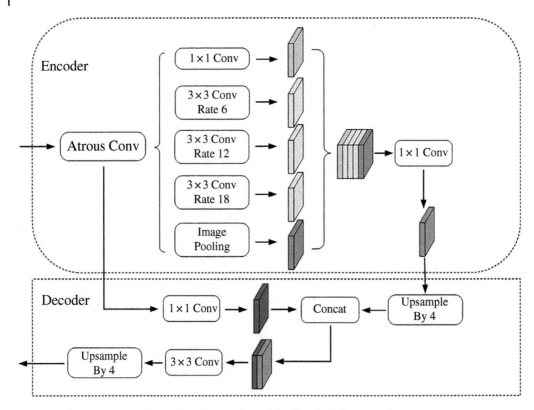

Figure 8.6 An overview of the DeepLabV3+ network.

Table 8.2 Results on Cityscapes testing set.

Methods	Backbone	MIoU (%)
DeepLabV3+ [29]	ResNet101	74.2
DANet [32]	ResNet101	74.5
SSN (Ours)	**ResNet101**	**75.1**

The specific experimental results are shown in Table 8.2. It can be seen that when using ResNet101 as the backbone network, the ASPP module that is used to capture the multi-scale features and fuse them effectively can better understand the traffic scene.

8.4 Self-Attention Network for Scene Segmentation

Although the method based on multi-scale feature fusion helps to capture objects of different scales, it cannot exploit the relationship between objects or contents in the global view, which is also the key to traffic scene semantic segmentation. To address this problem, the current work focuses on capturing feature correlations in different dimensions by improving the self-attention mechanism originated from natural language processing [32]. The self-attention mechanism can capture the spatial dependencies of any two positions in feature maps and thus obtains long-range context dependence.

8.4.1 Non-local attention Module

In the previous part, the basic unit of the convolutional neural network for semantic segmentation mentioned is the convolution layer, and the convolution operation is used to capture local spatial information. The method based on multi-scale feature fusion only obtains a wider range of receptive field yet cannot obtain long-range dependencies between effective features. Inspired by the non-local information in computer vision and self-attention information in natural language processing, researchers have proposed the non-local attention [24] to capture long-range dependencies to model the relationship between two pixels with a certain distance on an image. When calculating the response of a location, the non-local attention considers the weighting of all location features—spatial, temporal, and spatiotemporal. This structure is plug-and-play and thus can be used in many network structures. The non-local operation can be expressed as:

$$y_i = \frac{1}{C(x)} \sum_{\forall j} s_{cp}(x_i, x_j) g_f^j(x_j), \tag{8.1}$$

where x is the input feature map, i and j represent the output position index, and the s_{cp} function calculates the similarity of the corresponding positions of i and j. The g_f^j function computes a representation of the feature map at position j. y_i is the output feature map normalized by the response factor $C(x)$. Non-local attention can easily be used in convolutional layers, which can be plugged into shallow or deep layers of the network. This allows us to construct richer structures by combining non-local and local image information.

The architecture of a non-local module is shown in Figure 8.7, which can be easily plugged into the existing network structure. The module can be defined as:

$$z_i = W_z y_i + x_i, \tag{8.2}$$

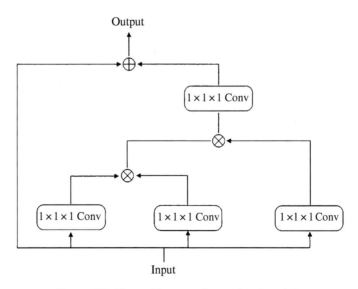

Figure 8.7 The architecture of a non-local module.

where z_i represents the output feature map, $W_z y_i$ is the $1 \times 1 \times 1$ convolution in Figure 8.7, and x_i represents the original input feature map.

8.4.2 Dual Attention Module

Non-local information can effectively solve the problem of missing global information in convolutional neural networks, but it shows limited capability when capturing the long-range dependencies of features in channel dimensions. In order to extend the non-local information to the two dimensions of spatial and channel, DANet proposes a dual attention model, which captures the feature dependencies in the spatial dimension and channel dimension respectively based on the self-attention mechanism [32] and conducts comparative experiments with other methods [16, 20, 27, 33–37] on the Cityscapes data set. Specifically, DANet integrates two parallel attention modules to FCN; one is a positional attention module, and the other is a channel attention module. The semantic dependencies in spatial and channel dimensions are modeled separately. For the position attention module, a self-attention mechanism is introduced to capture the spatial dependency between any two positions in feature maps. The features of all positions are aggregated and updated by weighted summation. The weight mentioned above is determined by the similarity of the features corresponding to the two positions. For the channel attention module, a similar self-attention mechanism is used to capture the channel dependencies between any two different channel feature maps, and the weighted sum of all channel maps is used to update each channel map. Finally, the outputs of these two attention modules are fused to further enhance the feature representation.

For the traffic scene segmentation task, the modeling of spatial long-range information is very important. However, the continuous local convolution employed by the encoder-decoder structure apparently ignore the aggregation of spatial global context information, which may lead to the poor classification results. For this issue, the position attention module is presented in [32]. The module can capture rich image contextual information by encoding the spatial long-range information into local features. Thus, the introduction of the position attention module is helpful to enhance feature representation ability.

As shown in Figure 8.8, giving an input feature map $A \in R^{C \times H \times W}$, two feature maps B and C are generated by convolutional layers, where $\{B, C\} \in R^{C \times H \times W}$, and the spatial attention map $S \in R^{N \times N}$ is then obtained. After that, the attention map is assigned to the transformed form of the original feature map to obtain the output feature map $E \in R^{C \times H \times W}$, and the whole process can be expressed as follows:

$$s_{ji} = \frac{\exp\left(B_i \cdot C_j\right)}{\sum_{i=1}^{N} \exp\left(B_i \cdot C_j\right)}, \tag{8.3}$$

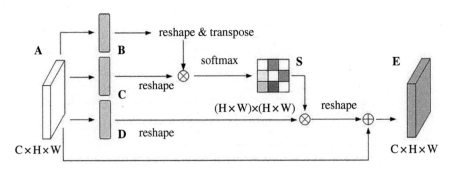

Figure 8.8 The detail of position attention module.

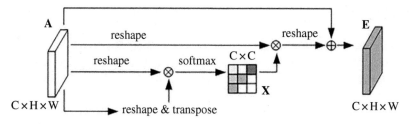

Figure 8.9   The channel attention module.

$$E_i = \alpha \sum_{i=1}^{N} \left(S_{ji} D_i \right) + A_j, \tag{8.4}$$

where the scale coefficient α is initialized to 0, and more weights are gradually assigned during the learning process. From Eq. (8.4), it can be inferred that the output E at each location is the weighted sum of all location features and the original features. Thus, it aggregates the global context focused on spatial attention. Similar semantic features achieve mutual enhancement, thereby improving intra-class compactness and semantic consistency.

Meanwhile, each feature map in high-level features can be viewed as a class-specific response, and different semantic responses are interrelated. By exploiting the interdependencies between channel mappings, interdependent feature mappings can be emphasized, and semantics-specific feature representations can be also improved. Therefore, by analogy with positional attention, the channel attention module is constructed to model explicitly the interdependence between channels.

As shown in Figure 8.9, unlike positional attention, there is no need to use convolution for feature embedding since the original channel mapping needs to be maintained, and the channel attention map $X \in R^{C \times C}$ can be computed directly from the input feature map $A \in R^{C \times H \times W}$ as follows:

$$x_{ji} = \frac{\exp\left(A_i \cdot A_j\right)}{\sum_{i=1}^{C} \exp\left(A_i \cdot A_j\right)}, \tag{8.5}$$

$$E_j = \beta \sum_{i=1}^{C} \left(x_{ji} A_i \right) + A_j, \tag{8.6}$$

where β is a learnable parameter, the initialization value is 0, and it is gradually assigned to larger weights by iterative learning. The resulting feature E for each channel is the weighted sum of all channel features and the original features. The modeling of long-range semantic dependencies between feature maps is realized, which helps to improve the discriminative capability of features, thus enhancing the feature representation of the network.

In order to utilize fully the long-range contextual information, the dual-attention module aggregates the features of these two attention modules. This module is simple yet effective and can be directly plugged into any backbone to effectively enhance the feature representation and requires less memory and computational costs.

8.4.3   Criss-Cross Attention

The proposed dual-attention module can be easily plugged into any semantic segmentation network benchmark model, and the feature representation can be effectively enhanced by aggregating contextual information at different locations by learning the generated attention maps [38]. However, these attention-based methods need to generate huge attention maps to measure the relationship of each pixel pair, which leads to high time and space complexity since the input feature maps in semantic segmentation tasks always has high resolution. Therefore, the method based on self-attention suffers from high computational complexity and occupies a large amount of GPU memory.

To address the above problems, the CCNet [39] uses dense attention instead of continuous sparse attention in the non-local neural network [24]. While ensuring generality, the researchers use two consecutive criss-cross attention modules, where each module has only sparse connections for each location in feature maps. The criss-cross attention module aggregates image contextual information in both horizontal and vertical directions. The decomposition strategy can greatly reduce the computational cost of standard attention module.

As shown in Figure 8.10, given an input feature map $H \in R^{C \times W \times H}$, two feature maps Q and K are generated using two 1×1 convolutions, where $\{Q, K\} \in R^{C' \times W \times H}$. A vector $Q_u \in R^{C'}$ can be obtained for each position u in the spatial dimension of Q. Furthermore, we can extract a set of feature vectors set $\Omega_u \in R^{(H+W-1) \times C'}$ from each position species of K corresponding to u. $\Omega_{i,u} \in R^{C'}$ denotes the i-th element of the set Ω_u. Thus, the feature correlation degree d can be defined as:

$$d_{i,u} = Q_u \Omega_{i,u}^T. \tag{8.7}$$

Meanwhile the 1×1 convolution is performing on the input feature map $H \in R^{C \times W \times H}$ to generate $V \in R^{C \times W \times H}$, the set $\Phi_u \in R^{(H+W-1) \times C}$ corresponding to the same row or column obtained with position u can be obtained in V, then the feature vector H'_u in the output feature map $H' \in R^{C \times W \times H}$ at position u can be defined as:

$$H'_u = \sum_{i=0}^{H+W-1} A_{i,u} \Phi_{i,u} + H_u. \tag{8.8}$$

where A is an attention map and $A_{i,u}$ is a scalar value at channel i and position u in A. The feature representation is enhanced by fusing contextual information into the local features H by the criss-cross module. Thus, it has extensive contextual information and selective feature enhancement based on spatial attention and requires less computation and memory overhead compared to other self-attention modules.

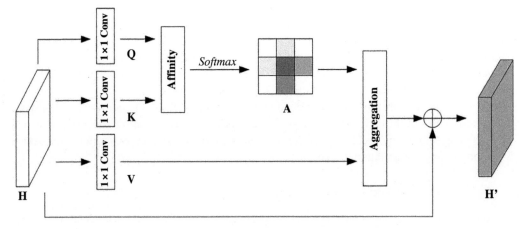

Figure 8.10 The detail of criss-cross attention module.

8.4.4 Multi-scale Non-local Module

For traffic scene semantic segmentation, as the spatial long-range relationship is a key factor that decides the final segmentation accuracy, the non-local attention is popular for improving network structure. However, it is unreasonable to apply directly the non-local module to a segmentation network due to the loss of multi-scale information that is important for segmentation tasks. To this end, we combine the advantages of non-local module and multi-scale feature fusion, and thus propose a multi-scale non-local module using multi-scale parallel sampling module (MPS).

The architecture of the multi-scale non-local module is shown in Figure 8.11. In this module, we can see that the structure of the multi-scale non-local module is similar to the vanilla non-local module. The difference between them is the introduction of MPS. First, the input feature maps X is divided into three branches, Q, K, and V, where $\{Q, K, V\} \in \mathbb{R}^{C \times H \times W}$. Then, the feature maps X are proceeded by the MPS module. Finally, the output matrix $C \times S$ and its transpose $S \times C$ are multiplied, and the substantial steps are same as in a non-local operation.

In the MPS module, we choose the average-pooling instead of the max-pooling since the former can provide better global relationship of images. Firstly, the MSP module employs four scales ($n = 2, 4, 8$, and 16) for parallel sampling on branches Q and K. Four different-scale feature maps are then obtained. To achieve feature fusion, these feature maps are reshaped to $\mathbb{R}^{C \times Z}$, where $Z = (H/n) \times (W/n)$ is the number of pixels. The result outputted by MSP is the concatenation of these four reshaped maps.

According to the above analysis, the proposed multi-scale non-local module is clearly different from the previous attention mechanism. On the one hand, the proposed multi-scale non-local module can incorporate multi-scale spatial information into channel attention. On the other hand, it can effectively improve the performance of the network while reducing the number of parameters.

8.4.5 Experiments

To evaluate the self-attention-based convolutional neural networks proposed in this chapter, this section conducts comprehensive experiments on the Cityscapes data set [31], and the experiments show that the multi-scale feature fusion network and the self-attention network achieve advanced performance on the Cityscapes data set. This section will first introduce the data set and experimental details and then show the experimental results on the Cityscapes data set.

We implement the proposed method based on PyTorch with a learning rate decay policy, where the initial learning rate is set to 0.01. We used SGD optimization algorithm, where the momentum and weight decay coefficients are set to 0.9 and 0.0001, the batch size is set to 8, and the iteration

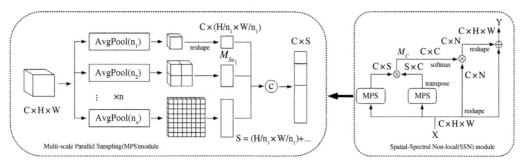

Figure 8.11 The detail of multi-scale non-local module.

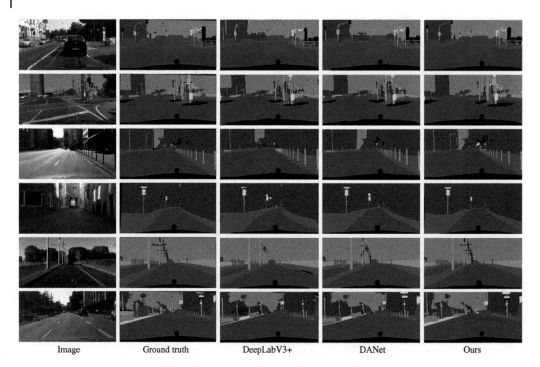

| Image | Ground truth | DeepLabV3+ | DANet | Ours |

Figure 8.12 Traffic scene segmentation results using different approaches.

number is set to 30 000. For data enhancement, random crop is used during the training, and the size is 768×768.

Meanwhile, the comparison with existing deep learning methods based on convolutional neural networks [29, 32] was carried out, and the detailed results of the MIoU are shown in Table 8.2. It can be seen that our method shows obvious advantages compared to others. The experimental results are shown visually in Figure 8.12.

8.5 Discussion and Summary

This chapter reviews the methods of traffic scene image segmentation based on deep learning. We summarize and analyzes from the typical works and the latest methods, focuses on the latest semantic segmentation methods based on multi-scale feature fusion network and self-attention network. Then we conducted experiments and compared their performance. The advantages and disadvantages of the two methods are compared from two aspects: IoU metric and segmentation result visualization. Although significant progress has been made in the research of traffic scene semantic segmentation, there is still a long way to go to make the computer vision methods recognize the traffic environment as accurately as human beings. Based on the review of this chapter, there are still the following challenges in the field of traffic scene image segmentation.

8.5.1 Network Architecture Search

Due to the continuous development of convolutional neural networks, the design of network architectures for traffic scene image segmentation tasks has become increasingly mature, and it has become difficult to efficiently obtain the required network structure for the task by relying on expert a priori knowledge alone. In order to further investigate and develop more efficient network model architectures, the neural architecture search (NAS) technique [40, 41] has been proposed and gradually developed in recent years. It is used to obtain the network structure with optimal performance in a given search space by an optimization algorithm [42]. For image semantic segmentation tasks, more complex and diverse network architectures bring more challenges to NAS techniques. Especially for semantic segmentation networks of traffic scenes images, the complex scene objectives lead to a network structure that requires more functional modules such as ASPP and non-local attention module. At the same time, the lack of interpretability of network architectures based on NAS technology makes it difficult for subsequent researchers to explore in this area. Therefore, it will be an opportunity and challenge for future research to better understand NAS technology and extend it to the task of semantic segmentation of traffic scenes.

8.5.2 Compact Networks

The successful application of convolutional neural networks for image semantic segmentation of traffic scenes tasks, however, depends on the large computational and memory consumption required by their complex network structure, which is highly detrimental to industrial application deployment. Therefore, more and more researchers tend to compress and accelerate image segmentation networks by designing compact networks [43, 44]. The compact network aims to simplify the basic operation of convolution to reduce overhead while maintaining accuracy as much as possible. In the spatial dimension, it is usually intuitive to decompose a convolution kernel with a larger receptive field into a superposition of smaller convolution kernels. On this basis, more simple and effective improvements are derived. Asymmetric convolution [45] and atrous convolution [25] use limited parameters and computation to make the vanilla convolution have receptive fields of any size. Deformable convolution [46] can be regarded as a general form of atrous convolution, which makes the convolution operation more flexible with fixed parameters and computation amount by learning the convolution bias. In the channel dimension, more work focuses on the study of channel aggregation topology. Grouped convolution [47] effectively reduces the parameter redundancy caused by too many channels in the deep feature space by grouping the input channels and performing conventional convolutions respectively. Depth-wise separable convolution [30] can be regarded as the extreme form of grouping convolution, in which the convolution is carried out channel by channel, and then the small-size convolution kernel is used for channel dimension information fusion for the generated feature map, which greatly reduces the parameter redundancy and ensures the aggregation of information between channels. Therefore, how to improve the speed of the semantic segmentation network model on the premise of ensuring the segmentation accuracy will also become a research hot topic in the field of semantic segmentation.

8.5.3 Vision Transformer

In order to obtain better global information of images and to improve the feature representation of neural network models, many researchers have introduced self-attention from natural language processing to computer vision and achieved equal or better performance than convolutional neural networks. The neural network that uses this basic unit is called transformer [48], which was first applied in the field of natural language processing and achieved SOTA performance on two machine translation tasks in WMT 2014 for English to German and WMT 2014 for English to French. However, transformer-based models in computer vision are limited to the image classification domain only [49, 50]. There is still much room for improvement of the current traffic scene interpretation tasks. At the same time, the efficiency of transformers in computer vision needs to be improved. The development of efficient vision transformers, especially those with high performance and low resource cost, determines whether their models can be applied to practical engineering applications [51]. Therefore, determining how to achieve a better balance between accuracy and efficiency is a meaningful topic for future vision transformer research.

References

1 Bing, Z.H.U., Pei-xing, Z.H.A.N.G., Jian, Z.H.A.O. et al. (2019). Review of scenario-based virtual validation methods for automated vehicles. *China J. Highway Transport* **32** (6): 1.

2 Janai, J., Güney, F., Behl, A., and Geiger, A. (2020). Computer vision for autonomous vehicles: problems, datasets and state of the art. *Foundations Trends Computer Graphics Vision* **12** (1–3): 1–308.

3 Chuang, K.S., Tzeng, H.L., Chen, S. et al. (2006). Fuzzy c-means clustering with spatial information for image segmentation. *Comput. Med. Imaging Graph.* **30** (1): 9–15.

4 Ng, H.P., Ong, S.H., Foong, K.W.C. et al. (2006). Medical image segmentation using k-means clustering and improved watershed algorithm. In: *2006 IEEE southwest symposium on image analysis and interpretation*, 61–65. IEEE.

5 Van Den Heuvel, M., Mandl, R., and Hulshoff Pol, H. (2008). Normalized cut group clustering of resting-state FMRI data. *PLoS One* **3** (4): e2001.

6 LeCun, Y., Bottou, L., Bengio, Y., and Haffner, P. (1998). Gradient-based learning applied to document recognition. *Proc. IEEE* **86** (11): 2278–2324.

7 Krizhevsky, A., Sutskever, I., and Hinton, G.E. (2012). Imagenet classification with deep convolutional neural networks. *Adv. Neural Inform. Process. Syst.* **25**: 1097–1105.

8 Deng, J., Dong, W., Socher, R. et al. (2009). Imagenet: A large-scale hierarchical image database. In: *2009 IEEE Conference on Computer Vision and Pattern Recognition*, 248–255. IEEE.

9 Simonyan, K. and Zisserman, A. (2014). Very deep convolutional networks for large-scale image recognition. arXiv preprint *arXiv:1409.1556*.

10 Szegedy, C., Liu, W., Jia, Y. et al. (2015). Going deeper with convolutions. In: *Proceedings of the IEEE Conference on Computer Vision and Pattern Recognition*, 1–9.

11 He, K., Zhang, X., Ren, S., and Sun, J. (2016). Deep residual learning for image recognition. In: *Proceedings of the IEEE conference on computer vision and pattern recognition*, 770–778.

12 Howard, A. G., Zhu, M., Chen, B. et al. (2017). Mobilenets: Efficient convolutional neural networks for mobile vision applications. arXiv preprint *arXiv:1704.04861*.

13 Zhang, X., Zhou, X., Lin, M., and Sun, J. (2018). Shufflenet: An extremely efficient convolutional neural network for mobile devices. In: *Proceedings of the IEEE Conference on Computer Vision and Pattern Recognition*, 6848–6856.

14 Han, K., Wang, Y., Tian, Q. et al. (2020). Ghostnet: More features from cheap operations. In: *Proceedings of the IEEE/CVF Conference on Computer Vision and Pattern Recognition*, 1580–1589.

15 Long, J., Shelhamer, E., and Darrell, T. (2015). Fully convolutional networks for semantic segmentation. In: *Proceedings of the IEEE Conference on Computer Vision and Pattern Recognition*, 3431–3440.

16 Badrinarayanan, V., Kendall, A., and Cipolla, R. (2017). Segnet: a deep convolutional encoder-decoder architecture for image segmentation. *IEEE Trans. Pattern Anal. Mach. Intell.* **39** (12): 2481–2495.

17 Ronneberger, O., Fischer, P., and Brox, T. (2015). U-net: Convolutional networks for biomedical image segmentation. In: *International Conference on Medical image computing and computer-assisted intervention*, 234–241. Cham: Springer.

18 Lin, G., Milan, A., Shen, C., and Reid, I. (2017). Refinenet: Multi-path refinement networks for high-resolution semantic segmentation. In: *Proceedings of the IEEE conference on computer vision and pattern recognition*, 1925–1934.

19 Paszke, A., Chaurasia, A., Kim, S., and Culurciello, E. (2016). Enet: A deep neural network architecture for real-time semantic segmentation. arXiv preprint *arXiv:1606.02147*.

20 Zhao, H., Shi, J., Qi, X. et al. (2017). Pyramid scene parsing network. In: *Proceedings of the IEEE Conference on Computer Vision and Pattern Recognition*, 2881–2890.

21 Hu, J., Shen, L., and Sun, G. (2018). Squeeze-and-excitation networks. In: *Proceedings of the IEEE Conference on Computer Vision and Pattern Recognition*, 7132–7141.

22 Chen, L.C., Yang, Y., Wang, J. et al. (2016). Attention to scale: Scale-aware semantic image segmentation. In: *Proceedings of the IEEE Conference on Computer Vision and Pattern Recognition*, 3640–3649.

23 Zhao, H., Zhang, Y., Liu, S. et al. (2018). Psanet: Point-wise spatial attention network for scene parsing. In: *Proceedings of the European Conference on Computer Vision (ECCV)*, 267–283.

24 Wang, X., Girshick, R., Gupta, A., and He, K. (2018). Non-local neural networks. In: *Proceedings of the IEEE Conference on Computer Vision and Pattern Recognition*, 7794–7803.

25 Yu, F., Koltun, V., and Funkhouser, T. (2017). Dilated residual networks. In: *Proceedings of the IEEE Conference on Computer Vision and Pattern Recognition*, 472–480.

26 Chen, L. C., Papandreou, G., Kokkinos, I. et al. (2014). Semantic image segmentation with deep convolutional nets and fully connected crfs. arXiv preprint *arXiv:1412.7062*.

27 Chen, L.C., Papandreou, G., Kokkinos, I. et al. (2017). Deeplab: semantic image segmentation with deep convolutional nets, atrous convolution, and fully connected crfs. *IEEE Trans. Pattern Anal. Mach. Intell.* **40** (4): 834–848.

28 Chen, L. C., Papandreou, G., Schroff, F., and Adam, H. (2017). Rethinking atrous convolution for semantic image segmentation. arXiv preprint *arXiv:1706.05587*.

29 Chen, L.C., Zhu, Y., Papandreou, G. et al. (2018). Encoder-decoder with atrous separable convolution for semantic image segmentation. In: *Proceedings of the European Conference on Computer Vision (ECCV)*, 801–818.

30 Chollet, F. (2017). Xception: Deep learning with depthwise separable convolutions. In: *Proceedings of the IEEE Conference on Computer Vision and Pattern Recognition*, 1251–1258.

31 Cordts, M., Omran, M., Ramos, S. et al. (2016). The cityscapes dataset for semantic urban scene understanding. In: *Proceedings of the IEEE Conference on Computer Vision and Pattern Recognition*, 3213–3223.

32 Fu, J., Liu, J., Tian, H. et al. (2019). Dual attention network for scene segmentation. In: *Proceedings of the IEEE/CVF Conference on Computer Vision and Pattern Recognition*, 3146–3154.

33 Peng, C., Zhang, X., Yu, G. et al. (2017). Large kernel matters--improve semantic segmentation by global convolutional network. In: *Proceedings of the IEEE conference on computer vision and pattern recognition*, 4353–4361.

34 Wang, P., Chen, P., Yuan, Y. et al. (2018). Understanding convolution for semantic segmentation. In: *2018 IEEE Winter Conference on Applications of Computer Vision (WACV)*, 1451–1460. IEEE.

35 Wu, Z., Shen, C., and Van Den Hengel, A. (2019). Wider or deeper: revisiting the resnet model for visual recognition. *Pattern Recogn.* **90**: 119–133.

36 Yu, C., Wang, J., Peng, C. et al. (2018). Bisenet: Bilateral segmentation network for real-time semantic segmentation. In: *Proceedings of the European conference on computer vision (ECCV)*, 325–341.

37 Yang, M., Yu, K., Zhang, C. et al. (2018). Denseaspp for semantic segmentation in street scenes. In: *Proceedings of the IEEE Conference on Computer Vision and Pattern Recognition*, 3684–3692.

38 Zhang, H., Dana, K., Shi, J. et al. (2018). Context encoding for semantic segmentation. In: *Proceedings of the IEEE conference on Computer Vision and Pattern Recognition*, 7151–7160.

39 Huang, Z., Wang, X., Huang, L. et al. (2019). Ccnet: Criss-cross attention for semantic segmentation. In: *Proceedings of the IEEE/CVF International Conference on Computer Vision*, 603–612.

40 Zoph, B., and Le, Q.V. (2016). Neural architecture search with reinforcement learning. arXiv preprint *arXiv:1611.01578*.

41 He, X., Zhao, K., and Chu, X. (2021). AutoML: a survey of the state-of-the-art. *Knowledge-Based Syst.* **212**: 106622.

42 Ha, H., Rana, S., Gupta, S. et al. (2019). Bayesian optimization with unknown search space. *Adv. Neural Inform. Process. Syst.* **32**: 11795–11804.

43 Wang, X. and Stella, X.Y. (2021). Tied block convolution: leaner and better CNNs with shared thinner filters. In: *Proceedings of the AAAI Conference on Artificial Intelligence*, vol. **35**, No. 11, 10227–10235.

44 Bhalgat, Y., Zhang, Y., Lin, J.M., and Porikli, F. (2020). Structured convolutions for efficient neural network design. *Adv. Neural Inform. Process. Syst.* **33**: 5553–5564.

45 Ding, X., Guo, Y., Ding, G., and Han, J. (2019). Acnet: Strengthening the kernel skeletons for powerful cnn via asymmetric convolution blocks. In: *Proceedings of the IEEE/CVF International Conference on Computer Vision*, 1911–1920.

46 Dai, J., Qi, H., Xiong, Y. et al. (2017). Deformable convolutional networks. In: *Proceedings of the IEEE International Conference on Computer Vision*, 764–773.

47 Xie, S., Girshick, R., Dollár, P. et al. (2017). Aggregated residual transformations for deep neural networks. In: *Proceedings of the IEEE conference on computer vision and pattern recognition*, 1492–1500.

48 Vaswani, A., Shazeer, N., Parmar, N. et al. (2017). Attention is all you need. In: *Advances in Neural Information Processing Systems*, 5998–6008.

49 Dosovitskiy, A., Beyer, L., Kolesnikov, A. et al. (2020). An image is worth 16x16 words: Transformers for image recognition at scale. arXiv preprint *arXiv:2010.11929*.

50 Liu, Z., Lin, Y., Cao, Y. et al. (2021). Swin transformer: Hierarchical vision transformer using shifted windows. arXiv preprint *arXiv:2103.14030*.

51 Liao, H., Tu, J., Xia, J., and Zhou, X. (2019). Davinci: A scalable architecture for neural network computing. In: *2019 IEEE Hot Chips 31 Symposium (HCS)*, 1–44. IEEE Computer Society.

9

Image Segmentation for Medical Analysis

This chapter mainly studies image segmentation in medical analysis. Firstly, this chapter begins with an introduction and then presents recent work related to medical image segmentation. Then two representative medical image segmentation methods are introduced. Finally, the methods for medical image segmentation are discussed and summarized.

9.1 Introduction

Medical image segmentation aims to make anatomical or pathological structure changes clearer in images; it often plays a key role in computer-aided diagnosis and smart medicine due to the great improvement in diagnostic efficiency and accuracy. Popular medical image segmentation tasks include liver and liver-tumor segmentation [1, 2], brain and brain-tumor segmentation [3, 4], optic disc segmentation [5, 6], cell segmentation [7, 8], and lung segmentation and pulmonary nodules [9, 10]. With the development and popularization of medical imaging equipment, x-ray, computed tomography (CT), magnetic resonance imaging (MRI), and ultrasound have become four important image-assisted means to help clinicians diagnose diseases, to evaluate prognosis, and to plan operations in hospitals. Despite the fact that these ways of imaging have advantages as well as disadvantages, they are useful for the medical examination of different parts of human body in practical applications.

Since medical image segmentation algorithms can help doctors improve the level of disease diagnosis and formulate better treatment plans, it is of great significance to study the medical image segmentation methods. However, there are some problems in medical images, such as blurred edges, large noise, and variable target shapes and sizes. In order to solve these problems, a large number of researchers have conducted in-depth research on medical image segmentation and proposed many methods. These methods are mainly divided into traditional methods and deep learning methods. Traditional segmentation methods are usually semiautomatic and mainly rely on model-driven image segmentation algorithms, such as region growth [11], active contour model [12], graph cutting [13], and shape statistical model [14]. Traditional image segmentation methods rely on the bottom features extracted manually. Such semiautomatic methods can achieve accurate medical image segmentation by simply marking the foreground and background, and don't need iterative operations. Therefore, image segmentation results are easily influenced by various factors. With the development of deep learning techniques, end-to-end medical image segmentation has attracted the attention of researchers. As deep learning can learn high-layer semantic features

Image Segmentation: Principles, Techniques, and Applications, First Edition. Tao Lei and Asoke K. Nandi.
© 2023 John Wiley & Sons Ltd. Published 2023 by John Wiley & Sons Ltd.

of liver from abdominal images, convolutional neural networks (CNNs) based on deep learning can provide excellent segmentation results. Currently, there are two types of popular deep CNNs used for liver segmentation, the first is 2D networks such as U-Net [7], CE-Net [15], U-Net++ [16], and mU-Net [17], and the second is 3D networks such as 3D U-Net [18] and V-Net [19].

9.2 Related Work

In order to help clinicians make accurate diagnoses, it is necessary to segment some key objects in medical images and extract features from the segmented regions. Some mainstream methods have important research values, and we divide them into two categories: traditional and deep learning.

9.2.1 Traditional Approaches for Medical Image Segmentation

Early approaches to medical image segmentation often depended on edge detection, template matching techniques, statistical shape models, active contours, and machine learning, among other techniques. Zhao et al. [20] proposed a new mathematical morphology edge detection algorithm for lung CT images. Lalonde et al. [21] applied Hausdorff-based template matching to disc inspection, and Chen et al. [22] also employed template matching to perform ventricular segmentation in brain CT images. Tsai et al. [23] proposed a shape-based approach using horizontal sets for 2D segmentation of cardiac MRI images and 3D segmentation of prostate MRI images. Li et al. [24] used the activity profile model to segment liver tumors from abdominal CT images, while Li et al. [25] proposed a framework for medical body data segmentation by combining level sets and support vector machines (SVMs). Held et al. [26] applied Markov random fields (MRF) to brain MRI image segmentation.

9.2.2 Deep Learning for Medical Image Segmentation

Although a large number of approaches have been reported, and they are successful in certain circumstances, image segmentation is still one of the most challenging topics in the field of computer vision due to the difficulty of feature representation. In particular, it is more difficult to extract discriminating features from medical images than normal RGB images since the former often suffers from problems of blur, noise, low contrast, and other issues. Due to the rapid development of deep learning techniques [27], medical image segmentation will no longer require handcrafted features, and CNNs successfully achieve hierarchical feature representation of images, and thus become the hottest research topic in image processing and computer vision. As CNNs used for feature learning can be insensitive to image noise, blur, contrast, and so forth, they provide excellent segmentation results for medical images. Among them, the encoder-decoder structure based on deep learning is the most popular method of image segmentation because these methods have better segmentation results, such as the fully convolutional network (FCN) [28], U-Net [7], and DeepLab V3 [29]. In these structures, an encoder is often used to extract image features while a decoder is often used to restore extracted features to the original image size and output the final segmentation results.

These networks usually adopt multilevel encoder-decoder structures, and the encoder and decoder are often composed of a large number of standard convolutional or deconvolutional layers. In addition, there is a residual or long-range connection between encoders and decoders. This kind of design can automatically remove insignificant features and maintain interesting features through the contraction and expansion paths; it can also achieve the fusion of low-level and

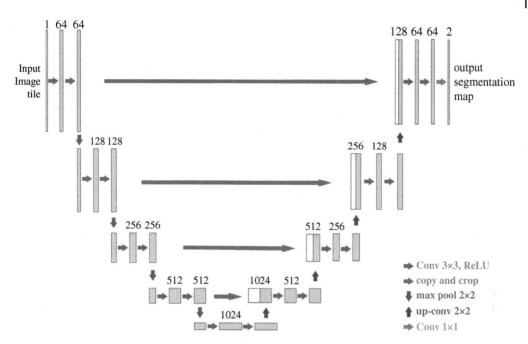

Figure 9.1 The U-Net architecture [7] / with permission of Springer Nature

high-level features. Figure 9.1 shows the U-Net architecture. Compared with FCN, U-Net, proposed by Ronneberger et al., obtains great success for medical image segmentation, since the encoder and decoder of U-Net are perfectly symmetrical, and upsampling gradually makes it possible to obtain finer segmentation results. Since then, researchers have focused on the improvements of U-Net [30, 31]. The most common way is to use the backbone of classic CNNs with pretrained parameters such as VGG [32], ResNet [33], DenseNet [34], and GhostNet [35] to replace the encoder, achieving transfer learning [36]. The other popular way of improving U-Nets is to add attention mechanisms between encoders and decoders to focus on interesting regions, such as attention U-Net [37] and RA-UNet [38].

To further exploit potentially useful information in feature maps, R2-UNet [39] introduces recurrent convolution, which is able to extract features using the same layer many times. UNet++ [16] employs U-Nets with different depths instead of long-range connections to avoid the rough fusion of low-level and high-level features. Furthermore, mU-Net [17] adds a residual path with deconvolution and activation operations to the skip connection of U-Net, and it thus achieves better segmentation results for small targets in medical images. CE-Net [15] and MSB-Net [40] employ multiscale feature fusion to enhance feature representation of networks. These improved 2D networks not only show better performance in medical image segmentation but also achieve simpler design of data augmentation schemes while keeping a lower memory requirement than 3D networks. However, they cannot capture the spatial information along the z-axis due to the employment of 2D convolution kernels, which may degrade the performance in volumetric segmentation.

To extract the spatial information along the third dimension, Ji et al. [41] employed 3D convolution kernels to achieve a 3D CNN, which makes it possible to process 3D volume data directly. Based on the 3D CNN and U-Net, Cicek and Milletari et al. proposed 3D U-Net [18] and V-Net [19], respectively. Figure 9.2 shows the V-Net architecture. The V-Net applies 3D convolutions together

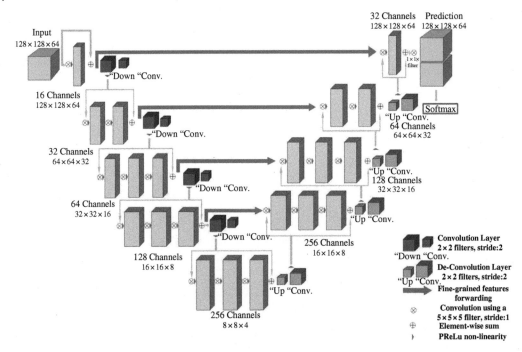

Figure 9.2 The V-Net architecture.

with residual connection to the feature encoder stage and deepens the network depth to obtain better segmentation results than 3D U-Net. Furthermore, by introducing the strategy of depth supervision, both Med3D [42] and 3D DSN [43] achieve faster and more accurate segmentation of volumetric medical images. More applications of 3D CNNs can be seen in [44]. Although these 3D networks can simultaneously explore the spatial information of inter-slice and inner-slice, these networks suffer from some new problems such as more parameters, high memory usage, and much narrower reception fields than 2D networks. To combine the advantages of 2D and 3D networks, researchers proposed H-DenseUNet [45]. This network firstly uses a 2D network to extract image features and perform segmentation tasks on a slice-by-slice basis. The pixel-wise probabilities produced by the 2D network are then concatenated with the original 3D volume and fed into a 3D network for a refinement. The H-DenseUNet finally achieves excellent liver and liver-tumor segmentation. In addition, Vu et al. [46] applied the overlay of adjacent slices as input to the central slice prediction and then fed the obtained 2D feature maps into a standard 2D network for model training. Although these pseudo-3D approaches can segment objects from 3D volume data, they only obtain limited accuracy improvement due to the utilization of local temporal information. Compared to pseudo-3D networks, hybrid cascading 2D and 3D networks are more popular for medical image segmentation.

Although codec network structures based on deep learning (U-Net, U-Net++, Mu net, CE net, 3D U-Net, V-Net) are widely used in medical image segmentation because of their excellent performance, there are still two challenges. The first is the data challenge. Due to the high requirements of medical image data annotation for doctors and because of how time-consuming they are, there is a lack of data samples in the medical field, and most networks are based on small samples. Therefore, it is of great significance to effectively extract the small sample features of medical images.

The second is the model challenge. Although the above network has achieved high segmentation accuracy, with the increase of network complexity, its larger number of parameters and greater memory usage makes it difficult to deploy the model in edge devices to achieve effective value. In order to solve the above two problems, this chapter introduces the following two medical image segmentation methods.

9.3 Lightweight Network for Liver Segmentation

Although deep learning can achieve better end-to-end liver segmentation, it causes some new problems that limit the clinical deployment of deep learning. As liver slices constitute volumetric data, it is difficult to utilize 3D spatial information of liver slices when a 2D CNN is used for liver segmentation. Compared with 2D CNNs, 3D CNNs can utilize the spatial information among neighboring liver slices effectively. Therefore, they achieve better segmentation results. V-Net [19] is a very popular 3D fully CNN in medical image segmentation. It has a symmetric structure, and it is composed of an encoder and a decoder. The encoder is used to extract useful features from input data, and the decoder is used to reconstruct the features to obtain the final segmentation result. Unfortunately, these 3D CNNs require a large number of network parameters and have a high computational cost. Researchers usually use image patch or image zooming to train these 3D CNNs, which is a trade-off between segmentation performance and hardware resource requirements. For example, V-Net employs long-range skip connections between encoder and decoder at symmetric layers to fuse low-layer features and high-layer features together, which improves the final predictions. Both encoder and decoder depend on vanilla 3D convolution. The design of V-net causes large memory usage and high computational cost. Therefore, how to remove redundant parameters and reduce the computational cost effectively of 3D CNNs are important when we extend 3D CNNs to practical clinical applications.

To address these issues mentioned above, we present a lightweight V-Net (LV-Net) for liver segmentation. The presented network is more practical since it requires less memory usage while maintaining liver segmentation accuracy. Experiments demonstrate that the presented LV-Net is superior to popular CNNs since it provides better segmentation results with lower memory usage. This is based on the recently published article by Lei et al. [47]. Specifically, the presented LV-Net, shown in Figure 9.3, has two advantages: (i) the LV-Net only requires low memory usage since it removes a large number of redundant parameters existing in the V-Net, and (ii) the LV-Net provides better segmentation results than popular 2D and 3D networks due to the employment of deep supervision.

9.3.1 Network Compression

According to Figure 9.3, we use an inverted residual bottleneck (IRB) block [48] instead of vanilla convolution to construct encoders and decoders of LV-Net. The IRB block is composed of depthwise convolution and pointwise convolution. Figure 9.4a shows the detailed architectures of the IRB block.

In the V-Net, $5 \times 5 \times 5$ convolutional kernels are used to extract spatial-dimension features and channel-dimension features. But in an IRB block, the input feature maps are firstly expanded on channels via the operation of $1 \times 1 \times 1$ pointwise convolution; secondly, the operation of $5 \times 5 \times 5$ depthwise convolution is used to extract spatial-dimension features; thirdly, the $1 \times 1 \times 1$ pointwise convolution is used to squeeze feature channels. Finally, the composition of residual feature maps

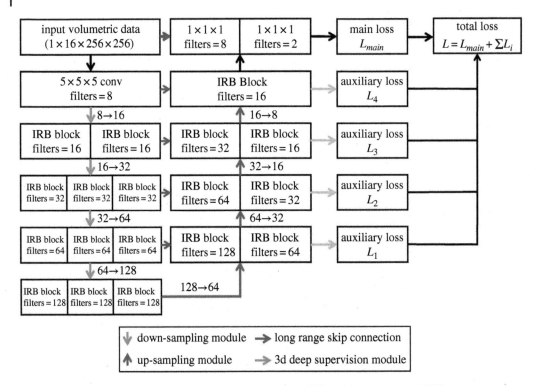

Figure 9.3 The detailed structure of LV-Net [55] / with permission of IEEE.

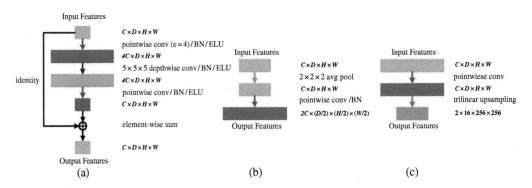

Figure 9.4 The modules in LV-Net [55] / with permission of IEEE. (a) IRB block. (b) Downsampling module. (c) 3D supervision module.

and the input feature maps is considered as the final output feature maps. Compared with the bottleneck block of ResNet [33], both the entrance and exit of the IRB block are narrow, but the middle part of the IRB block is wide. There are two advantages of the design of IRB block: (i) the IRB block can extract more features from input feature maps at the input stage, and (ii) the IRB block can remove redundant features by squeezing the channel dimension of output feature maps at the output stage. Besides, the width of the middle part of IRB block is decided by an expansion rate ε. The parameter ε is an important hyperparameter that can adjust the model capacity of the network.

Tuning the ε reasonably can avoid overfitting and make LV-Net fit other segmentation tasks flexibly. Due to the limitation of GPU memory, in this chapter, we fix ε to 4 to get the best performance.

As the vanilla convolution achieves the united mapping of feature maps on spatial correlations and cross-channel correlation, spatial features and channel features are often coupled together at the output feature maps of a vanilla convolution, which limits the feature extraction of subsequent convolutional layers. The IRB block is a variant of depthwise separable convolution, and it has stronger capability of feature extraction since the IRB block can overcome the coupling between spatial features and channel features via depthwise convolution and pointwise convolution. Besides, as the pointwise convolution and depthwise convolution employ $1 \times 1 \times 1$ kernels and $k \times k \times k$ kernels, respectively, the IRB block reduces of the number of network parameters. For Figure 9.4b, Table 9.1 shows the comparison of the number of network parameters between one IRB block and one vanilla convolution layer, where $k \times k \times k$ is the size of a kernel, $k = 5$, C is the number of channels = 64, and $\epsilon = 4$. It is clear that the IRB block requires fewer parameters (12.65%) than a convolution layer of V-Net.

To compress the size of V-Net further, we use the composition of pointwise convolution and average pooling instead of vanilla convolution in the stage of downsampling and use pointwise convolution and trilinear interpolation instead of deconvolution in the upsampling stage. As there is no trainable parameter in pooling layers and trilinear interpolation layers, the presented down- and upsampling module can reduce the number of network parameters. Figure 9.4b shows the detailed architecture of our downsampling module, where the parameter number of LV-Net is $2C^2$ while for V-Net it is $16C^2$ at the stage of downsampling, and LV-Net is $C^2/2$ while V-Net is $4C^2$ at the stage of upsampling.

9.3.2 3D Deep Supervision

In the training phase of deep neural networks, gradient vanishing is a notorious problem due to the difficulty of transmitting gradient value to shallow layers, which makes it more difficult to train deep networks. In particular, this will be worse when we use a 3D CNN including a large number of parameters to train a small data set [43]. To address the issue, we present a novel strategy that integrates a deep supervision mechanism into the decoder, as shown in Figure 9.3. We can see that a branch network is considered as a constraint after each decoder stage, and the branch is able to inject gradient values from different losses, which can avoid the problem of gradient vanishing. Figure 9.4c shows the detailed architecture of 3D deep supervision, where the pointwise convolution is firstly used for the input feature maps, then the trilinear interpolation is used for upsampling,

Table 9.1 Comparison of the efficiencies of different networks [55] / with permission of IEEE.

Models	IRB block	Vanilla 3D Convolution
Expansion pointwise Convolution	$(1 \times 1 \times 1 \times C) \times C \times \varepsilon$	
Depth-wise convolution	$(k \times k \times k \times 1) \times C \times \varepsilon$	$(k \times k \times k) \times C \times C$
Contraction pointwise Convolution	$(1 \times 1 \times 1 \times C \times \varepsilon) \times C$	
Total	$k^3 C\epsilon + 2C^2\varepsilon$	$k^3 C^2$
	64 768(12.65%)	512 000(100%)

and finally a softmax layer is used for computing the probability map of the segmentation result. Here, we use a cross-entropy loss function to estimate the difference between the final feature maps and labels.

We define the loss function from the end of decoder as the main loss function denoted by L_{main},

$$L_{main}(X, W) = \sum_{x_i \in X} - logp(t_i \mid x_i; W), \tag{9.1}$$

where X denotes training samples, W denotes the parameters of backbone network, t_i is the label of x_i, $x_i \in X$. Besides, there are also four auxiliary loss functions denoted by L_1, L_2, L_3, and L_4, respectively,

$$L_i(X, W_i, \hat{w}_i) = \sum_{x_i \in X} - logp(t_i \mid x_i; W_i; \hat{w}_i), \tag{9.2}$$

where W_i denotes parameters of backbone networks, and $\hat{w}_i$ denotes the parameters of pointwise convolution in deep supervision block. According to Eqs. (9.1) and (9.2), we present the final loss function of LV-Net

$$L = L_{main} + \sum_{i=1}^{4} \eta_i L_i + \lambda \left(||W||^2 + \sum_{i=1}^{4} ||\hat{w}_i||^2 \right), \tag{9.3}$$

where η_i represents the weight of the i-th auxiliary loss function. Here the third term of Eq. (9.3) is the weight decay, where λ denotes the decay coefficient.

As different convolutional layers of decoder have different contributions on the final loss function, the value of the balancing weight η_i is usually set empirically. Generally, the deeper a network is, the wider the perceptive field is, and the representation capability of feature is stronger, which means the output of decoder at deep layers is more important than the output of shallow layers. According to this principle, we set $\eta_1 = 0.2$, $\eta_2 = 0.4$, $\eta_3 = 0.6$, and $\eta_4 = 0.8$, respectively.

Based on the analysis above, the introduction of deep supervision has two advantages: (i) it improves the training efficiency of a network, i.e. it can speed up the convergence of the network; and (ii) it can help the network to learn more complex and useful features leading to high segmentation performance.

9.3.3 Experiment

To evaluate the performance of the presented LV-Net on liver segmentation tasks, we consider the MICCAI 2017 Liver Tumor Segmentation Challenge (LiTS) data set [49] as experimental data. The LiTS includes 131 labeled 3D CT scans, where the resolution in-plane ranges from 0.55 to 1.0 mm, and slice spacing ranges from 0.45 to 6.0 mm. In our experiments, we randomly choose 90 and 10 volume data to construct a training set and validation set, respectively. The other 31 volume data are considered as test set. Experiments are performed on a workstation with Intel core i9 9900X at 3.5GHz, 128 GB RAM, double NVIDIA GeForce RTX 2080Ti GPU, Windows 10 Pro, and PyTorch 1.2.

9.3.3.1 Data Set Preprocessing

In this experiment, image preprocessing includes three stages: truncating the range of image intensity values, scaling the slice, and normalizing the grayscale value of slice. The first stage is used to enhance the liver area and remove irrelevant details, which can achieve better feature learning and thus improve the segmentation effect. Here we set the range of [−200, 200] Hounsfield units (HU). The second stage is used to reduce the memory requirement of the hardware environment. Here we choose sequential 16 slices and resize each slice from 512×512 to 256×256. The last stage is used for the normalization of input samples, which is a key factor that affects the final segmentation performance. Here, we use mean value and variance to normalize input data.

9.3.3.2 Training

We set the values of hyperparameters to train LV-Net. The batch size is set to 4. The initial learning rate is 0.001, and it multiplies 0.9 at the end of each epoch. We use cross-entropy loss and adaptive moment estimation (ADAM) [50] to optimize the network, and weight decay is set to 1e−5. The total loss is the weighted sum of main loss and auxiliary losses. In addition, ELU [51] is considered as the activate function, which can not only boost up the traning speed but also bring better generalization performance than ReLU. Batch normalization is performed after each convolutional layer. The validation set mentioned above is used to check whether the model is overfitting or not at the end of each epoch. Once the model achieves the best performance on validation set, which often happens after about 20 epochs of training, the training stops and the model parameters are saved for further evaluation.

9.3.3.3 Evaluation and Results

We use five metrics to evaluate comprehensively the segmentation quality of each network in this experiment. They are DICE per case (DICE), volume overleap error (VOE), relative volume difference (RVD), average symmetric surface distance (ASSD [mm]), and maximum symmetric surface distance (MSSD [mm]). Note that a perfect segmentation means that the value of DICE score is 1, while the value of each of VOE, RVD, ASSD, and MSSD score is 0.

Table 9.2 presents the segmentation performance on the test set using U-Net [7], CE-Net [15], 3D U-Net [18], V-Net [19], and LV-Net. It is clear that our LV-Net achieves an average DICE of 0.954, an average VOE of 0.086, an average RVD of 0.016, an average ASSD of 1.871 mm, and an average MSSD of 29.496 mm. Except for the value of RVD, which is slightly lower than the value provided by 3D U-Net in the first place, the remaining metrics of LV-Net are higher than the ones for comparable networks. LV-Net shows the best segmentation performance on the LiTS data set.

We also count the quantities of trainable parameters and computational costs of networks above, as shown in Table 9.3. Compared with 2D CNNs, 3D CNNs obtain a certain increase in segmentation performance, but they require more memory usage and have a high computational cost. The presented LV-Net overcomes the drawbacks of 3D CNNs due to the utilization of depth separable convolution and the design of down- and upsampling modules. Consequently, LV-Net is significantly ahead of other 2D CNNs and 3D CNNs on the number of trainable parameters,

Table 9.2 Quantitative evaluation results of different networks on the liver segmentation testing set [55] / with permission of IEEE.

Models	DICE	VOE	RVD	ASSD (mm)	MSSD (mm)
U-Net [7]	0.9399	0.1114	0.0322	5.7985	123.5763
CE-Net [15]	0.9404	0.1103	0.0619	4.1162	115.4076
3D U-Net [18]	0.9400	0.1113	0.0142	2.6173	36.4352
V-Net [19]	0.9426	0.1065	0.0192	2.4887	38.2826
LV-Net	0.9543	0.0856	0.0156	1.8705	29.4960

Table 9.3 Comparison of the efficiencies of different networks [55] / with permission of IEEE.

Models	Trainable parameters	Operation (GFLOPs)	Storage usage (MB)
U-Net [7]	13 394 242	123.96	51.15
CE-Net [15]	29 003 668	35.78	110.77
3D U-Net [18]	16 320 322	1032.80	62.27
V-Net [19]	65 173 903	516.12	248.69
LV-Net	1 659 282	58.07	6.56

computational cost, and storage usage. For example, the number of trainable parameters of LV-Net is only 2.55% of the vanilla V-Net, the computational cost is 11.25%, and the storage usage is 2.64%.

9.4 Deformable Encoder–Decoder Network for Liver and Liver-Tumor Segmentation

Deep CNNs have been widely used for medical image segmentation due to their superiority in feature learning. Although the networks mentioned above can perform end-to-end liver and liver-tumor segmentation well, the use of vanilla convolution limits further improvements of segmentation accuracy. Since standard convolution kernels have a regular sampling grid, they are unable to capture accurately liver and liver-tumor features with variable shapes in different slices. Besides, some improved networks such as CE-Net [15] and MSB-Net [40] employ multi-scale feature fusion to enhance feature representation of networks, but many network branches lead to the requirement of more parameters. To address these issues, we present a deformable encoder–decoder network (DefED-Net) to improve liver and liver-tumor segmentation. This is based on the recently published article by Lei et al. [52]. The presented DefED-Net includes following advantages:

1) Feature extraction layers of the DefED-Net are constructed by using the deformable convolution with residual design. The design can more effectively extract the spatial context information of images while maintaining high-level features.
2) The feature fusion module of the DefED-Net depends on a ladder-atrous-spatial-pyramid-pooling (Ladder-ASPP), which employs multi-scale dilated convolution kernels using variable dilation rate to obtain better spatial context information.
3) The DefED-Net provides higher segmentation accuracy for liver and liver-tumor than state-of-the-art approaches, and it requires less memory usage due to the employment of depth separable convolution.

Specifically, we present a DefED-Net and apply it to liver and liver-tumor segmentation. Figure 9.5 shows the architecture of the DefED-Net. As can be seen from Figure 9.5, the DefED-Net is an enhanced U-net, and it is composed of three parts including an encoder, a middle processing module, and a decoder. In contrast with the U-net, the DefED-Net employs deformable convolution with residual structure to generate feature maps. Moreover, the original image is concatenated with outputs at different layers of the decoder to obtain better feature representation. Different from general pyramid pooling (PP) modules [15, 53], we design a better feature fusion module, namely Ladder-ASPP, and apply it to our DefED-Net. Although the Ladder-ASPP adopts

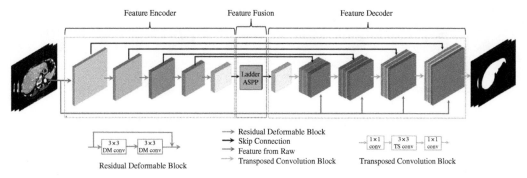

Figure 9.5 The architecture of the presented DefED-Net [56] / with permission of IEEE. Firstly, the feature encoder employs deformable convolution using the residual connection. Secondly, the Ladder-ASPP block is used to extract richer context information. Finally, both the skip connection and the dense connection of original images are used for the fusion of feature maps in decoder.

the way of dense connection, it only requires smaller memory usage due to the utilization of the depth separable convolution. It is worth mentioning that the DefED-Net is designed in 2D domain.

9.4.1 Deformable Encoding

Although a large number of improved U-Nets have been proposed for medical image segmentation, they provide limited segmentation accuracy for livers and liver-tumors in CT images. Here are two reasons that limit the performance of U-Nets. First, convolutional kernels with fixed geometric structures are employed by the U-Nets, which ignores the shape information of objects in an image. Second, the operation of polling and strided convolution leads to the loss of spatial context detail information.

To illustrate the first reason, we present an example of image filtering, shown in Figure 9.6, which demonstrates that the morphological opening filter is able to smooth noise effectively by employing different structuring elements (SEs). However, these filtering results depend on the choice of SEs. Figure 9.6b shows that a circular SE is useful for preserving the details of circular objects, and Figure 9.6c shows that a square SE is effective for square objects. Similarly, Figure 9.6d shows that a linear SE can maintain the details of linear objects. Therefore, it is better to adopt multiple different SEs for an image including many different objects. In other words, we should consider adaptive filters that can obtain better filtering effect due to the consideration of geometrical shape information of objects. In addition, the design of convolution kernels also plays the same important role for CNNs. In practical applications, researchers often employ fixed-shape square convolutional kernels to perform feature learning such as U-Net, PSP-Net, and CE-Net. Since convolution kernels with fixed shape show weak ability for the extraction of image contextual information, these aforementioned networks only provide tolerable accuracy for liver and liver-tumor segmentations. Instead, we use deformable convolution kernels to extract richer geometry information of the liver and liver tumor, which can better accommodate the irregular shape of liver and liver tumor and lead to better segmentation results. In Figure 9.7, the deformable convolution shows better adaption for liver in a CT image than the vanilla convolution. In fact, Sun et al. [54] have started to explore the utilization of deformable convolution on automatic segmentation networks for gastric cancer, and their proposed network achieves better segmentation results than vanilla U-Net [7] and ResU-Net [55].

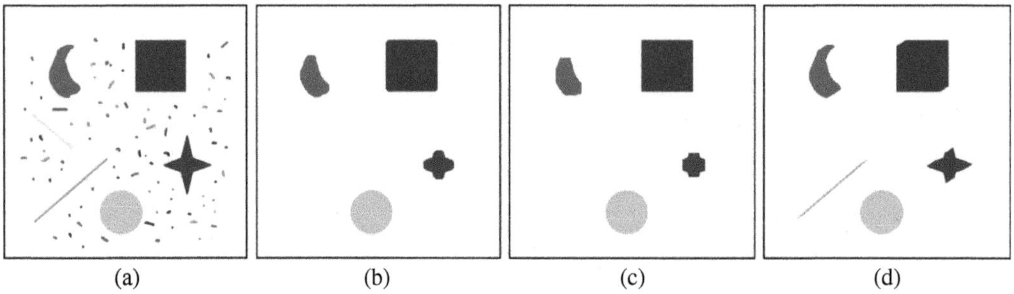

Figure 9.6 Image filtering using a morphological opening filter with different structuring elements [56] / with permission of IEEE (a) The original image. (b) The SE is a disk of size 20 × 20. (c) The SE is a square of size 20 × 20. (d) The SE is a line that the length is 20 and the orientation is $1/6\pi$.

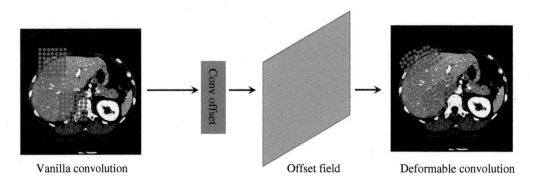

Vanilla convolution   Offset field   Deformable convolution

Figure 9.7 Comparison of vanilla convolution and deformable convolution for liver segmentation [56] / with permission of IEEE. In contrast with the standard convolution, the deformable convolution requires offset locations for each sampling location.

The deformable convolution is able to provide convolutional kernels with arbitrary shapes by learning offset locations and thus adaptively decide scales of receptive field with fine localization. Therefore, the DefED-Net possesses better capability of modeling geometric transformations than common U-Nets because of the employment of a deformable convolution. However, the implementation of deformable convolution is more complex than vanilla convolution since additional spatial offset locations are limited. Based on learned offset locations, the convolution kernels can achieve the deformation of different scales, shapes, and orientations. Figure 9.7 illustrates the principle of deformable convolution on liver segmentation.

In practical applications, a deformable convolution is composed of four layers: a convolutional layer, a convolutional offset layer, a batch normalization layer, and an activation layer. The principle of deformable convolution is given as follows. Let x and y be the input and the output feature map, respectively. L denotes a regular grid in 2D domain.

When performing the convolution operation on x using L, the output is denoted by:

$$y(e_0) = \sum_{e_n \in L} w(e_n) \times x(e_0 + e_n),$$

(9.4)

where w denotes the weight, e_0 denotes the location of a pixel, and e_n denotes the location of neighboring pixels falling into L. If we perform the deformable convolution on x, the output can be represented by:

$$\tilde{y}(e_0) = \sum_{e_n \in \tilde{L}} w(e_n) \times x(e_0 + e_n + \Delta e_n), \tag{9.5}$$

where $\tilde{L}$ is the deformation result of L. Compared to L, $\tilde{L}$ is an irregular grid including offset locations Δe_n.

The offset Δe_n is usually a float number, and the sampling position of the deformable convolution becomes irregular, so a bilinear interpolation is used to perform the process of determining the pixel value of the final sampling position.

Suppose the coordinates of the sampled position after offset is (a, b), which we convert into four integers $floor(a)$, $ceil(a)$, $floor(b)$ and $ceil(b)$. By integrating these four integers, we obtain four pairs of coordinates $(floor(a), floor(b))$, $(floor(a), ceil(b))$, $(ceil(a), floor(b))$, $(ceil(a), ceil(b))$. The $floor(\cdot)$ operation is rounding down and the $ceil(\cdot)$ operation is rounding up. Each of these four pairs of coordinates corresponds to the pixel values $p, q, m_{and} n$ in the feature map, so we use bilinear interpolation of them to calculate the pixel value of (a, b). The pixel value $x(e)$ at the final sampling location is defined as:

$$
\begin{aligned}
x(e) = {} & \frac{p}{(ceil(a) - floor(a)) \times (ceil(b) - floor(b))} \times (ceil(a) - a) \times (ceil(b) - b) \\
& + \frac{m}{(ceil(a) - floor(a)) \times (ceil(b) - floor(b))} \times (a - floor(a)) \times (ceil(b) - b) \\
& + \frac{q}{(ceil(a) - floor(a)) \times (ceil(b) - floor(b))} \times (ceil(a) - a) \times (b - floor(b)) \\
& + \frac{n}{(ceil(a) - floor(a)) \times (ceil(b) - floor(b))} \times (a - floor(a)) \times (b - floor(b)).
\end{aligned}
\tag{9.6}
$$

For instance, if the coordinates we got from the sampling position are (2.2, 4.6), then its nearest pixel set is {(2, 4), (2, 5), (3, 4), (3, 5)}. Therefore, in the actual program calculation, we will use the bilinear interpolation of {(2, 4), (2, 5), (3, 4), (3, 5)} pixels for the sampled location (2.2, 4.6) pixels.

As shown in Figure 9.7, the offset is obtained by applying a convolutional layer. Note that the convolutional layer to obtain the offset needs to have the same spatial resolution and dilation rate as the convolutional layer to extract the features in the offset feature map. For each layer of the deformable convolution, when the input of a convolutional layer is a feature map with C channels, the corresponding offset map includes $2C$ channels in this convolutional layer because each channel includes two offset maps in the x and y directions, separately. Note that the offset map of the output has the same spatial resolution as the input map in a convolutional layer. During training, the offset can be learned through the back-propagation of Eqs. (9.6). After the pixel values of all sampled positions are obtained, a new feature map will be generated. Although the deformable convolution is superior to vanilla convolution due to the employment of convolutional kernels with flexible shape, it can be further improved by using multi-scale convolutional kernels instead of single-scale kernels. For liver segmentation task, a large convolutional kernel is better than small ones for capturing coarse liver areas. However, a small convolutional kernel is more useful for obtaining accurate contour details. Therefore, here we use a large convolutional kernel of 7×7 for the first deformable convolution layer while using a small convolutional kernel of 3×3 for subsequent layers. The presented multi-scale deformable convolution is able to achieve better feature representation than single-scale deformable convolution and thus leads to better liver and liver-tumor segmentation results due to more accurate liver and liver-tumor contours. In addition, the residual design is integrated in the presented deformable encoder to avoid vanishing gradients and to speed up the convergence of networks.

9.4.2 Ladder-ASPP

Both PP and atrous spatial pyramid pooling (ASPP) are two popular ways for encoding context information due to wider receptive fields than standard pooling. Since PP directly performs pooling operation using multi-scale pooling kernels, it often causes irreversible information loss leading to poor segmentation results for small objects such as a liver tumor. However, ASPP performs atrous convolution using multiple dilation convolutional kernels instead of multi-scale pooling kernels. Compared with PP, ASPP provides better context information since atrous convolution is superior to the pooling operation for the preservation of detailed information. However, ASPP still faces two challenges in practical applications. The first one is that the fixed dilation rate is used for ASPP, which causes a gridding effect, as shown in Figures 9.8a–c; some pixels falling into receptive fields

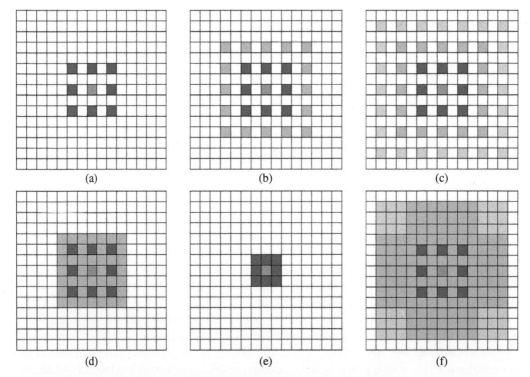

Figure 9.8 Comparison of standard atrous convolution and the atrous convolution with variable dilation rate [56] / with permission of IEEE. (a) Convolutional kernel: 3 × 3, rate = 2. (b) The cascade of two convolutional kernels: 3 × 3, rate = 2. (c) The cascade of three convolutional kernels: 3 × 3, rate = 2. (d) Convolutional kernel: 3 × 3, rate = 1. (e) The cascade of two convolutional kernels: 3 × 3, rate = 1, 2. (f) The cascade of three convolutional kernels: 3 × 3, rate = 1, 2, and 3. Note that although (c) and (f) have similar receptive fields 13 × 13, (c) has 70%-pixel loss compared to f.

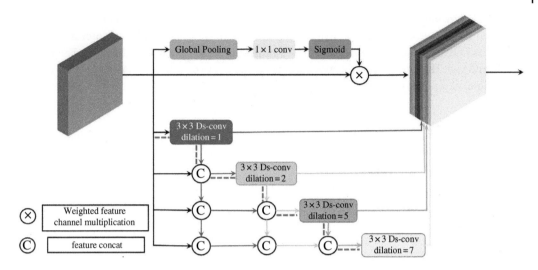

Figure 9.9 The architecture of the Ladder-ASPP [56] / with permission of IEEE. The output feature maps are concatenated by two parts. The first one is the output from global pooling and the second one is the densely connected feature fusion liking ladder.

cannot take part in the convolutional operation. The second one is that ASPP ignores the global context information. To address these issues, we present a novel Ladder-ASPP, as shown in Figure 9.9.

The standard atrous convolution easily leads to the loss of spatial detailed information. To overcome the drawback, we use variable dilation rate instead of fixed dilation rate, which leads to better receptive fields. It is clear that each pixel in the receptive fields is covered as shown in Figures 9.8d–f. Therefore, atrous convolution with variable dilation rate can overcome the gridding effect caused by the standard atrous convolution.

Based on atrous convolution with variable dilation rate, we design a Ladder-ASPP to improve context encoding. Figure 9.9 shows the architecture of the Ladder-ASPP.

First, the Ladder-ASPP employs a variable dilation rate to achieve atrous convolution that was mentioned previously.

Second, the Ladder-ASPP uses a dense ladder connection that is helpful for ASPP to achieve better feature fusion. However, the dense connection easily leads to the increase of the number of parameters and high memory requirement. To reduce the number of parameters to obtain a lightweight network, we introduce depthwise separable convolution (DSC) [56] to Ladder-ASPP. Compared to the standard convolution in which spatial features and channel features are often coupled together, the DSC can achieve the decoupling computation between spatial features and channel features leading to the requirement of fewer parameters.

It is well known that the standard convolution requires parameters $D_K \times D_K \times M \times N$, where M is the dimension of input feature maps, N is the dimension of output feature maps, and D_K is the space resolution of convolution kernels. In the DSC, the depthwise convolution only requires parameters $D_K \times D_K \times 1 \times M$, and the pointwise convolution only requires parameters $1 \times 1 \times M \times N$. Therefore, the number of parameters of DSC is $1/(N \times D_K^2)$ of the standard convolution. Here, the presented Ladder-ASPP employs four kernels of size 3×3. Consequently, the Ladder-ASPP only requires 36% of the parameters required without using DSC.

Finally, to improve feature representation of ASPP, the global pooling is integrated into Ladder-ASPP since it can achieve the priority of channels including more important information. We can see from Figure 9.9 that the information hidden in both space dimension and channel dimension is exploited simultaneously. The final feature maps fuse both the global and local information.

To illustrate the presented Ladder-ASPP, let Y be the output feature map, y_1 be the global pooling result, and y_2 be the output from the module of ladder autrous convolution. It is clear that $Y = y_1 + y_2$. The global pooling result y_1 is defined as:

$$y_1 = N_{or}[C_1[GP_S(x)]] \times x, \tag{9.7}$$

where x is the feature map obtained from the feature encoder, followed by global pooling denoted by $GP_S(x)$, C_1 represents the weight of each feature channel through a 1×1 convolution, and N_{or} is the normalization of feature weight.

In our Ladder-ASPP, we adopt variable dilation rate, i.e. 1, 2, 5, and 7. Let $G_{K^p,\varepsilon}$ be the output of densely connected PP, where K^p is the level of pyramid and ε is dilation rate; we get:

$$y_2 = G_{1,1}\left(x^{(1)}\right) \oplus G_{2,2}\left(x^{(2)}\right) \oplus G_{3,5}\left(x^{(3)}\right) \oplus G_{4,7}\left(x^{(4)}\right), \tag{9.8}$$

where,

$$\begin{cases} x^{(1)} = x \\ x^{(2)} = x^{(1)} \oplus G_{1,1}\left(x^{(1)}\right) \\ x^{(3)} = x^{(2)} \oplus G_{2,2}\left(x^{(2)}\right) \\ x^{(4)} = x^{(3)} \oplus G_{3,5}\left(x^{(3)}\right) \end{cases}, \tag{9.9}$$

and $\oplus$ denotes the concatenation operation.

According to Eqs. (9.7)–(9.9), we can see that the output from Ladder-ASPP includes richer information than the original input. The Ladder-ASPP can help our DefED-Net to achieve better segmentation results due to the exploitation of significant spatial information.

9.4.3 Loss Function

Our framework is an end-to-end deep learning system. As illustrated in Figure 9.5, we need to train the proposed method to predict each pixel as foreground or background, which is a pixel-wise classification problem. Cross-entropy is one of the most popular loss functions, and it is defined as:

$$L_{cross} = -(p \times \log(\hat{p}) + (1-p) \times \log(1-\hat{p})) \tag{9.10}$$

where p and $\hat{p}$ are the ground truth and predicted segmentation, respectively.

However, the tumor often occupies a small region in an image. The cross-entropy loss is not optimal for such tasks. It is worth noting that the Dice loss [19] is suitable for uneven samples. This metric is essentially a measure of overlap between a segmentation result and corresponding ground truth. The Dice loss is defined as:

$$L_{dice} = 1 - \frac{2\langle p, \hat{p} \rangle}{|p|_1 + |\hat{p}|_1} \tag{9.11}$$

where $p \in (0, 1)$, $0 \le \hat{p} \le 1$, p and $\hat{p}$ are the ground truth and predicted segmentation, respectively, and $\langle p, \hat{p} \rangle$ denotes dot product.

However, the use of the Dice loss easily influences the back-propagation and leads to a training difficulty. Therefore, the final loss function is defined as a combination of both losses:

$$L_{loss} = L_{cross} + L_{dice} \tag{9.12}$$

9.4.4 Postprocessing

Generally, the task of liver and liver-tumor segmentation aims to obtain a binary image where the foreground denotes liver and liver tumor, and the background denotes other areas. Based on the aforementioned analysis, we can obtain a coarse segmentation result for livers and liver tumors. However, the segmented image often includes a lot of small and isolated areas or some holes. In practical applications, binary image filtering is often used to remove false liver areas or fill holes within livers. For binary image filtering, morphological filters are very popular for the removal of small segmentation areas.

Although both classic morphological opening and closing operations can effectively improve binary segmentation results, they often smooth the boundaries of main objects as well. It is difficult to remove false objects while maintaining the boundary accuracy of real objects. For this problem, morphological reconstruction is an excellent tool, and it has been widely used for object extraction. Morphological reconstruction is able to achieve binary image filtering while maintaining the large objects unchanged. The operation requires to set the parameter of SEs. If the parameter is large, more small areas would be removed. On the contrary, fewer areas are removed in the case of small value of parameters. To address the issue, we present an adaptive morphological reconstruction to optimize liver and liver segmentation results from the DefED-Net.

We first compute the proportion of the maximal connected component in an image to the total area of the image. If the value is large, then a large SE will be adopted. Conversely, a small SE will be adopted when the value is small. Here, the SE is a disk, and its radius is denoted by r:

$$r = 30 \times round(R/(H \times W)) + 1 \tag{9.13}$$

where R denotes the area of the maximal connected component in the segmentation result, and H and W denote the height and width of the input image, respectively. Figure 9.10 shows postprocessing results using the presented adaptive morphological reconstruction. Note that it is unnecessary to make postprocessing for liver-tumor segmentation since the area of liver-tumors is generally small.

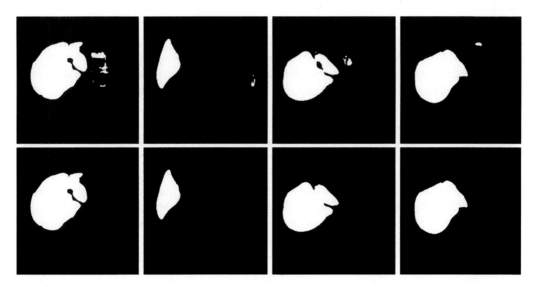

Figure 9.10 The post-processing results using adaptive morphological reconstruction [56] / with permission of IEEE. Top: segmentation results from the DefED-Net. Bottom: post-processing results.

9.4.5 Experiment

9.4.5.1 Data Set and Preprocessing

Two public contrast-enhanced CT scans data sets, Liver Tumor Segmentation Challenge (LiTS-ISBI2017) and the 3D Image Reconstruction for Comparison of Algorithm and DataBase (3Dircadb), are considered as experimental data. LiTS is a large data set, which contains 130 3D abdominal CT scans, where the image size is 512×512, slice thickness varies from 0.55 to 6 mm, and pixel spacing varies from 0.55 to 1 mm. 3DIRCADb is a small data set, which contains 22 3D data, where the image size is 512×512, slice thickness varies from 1 to 4 mm, pixel spacing varies from 0.56 to 0.86 mm, and slice number varies from 184 to 260. We constructed the training set and validation set using 90 patients (total 43 219 axial slices) and 10 patients (total 1500 axial slices), respectively. Then the other 30 patients (total 15 419 axial slices) are considered as the test set. For the 3DIR-CADb, it was split into 17 patients for training and 5 patients for test.

Medical CT axial slices are different from normal axial slices; the former is able to obtain a wider range of values from -1000 to 3000 than the latter, from 0 to 255. To remove interferences and enhance liver areas, we truncated the image intensity values of all scans of $[-200, 250]$ HU and performed the normalization on these scans. In our experiments, the given models are independently and separately performed for liver and liver-tumor segmentation.

9.4.5.2 Experimental Setup and Evaluation Metrics

All algorithms were implemented on a desktop PC with double NVIDIA GeForce RTX 2080 Ti with 11GBVRAM. The CNNs were performed and trained using the framework of PyTorch 1.3.0.

On the model training, we set the initial learning rate (lr) to 0.001 and define the decay strategy for learning rate during training as:

$$lr = lr \times (1 - i/t_i)^{0.9}, \tag{9.14}$$

where i denotes the number of iterations of this training, and t_i denotes the total number of iterations.

Note that the deformable convolution requires two learning rates compared to one for vanilla convolution. We set $lr_2 = lr \times 0.01$ for offset convolutional layers used for deformable convolution networks and used the Adam gradient descent with momentum to optimize the model.

Five popular evaluation metrics are used to measure the accuracy of segmentation results such as Dice score (DICE) [49], volumetric overlap error (VOE), RVD, ASSD, and root mean square symmetric surface distance (RMSD). The tumor burden of the liver is a measure of the fraction of the liver afflicted by cancer. In particular, as a metric, we measure the root mean square error (RMSE) in tumor burden estimates from lesion predictions. The value of DICE ranges from 0 to 1, and a perfect segmentation yields a DICE value of 1. In fact, DICE is one of the most important metrics in image segmentation evaluation. The VOE is the complement of the Jaccard coefficient, and thus a perfect segmentation yields a VOE value of 0. The RVD is an asymmetric metric, and a smaller value of RVD means a better segmentation result. Both ASD and RMSD are used to measure the surface distance between segmentation results and ground truths; the former is used to compute the average distance, but the latter is used to compute the maximal distance. Consequently, a better segmentation result corresponds to high values of DICE but low values of VOE, RVD, ASD, and RMSD. Note that we evaluate segmentation results based on 3D volumes.

9.4.5.3 Ablation Study

This chapter focuses on liver and liver-tumor segmentation. Two contributions are highlighted: One is that the deformable convolution is used instead of the vanilla convolution; the other is that Ladder-ASPP is integrated into the presented DefED-Net to improve the context information. To demonstrate the two contributions and the effectiveness of the DefED-Net, we conducted comprehensive experiments on both LiTS liver and liver tumor data sets.

Effectiveness of the deformable convolution: We analyzed the performance of deformable convolution (DC) and residual deformable convolution (RDC), respectively. Figure 9.11 shows the comparison of U-Net, U-Net + DC, and U-Net + RDC on liver segmentation. It is clear that both DC and RDC can help U-Net to focus on the interesting regions and remove irrelevant background information, but RDC can help the network converge faster and obtain more accurate edge predictions. In the third column of Figure 9.11, the feature maps provided by U-Net + RDC include less information that is unrelated with the liver. Consequently, U-Net obtains more fake liver regions than U-Net + RDC and U-Net + DC in the fifth column of Figure 9.11. Table 9.4 demonstrates the effectiveness of the first contribution. We can see that the utilization of DC effectively raises the segmentation accuracy of U-Net. The residual design not only speeds up the convergence of U-Net but also further improves segmentation accuracy.

Effectiveness of Ladder-ASPP: Both U-Net + ASPP and U-Net + Ladder-ASPP use the idea of context encoding to improve feature representation of networks. The difference is that Ladder-ASPP uses atrous convolution with variable dilation rate and dense connection to obtain better context information than ASPP. Experimental results in Table 9.4 consistently demonstrate that both ASPP and Ladder-ASPP can help U-Net to improve segmentation accuracy of livers, and the latter is superior to the former. Figure 9.12 shows the difference of segmentation results between prediction results and ground truths, where the foreground is the difference and the background is the same. It is clear that the prediction result obtained by U-Net + Ladder-ASPP is closer to the ground truth than U-Net. Furthermore, U-Net + ASPP only improves the representation capability of models

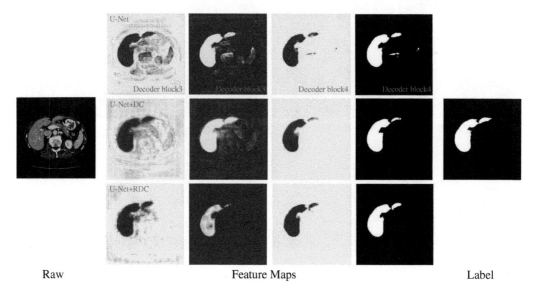

Figure 9.11 Comparison of feature maps generated by U-Net, U-Net + DC and U-Net + RDC, respectively [56] / with permission of IEEE.

Table 9.4 Comparison of ablation study on LITS test datasets [56] / with permission of IEEE.

Method	Liver DICE (%)	Tumor DICE (%)
U-Net [28]	93.99 ± 1.18	82.16 ± 6.26
U-Net + post-processing	94.25 ± 1.06	82.16 ± 6.26
U-Net + DC	95.23 ± 1.13	84.57 ± 6.21
U-Net + ASPP	94.30 ± 1.16	85.32 ± 6.08
U-Net + DenseASPP	95.32 ± 1.14	85.43 ± 6.02
U-Net + Ladder-ASPP	95.50 ± 1.09	86.72 ± 5.87
DefED-Net (without post-processing)	96.02 ± 1.04	$\mathbf{87.52} \pm \mathbf{5.32}$
DefED-Net	$\mathbf{96.30}^{a} \pm \mathbf{1.01}$	$\mathbf{87.52} \pm \mathbf{5.32}$

^{a} The best values are boldfaced.

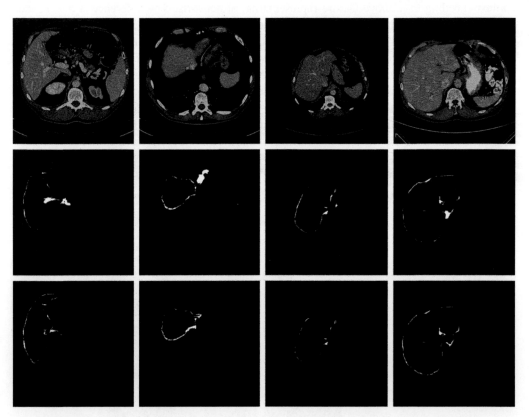

Figure 9.12 Difference of prediction results and ground truths [56] / with permission of IEEE. Top: Input images, Middle: U-Net results, and Bottom: U-Net + Ladder-ASPP results.

on the capture of spatial context information, which is unavailable for the optimization of channel dimension. Therefore, U-Net + Ladder-ASPP provides higher DICE than U-Net + ASPP, as shown in Table 9.4.

In addition, postprocessing is also useful for improving segmentation accuracy. Table 9.4 shows that RDC plays a more important role than ASPP and postprocessing for improving segmentation accuracy. The results further demonstrate that location information is more important than feature fusion for image segmentation. Figure 9.13 shows the comparison of segmentation boundaries, which further illustrates the ablation study. All these comparison results demonstrate the effectiveness of the deformable convolution, Ladder-ASPP, and postprocessing on liver and liver-tumor segmentation.

9.4.5.4 Experimental Comparison on Test Data Sets

To validate the superiority of the presented DefED-Net, six state-of-the-art networks used for liver and liver-tumor segmentation are considered as comparative approaches. These networks can be grouped into three categories: 2D networks, 3D networks, and hybrid networks with 2D and 3D, where 2D networks include U-Net, U-Net++, and CE-Net; 3D networks include 3D U-Net and V-Net; and hybrid networks include H-DenseUNet. Note that we do not give experimental results obtained by 3D U-Net and V-Net in Tables 9.4 and 9.5 due to high risk of overfitting on the 3DIR-CADb data set.

It is known that 3D networks can provide better segmentation results than 2D networks due to their exploitation of information between slices. Tables 9.5 and 9.6 demonstrate that both 3D U-Net

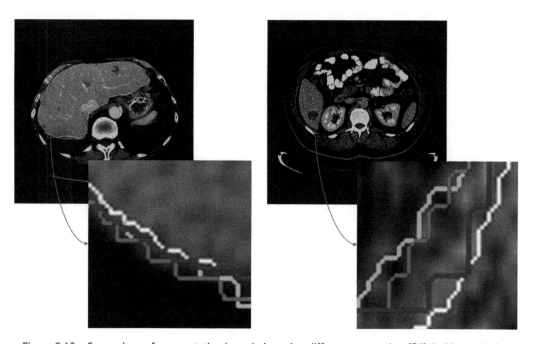

Figure 9.13 Comparison of segmentation boundaries using different approaches [56] / with permission of IEEE. The green denotes ground truth, the white denotes the result provided by U-Net, the purple denotes the result provided by CE-Net and the red denotes the result provided by DefED-Net.

Table 9.5 Quantitative scores of the liver segmentation results using different approaches on the LITS dataset [56] / with permission of IEEE.

Method	DICE (%)	VOE (%)	RVD (%)	ASD (mm)	RMSD (mm)
			LITS-Liver		
U-Net [7]	93.99 ± 1.23	11.13 ± 2.47	3.22 ± 0.20	5.79 ± 0.53	123.57 ± 6.28
U-Net++ [21]	94.01 ± 1.18	11.12 ± 2.37	2.36 ± 0.15	5.23 ± 0.45	120.36 ± 5.03
CE-Net [15]	94.04 ± 1.15	11.03 ± 2.31	6.19 ± 0.16	4.11 ± 0.51	115.40 ± 5.82
3D U-Net [18]	94.10 ± 1.06	11.13 ± 2.23	**1.42** ± 0.13	2.61 ± 0.45	**36.43** ± 5.38
V-Net [19]	94.25 ± 1.03	10.65 ± 2.17	1.92 ± **0.11**	2.48 ± 0.38	38.28 ± 5.05
H-DenseUNet [45]	96.10 ± 1.02	7.02 ± **2.00**	1.53 ± 0.12	1.56 ± 0.28	37.26 ± **3.64**
DefEnCe-Net	**96.30 ± 1.01**[a]	**6.88** ± 2.10	1.46 ± 0.12	**1.37 ± 0.23**	77.60 ± 4.26

[a] The best values are boldfaced.

Table 9.6 Quantitative scores of the liver segmentation results using different approaches on the LITS dataset [56] / with permission of IEEE.

Method	DICE (%)	VOE (%)	RVD (%)	ASD (mm)	RMSD (mm)	Tumor Burden RMSE
			LITS-Tumor			
U-Net [7]	82.16 ± 6.26	26.85 ± 16.21	3.54 ± 0.18	22.26 ± 0.30	155.15 ± 5.62	0.020
U-Net++ [21]	83.23 ± 6.36	26.03 ± 15.39	2.16 ± 0.17	21.36 ± 0.25	112.36 ± 4.89	0.017
CE-Net [15]	84.02 ± 6.15	25.62 ± 15.21	1.59 ± 0.17	20.79 ± 0.28	100.29 ± 5.23	0.018
3D U-Net [18]	85.13 ± 5.87	25.13 ± 15.02	1.23 ± 0.14	20.32 ± 0.27	62.36 ± 5.16	0.018
V-Net [19]	85.87 ± 5.42	24.52 ± 14.86	1.09 ± 0.15	19.23 ± 0.25	68.32 ± 4.52	0.017
H-DenseUNet [45]	86.23 ± 5.13[a]	**24.46 ± 13.25**	0.53 ± 0.13	18.83 ± **0.22**	**54.32 ± 4.32**	**0.015**
DefEnCe-Net	**87.52 ± 5.32**	23.85 ± 14.62	**0.52 ± 0.10**	**17.41** ± 0.28	64.25 ± 4.87	0.016

[a] The best values are boldfaced.

and V-Net provide higher segmentation accuracy than U-Net. However, CE-Net is superior to U-Net since it employs SPP to achieve feature fusion. In contrast with those networks mentioned above, H-DenseUNet provides better segmentation accuracy since it balances the advantages of both 2D networks and 3D networks. The presented DefED-Net provides the best quantitative scores (DICE, VOE, and ASD) among all available approaches. As the DefED-Net is a 2D network, it obtains lower values of RMSD than 3D networks such as 3D U-Net and V-Net. Since the 3DIRCADb is a small data set, 3D networks including a mountain of parameters easily lead to overfitting for the data set. Therefore, we only show the comparison results of U-Net, U-Net++, CE-Net, H-DenseU-Net, and DefED-Net in Tables 9.7 and 9.8, which demonstrates the DefED-Net outperforms the rest of the networks on the 3DIRCADb data set. DICE, VOE, and RVD are all overlap measures while ASD and RMSD are surface distance measures. The former focuses more on the interior of the

Table 9.7 Quantitative scores of the liver segmentation results using different approaches on the 3DIRCADb dataset [56] / with permission of IEEE.

Method	3DIRCADb-Liver				
	DICE (%)	VOE (%)	RVD (%)	ASD (mm)	RMSD (mm)
U-Net [7]	92.30 ± 1.27	11.78 ± 3.62	-2.83 ± 0.38	4.25 ± 1.56	75.67 ± 5.68
U-Net++ [17]	93.60 ± 1.29	10.36 ± 3.90	1.21 ± 0.23	3.87 ± 1.36	56.39 ± 4.76
CE-Net [20]	94.28 ± 1.22	10.02 ± 3.53	-1.80 ± 0.24	3.52 ± 1.25	30.29 ± 4.82
H-DenseUNet [45]	95.72 ± 1.14	9.88 ± 2.91	0.39 ± 0.12	2.85 ± 0.89	$\mathbf{9.63 \pm 3.95}$
DefEnCe-Net	$\mathbf{96.60 \pm 1.08}^{a}$	$\mathbf{5.65 \pm 2.81}$	$\mathbf{0.23 \pm 0.11}$	$\mathbf{2.61 \pm 0.84}$	12.76 ± 3.43

[a] The best two values are boldfaced.

Table 9.8 Quantitative scores of the liver-tumor segmentation results using different approaches on the 3DIRCADb dataset [56] / with permission of IEEE.

Method	3DIRCADb-Tumor					Tumor Burden RMSE
	DICE (%)	VOE (%)	RVD (%)	ASD (mm)	RMSD(mm)	
U-Net [7]	51.25 ± 8.28	50.75 ± 18.26	-1.11 ± 0.48	16.72 ± 0.92	130.54 ± 5.64	0.023
U-Net++ [21]	60.36 ± 7.36	46.72 ± 17.64	1.36 ± 0.78	14.76 ± 0.63	118.23 ± 5.64	0.018
CE-Net [20]	60.25 ± 7.18	40.36 ± 17.26	0.93 ± 0.32	12.42 ± 0.79	110.67 ± 4.92	0.020
H-DenseUNet [45]	65.47 ± 6.54^{a}	36.74 ± 12.86	$\mathbf{-0.74 \pm 0.18}$	12.21 ± 0.51	$\mathbf{32.52 \pm 3.28}$	**0.015**
DefEnCe-Net	$\mathbf{66.25 \pm 6.62}$	$\mathbf{34.28 \pm 13.43}$	0.81 ± 0.20	$\mathbf{11.21 \pm 0.63}$	70.05 ± 3.10	0.018

[a] The best values are boldfaced.

segmentation target, while the latter focuses more on the shape similarity of the segmentation target. It is important to note that the shape and size of liver tumors vary greatly among patients as well as in the same patient at different times compared to the liver, which make it more difficult to achieve fully automatic segmentation of liver tumors. Therefore, as recorded in Tables 9.3 and 9.5, the ASD and RMSD values for liver tumors are obviously larger than the values for livers.

Figure 9.14 shows the segmentation results from different approaches. First, from the segmentation results obtained by 2D networks, both U-Net and U-Net++ fail to identify large liver tumors, but CE-Net is successful in the first column of results. U-Net++ obtains poorer segmentation results than U-Net and CE-Net for small liver tumors as shown in the second column of results. It is clear that CE-Net is able to recognize a larger range of liver tumors due to the employment of an SPP module with multi-scale receptive fields. In the third column, both U-net and CE-Net obtain more false liver areas, but U-Net++ shows better performance for large liver target recognition because it has stronger generalization capability and more dense feature representation. In the fourth column of Figure 9.14, U-Net and U-Net++ are inaccurate in identifying liver boundaries, while CE-Net, which uses multiple atrous convolutional parallel modules, provides higher

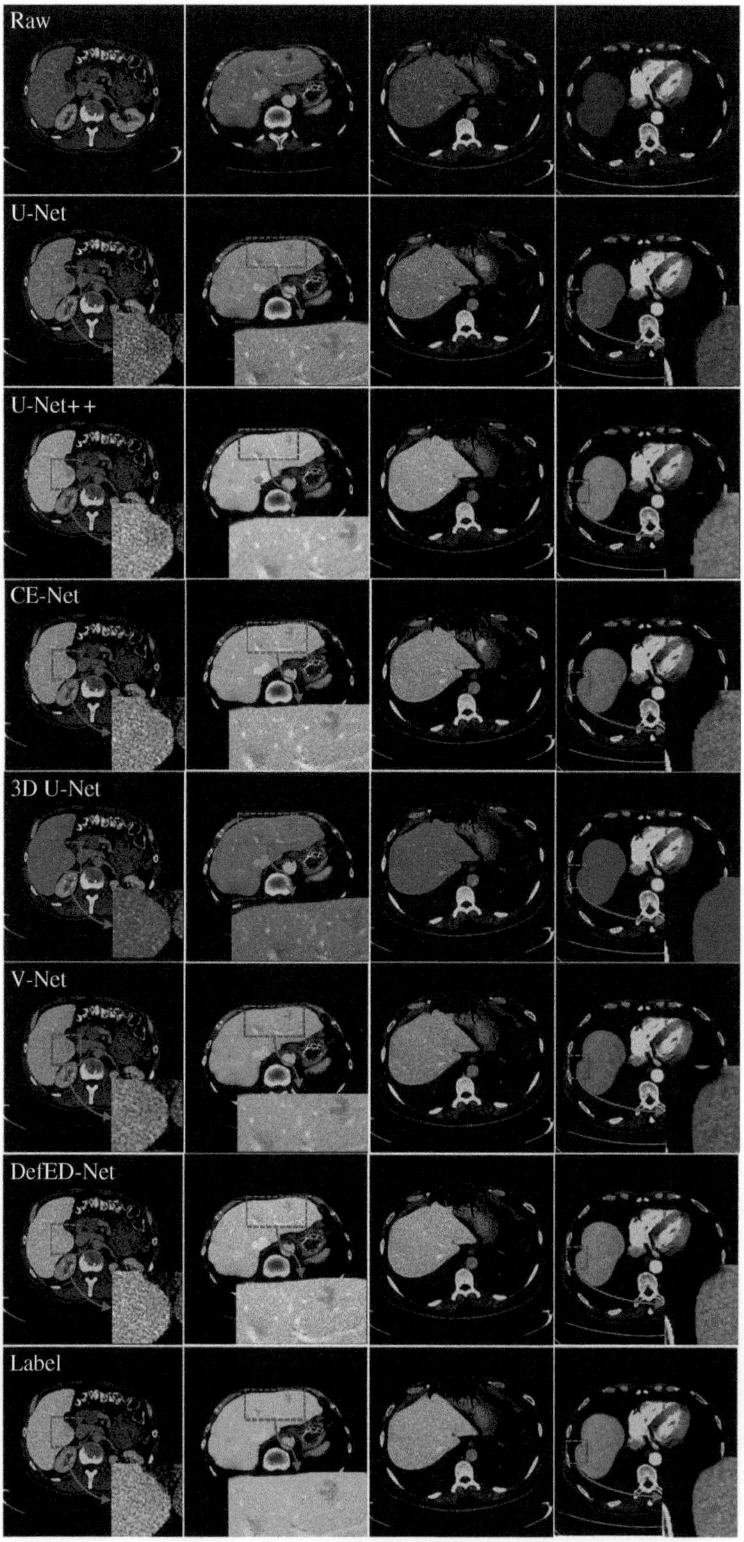

Figure 9.14 Liver and Liver-tumor segmentation results using different approaches [56] / with permission of IEEE.

accuracy for liver boundary detection. Second, it is well known that 3D networks can provide better segmentation results of liver and liver tumors than 2D networks as they can capture the temporal information of volumetic data. In both the first and second columns, it can be seen that the tumor boundaries obtained by the 3D network are clearer than results provided by 2D networks, and the tumor boundaries obtained by V-Net are clearer and more accurate than results obtained by 3D U-Net due to the utilization of feature extraction block with residual connection. In the third column, 3D networks are clearly superior to 2D networks since the former do not suffer from the problem of over-detection. Finally, it is evident from the results obtained by DefED-Net in the first and second columns that they provide more accurate segmentations of both liver and liver tumors than the above mentioned 2D and 3D networks. In addition, in the third and fourth columns, the DefED-Net focuses on the relevant liver region while suppressing the influence of surrounding organs; it thus provides smoother segmentation boundaries than comparable approaches. In general, Figure 9.14 shows that the DefED-Net achieves better feature encoding and context information extraction, which is helpful for improving the segmentation accuracy of liver and liver tumors.

9.4.5.5 Model-Size Comparison

We also counted the number of training parameters and computational costs of networks as shown in Table 9.9. Compared with 2D networks, 3D networks require much more memory and higher computational cost due to the employment of 3D convolutional kernels. The number of parameters of V-Net is greatly larger than the one of 3D U-Net since V-Net uses a deeper network structure than 3DU-Net, uses more convolutions, and uses residual connections. On the efficiency of models, the DefED-Net is similar to U-Net.

In fact, the DefED-Net adds a Ladder-ASPP block compared to U-Net. The Ladder-ASPP is a densely connected block, and thus it shows high computational complexity and requires a large number of parameters, as shown in Table 9.9. In this chapter, we utilize depth-separable convolution to decouple the operation of spatial dimension and channel dimension, which efficiently reduces the number of parameters. Thus, the added Ladder-ASPP is a very small block compared to the size of U-Net. Finally, the DefED-Net achieves excellent liver and liver-tumor segmentation with low computational cost.

Table 9.9 Comparison of the efficiencies of different networks [56] / with permission of IEEE.

Method	Operations (GFLOPs)	Parameters (M)	ModelSize (MB)
U-Net [7]	**123.96**[a]	**13 394 242**	**51.15**
U-Net++ [21]	**25.90**	**9 041 700**	**35.34**
CE-Net [20]	35.78	29 003 668	110.77
3D U-Net [18]	1032.80	16 320 322	62.27
V-Net [19]	516.12	65 173 903	248.69
H-DenseUNet [45]	547.81	35 580 976	139.19
DefEnCe-Net	222.44	14 529 959	56.96

[a] The best two values are boldfaced.

9.5 Discussion and Summary

Image segmentation based on deep learning is of great significance for medical image analysis to help doctors make diagnoses and treatment plans. However, there are two challenges in the field of medical image segmentation. One is the data challenge. Due to the different medical imaging equipment, the image noise is large and the contrast is low. Also medical image annotations are difficult and time-consuming for doctors, which leads to less data. Thus, most methods are based on small-sample learning. The second is the model challenge. Due to the requirements of accuracy, the networks are increasingly deeper, the number of model parameters is increasingly larger, and deployment and application are increasingly more difficult.

In order to solve the problem of small samples of medical images, we have presented DefED-Net, which can better extract image features. We have introduced deformable convolution into U-Nets to achieve better feature encoding. Furthermore, although ASPP is effective for improving the context information, atrous convolution and pooling lead to the loss of detailed information, so we have suggested the Ladder-ASPP for feature encoding and fusion. The Ladder-ASPP is superior to ASPP due to the dense connection and the atrous convolution with variable dilation rate. Finally, the presented DefED-Net provides the best liver segmentation results without increasing the size of models. Our studies also show that utilization of spatial information is more important than feature fusion via modifying the network architecture for liver and liver-tumor segmentations. Experiments demonstrate the advantages of the presented DefED-Net on improving segmentation accuracies and reducing model size for liver and liver-tumor segmentations.

In order to solve the problem that the deep learning model has too many parameters and is difficult to deploy, we mainly study the liver segmentation based on three-dimensional deep convolutional neural networks. We have presented a lightweight 3D network based on V-Net. The network adopts depth separation convolution and 3D depth supervision to reduce the memory requirements of 3D network and maintain the segmentation accuracy of the liver. Experiments show that the presented LV-Net can achieve higher segmentation accuracy and is much lighter than the popular 2D and 3D networks (such as U-Net, CE-Net, 3D U-Net, and V-Net). Therefore, the presented LV-Net is more suitable for clinical practice and can be easily extended to other medical volume segmentation tasks.

References

1 Li, W. (2015). Automatic segmentation of liver tumor in CT images with deep convolutional neural networks. *J. Computer Commun.* **3** (11): 146.

2 Vivanti, R., Ephrat, A., Joskowicz, L. et al. (2015). Automatic liver tumor segmentation in follow-up CT studies using convolutional neural networks. In: *Proc. Patch-Based Methods in Medical Image Processing Workshop*, **2**, 2.

3 Menze, B.H., Jakab, A., Bauer, S. et al. (2014). The multimodal brain tumor image segmentation benchmark (BRATS). *IEEE Trans. Med. Imaging* **34** (10): 1993–2024.

4 Cherukuri, V., Ssenyonga, P., Warf, B.C. et al. (2017). Learning based segmentation of CT brain images: application to postoperative hydrocephalic scans. *IEEE Trans. Biomed. Eng.* **65** (8): 1871–1884.

5 Cheng, J., Liu, J., Xu, Y. et al. (2013). Superpixel classification based optic disc and optic cup segmentation for glaucoma screening. *IEEE Trans. Med. Imaging* **32** (6): 1019–1032.

6 Fu, H., Cheng, J., Xu, Y. et al. (2018). Joint optic disc and cup segmentation based on multi-label deep network and polar transformation. *IEEE Trans. Med. Imaging* **37** (7): 1597–1605.

7 Ronneberger, O., Fischer, P., and Brox, T. (2015). U-net: Convolutional networks for biomedical image segmentation. In: *International Conference on Medical Image Computing and Computer-Assisted Intervention*, 234–241. Cham: Springer.

8 Song, T.H., Sanchez, V., Eidaly, H., and Rajpoot, N.M. (2017). Dual-channel active contour model for megakaryocytic cell segmentation in bone marrow trephine histology images. *IEEE Trans. Biomed. Eng.* **64** (12): 2913–2923.

9 Wang, S., Zhou, M., Liu, Z. et al. (2017). Central focused convolutional neural networks: developing a data-driven model for lung nodule segmentation. *Med. Image Anal.* **40**: 172–183.

10 Onishi, Y., Teramoto, A., Tsujimoto, M. et al. (2020). Multiplanar analysis for pulmonary nodule classification in CT images using deep convolutional neural network and generative adversarial networks. *Int. J. Comput. Assist. Radiol. Surg.* **15** (1): 173–178.

11 Wong, D., Liu, J., Fengshou, Y. et al. (2008). A semi-automated method for liver tumor segmentation based on 2D region growing with knowledge-based constraints. In: *MICCAI Workshop*, vol. **41**, No. 43, 159.

12 Bai, J.W., Li, P.A., and Wang, K.H. (2016). Automatic whole heart segmentation based on watershed and active contour model in CT images. In: *2016 5th International Conference on Computer Science and Network Technology (ICCSNT)*, 741–744. IEEE.

13 Chartrand, G., Cresson, T., Chav, R. et al. (2016). Liver segmentation on CT and MR using Laplacian mesh optimization. *IEEE Trans. Biomed. Eng.* **64** (9): 2110–2121.

14 Saito, A., Nawano, S., and Shimizu, A. (2016). Joint optimization of segmentation and shape prior from level-set-based statistical shape model, and its application to the automated segmentation of abdominal organs. *Med. Image Anal.* **28**: 46–65.

15 Gu, Z., Cheng, J., Fu, H. et al. (2019). Ce-net: context encoder network for 2d medical image segmentation. *IEEE Trans. Med. Imaging* **38** (10): 2281–2292.

16 Zhou, Z., Siddiquee, M.M.R., Tajbakhsh, N., and Liang, J. (2019). Unet++: redesigning skip connections to exploit multiscale features in image segmentation. *IEEE Trans. Med. Imaging* **39** (6): 1856–1867.

17 Seo, H., Huang, C., Bassenne, M. et al. (2019). Modified U-Net (mU-Net) with incorporation of object-dependent high level features for improved liver and liver-tumor segmentation in CT images. *IEEE Trans. Med. Imaging* **39** (5): 1316–1325.

18 Çiçek, Ö., Abdulkadir, A., Lienkamp, S.S. et al. (2016). 3D U-Net: learning dense volumetric segmentation from sparse annotation. In: *International Conference on Medical Image Computing and Computer-Assisted Intervention*, 424–432. Cham: Springer.

19 Milletari, F., Navab, N., and Ahmadi, S.A. (2016). V-net: Fully convolutional neural networks for volumetric medical image segmentation. In: *2016 Fourth International Conference on 3D Vision (3DV)*, 565–571. IEEE.

20 Yu-Qian, Z., Wei-Hua, G., Zhen-Cheng, C. et al. (2006). Medical images edge detection based on mathematical morphology. In: *2005 IEEE Engineering in Medicine and Biology 27th Annual Conference*, 6492–6495. IEEE.

21 Lalonde, M., Beaulieu, M., and Gagnon, L. (2001). Fast and robust optic disc detection using pyramidal decomposition and Hausdorff-based template matching. *IEEE Trans. Med. Imaging* **20** (11): 1193–1200.

22 Chen, W., Smith, R., Ji, S.Y. et al. (2009). Automated ventricular systems segmentation in brain CT images by combining low-level segmentation and high-level template matching. *BMC Med. Inform. Decis. Mak.* **9** (1): 1–14.

23 Tsai, A., Yezzi, A., Wells, W. et al. (2003). A shape-based approach to the segmentation of medical imagery using level sets. *IEEE Trans. Med. Imaging* **22** (2): 137–154.

24 Li, C., Wang, X., Eberl, S. et al. (2013). A likelihood and local constraint level set model for liver tumor segmentation from CT volumes. *IEEE Trans. Biomed. Eng.* **60** (10): 2967–2977.

25 Li, S., Fevens, T., and Krzyżak, A. (2004). A SVM-based framework for autonomous volumetric medical image segmentation using hierarchical and coupled level sets. In: *International Congress Series*, vol. **1268**, 207–212. Elsevier.

26 Held, K., Kops, E.R., Krause, B.J. et al. (1997). Markov random field segmentation of brain MR images. *IEEE Trans. Med. Imaging* **16** (6): 878–886.

27 Krizhevsky, A., Sutskever, I., and Hinton, G.E. (2012). Imagenet classification with deep convolutional neural networks. *Adv. Neural Inform. Process. Syst.* **25**: 1097–1105.

28 Long, J., Shelhamer, E., and Darrell, T. (2015). Fully convolutional networks for semantic segmentation. In: *Proceedings of the IEEE Conference on Computer Vision and Pattern Recognition*, 3431–3440.

29 Chen, L. C., Papandreou, G., Schroff, F., and Adam, H. (2017). Rethinking atrous convolution for semantic image segmentation. arXiv preprint *arXiv:1706.05587*.

30 Guo, Z., Li, X., Huang, H. et al. (2019). Deep learning-based image segmentation on multimodal medical imaging. *IEEE Trans. Radiat. Plasma Med. Sci.* **3** (2): 162–169.

31 De Sio, C., Velthuis, J.J., Beck, L. et al. (2020). R-UNet: leaf position reconstruction in upstream radiotherapy verification. *IEEE Trans. Radiat. Plasma Med. Sci.* **5** (2): 272–279.

32 Simonyan, K. and Zisserman, A. (2014). Very deep convolutional networks for large-scale image recognition. arXiv preprint *arXiv:1409.1556*.

33 He, K., Zhang, X., Ren, S., and Sun, J. (2016). Deep residual learning for image recognition. In: *Proceedings of the IEEE conference on computer vision and pattern recognition*, 770–778.

34 Huang, G., Liu, Z., Van Der Maaten, L., and Weinberger, K.Q. (2017). Densely connected convolutional networks. In: *Proceedings of the IEEE Conference on Computer Vision and Pattern Recognition*, 4700–4708.

35 Han, K., Wang, Y., Tian, Q. et al. (2020). Ghostnet: More features from cheap operations. In: *Proceedings of the IEEE/CVF Conference on Computer Vision and Pattern Recognition*, 1580–1589.

36 Conze, P.H., Kavur, A.E., Cornec-Le Gall, E. et al. (2021). Abdominal multi-organ segmentation with cascaded convolutional and adversarial deep networks. *Artif. Intell. Med.* **117**: 102109.

37 Oktay, O., Schlemper, J., Folgoc, L. L. et al. (2018). Attention u-net: Learning where to look for the pancreas. arXiv preprint *arXiv:1804.03999*.

38 Jin, Q., Meng, Z., Sun, C. et al. (2020). RA-UNet: a hybrid deep attention-aware network to extract liver and tumor in CT scans. *Front. Bioeng. Biotechnol.* **8**: 1471.

39 Alom, M.Z., Hasan, M., Yakopcic, C. et al. (2018). Recurrent Residual Convolutional Neural Network based on U-Net (R2U-Net) for Medical Image Segmentation.

40 Shao, Q., Gong, L., Ma, K. et al. (2019). Attentive CT lesion detection using deep pyramid inference with multi-scale booster. In: *International Conference on Medical Image Computing and Computer-Assisted Intervention*, 301–309. Cham: Springer.

41 Ji, S., Xu, W., Yang, M., and Yu, K. (2012). 3D convolutional neural networks for human action recognition. *IEEE Trans. Pattern Anal. Mach. Intell.* **35** (1): 221–231.

42 Chen, S., Ma, K., and Zheng, Y. (2019). Med3d: Transfer learning for 3d medical image analysis. arXiv preprint *arXiv:1904.00625*.

43 Dou, Q., Yu, L., Chen, H. et al. (2017). 3D deeply supervised network for automated segmentation of volumetric medical images. *Med. Image Anal.* **41**: 40–54.

44 Zhang, Y., Li, H., Du, J. et al. (2021). 3D multi-attention guided multi-task learning network for automatic gastric tumor segmentation and lymph node classification. *IEEE Trans. Med. Imaging* **40** (6): 1618–1631.

45 Li, X., Chen, H., Qi, X. et al. (2018). H-DenseUNet: hybrid densely connected UNet for liver and tumor segmentation from CT volumes. *IEEE Trans. Med. Imaging* **37** (12): 2663–2674.

46 Vu, M.H., Grimbergen, G., Nyholm, T., and Löfstedt, T. (2020). Evaluation of multislice inputs to convolutional neural networks for medical image segmentation. *Med. Phys.* **47** (12): 6216–6231.

47 Lei, T., Zhou, W., Zhang, Y. et al. (2020). Lightweight v-net for liver segmentation. In: *ICASSP 2020–2020 IEEE International Conference on Acoustics, Speech and Signal Processing (ICASSP)*, 1379–1383. IEEE.

48 Sandler, M., Howard, A., Zhu, M. et al. (2018). Mobilenetv2: Inverted residuals and linear bottlenecks. In: *Proceedings of the IEEE Conference on Computer Vision and Pattern Recognition*, 4510–4520.

49 Bilic, P., Christ, P. F., Vorontsov, E. et al. (2019). The liver tumor segmentation benchmark (lits). arXiv preprint *arXiv:1901.04056*.

50 Kingma, D.P. and Ba, J. (2014). Adam: A method for stochastic optimization. arXiv preprint *arXiv:1412.6980*.

51 Clevert, D.A., Unterthiner, T., and Hochreiter, S. (2015). Fast and accurate deep network learning by exponential linear units (elus). arXiv preprint *arXiv:1511.07289*.

52 Lei, T., Wang, R., Zhang, Y. et al. (2021). Defed-net: Deformable encoder-decoder network for liver and liver tumor segmentation. *IEEE Transactions on Radiation and Plasma Medical Sciences*.

53 Zhao, H., Shi, J., Qi, X. et al. (2017). Pyramid scene parsing network. In: *Proceedings of the IEEE Conference on Computer Vision and Pattern Recognition*, 2881–2890.

54 Sun, M., Zhang, G., Dang, H. et al. (2019). Accurate gastric cancer segmentation in digital pathology images using deformable convolution and multi-scale embedding networks. *IEEE Access* **7**: 75530–75541.

55 Xiao, X., Lian, S., Luo, Z., and Li, S. (2018). Weighted res-unet for high-quality retina vessel segmentation. In: *2018 9th International Conference on Information Technology in Medicine and Education (ITME)*, 327–331. IEEE.

56 Howard, A. G., Zhu, M., Chen, B., et al. (2017). Mobilenets: Efficient convolutional neural networks for mobile vision applications. arXiv preprint *arXiv:1704.04861*.

10

Image Segmentation for Remote Sensing Analysis

Image segmentation for remote sensing analysis has become a major topic of interest in the environmental remote sensing field due to the ever-increasing quantity of high spatial resolution (HSR) imagery acquired from satellites, airplanes, unmanned aerial vehicles (UAVs), and other platforms. It has also been widely used in resource and environmental surveys [1], environmental monitoring [2], urban expansion and change information acquisition [3], land use [4], fire area survey and disaster detection and assessment [5], and so forth. At the same time, with the development of aerospace technology, high-resolution remote sensing images have received extensive attention due to their richer feature information and varied applications. Therefore, image analysis based on high-resolution remote sensing images can interpret the information reflected in the images more intuitively and clearly. It further improves the feasibility of image segmentation for remote sensing analysis.

This chapter focuses on image segmentation for remote sensing analysis to solve change detection tasks. First, it begins with an introduction and review of recent work related to change detection. Then, two representative approaches for change detection tasks are introduced. Finally, it provides a discussion and summary on change detection based on image segmentation for remote sensing analysis.

10.1 Introduction

Change detection based on image segmentation is used to detect the changed information of the target area by analyzing the multi-temporal images acquired in different periods of the same geographical area. For example, landslides occur at various spatial and temporal scales in mountains, reshaping landscapes and altering local topography. Landslides can be caused by earthquakes, rainfalls, water-level change, storm waves, human activities, and so on. A sudden and rapid landslide event is often associated with fatalities, environmental degradation, damages to businesses, buildings, roads, public utilities, and so on. For instance, landslides cause damages of nearly US$1 billion in China and more than US$3 billion in Japan annually [6]. Since landslides seriously endanger people's property and safety, it is important to study automatic approaches of landslide identification and apply these approaches to mark the location and range of the landslide for mitigating potential harm, assisting post-disaster rescue, and rebuilding.

For a geological disaster such as landslides, landslide inventory mapping (LIM) focuses on outlining slide boundaries and neglecting the wealth of information revealed by internal deformation features. It is able to provide some significant information, e.g. the sizes, spatial distributions, number, and types of landslides for disaster relief strategy and hazard prevention. Thus, LIM is one of the most important features in landslide risk assessment. Researchers can obtain disaster information by analyzing the changed information of multi-temporal remote sensing images without entering the

Image Segmentation: Principles, Techniques, and Applications, First Edition. Tao Lei and Asoke K. Nandi.

disaster scene. More specifically, automatic approaches to landslide identification can quickly provide detailed landslide information including the landslide site, the range, and the degree of disaster without entering the disaster scene. Based on the landslide information obtained, rescue and reconstruction work can be carried out effectively. It can effectively reduce the degree of disasters and threats to personal safety, and improve work efficiency. As early techniques used for landslide identification rely on visual interpretation, they are highly labor-intensive and time-consuming [3]. Existing techniques always employ different change detection approaches that compute the difference between pre- and post-event images to achieve automatic landslide identification.

10.2 Related Work

So far, LIM relies on the visual interpretation of aerial photographs and intensive field surveys, which are highly labor-intensive and time-consuming for mapping of large areas. With the rapid progress of machine learning and remote sensing technologies, a large number of advanced approaches used for LIM have been proposed in recent years. Most of them depend on change detection that aims to detect the changed information of target areas by analyzing the multi-temporal images acquired at different times of the same geographical area. The popular ones can be roughly divided into four categories: approaches based on threshold segmentation [7–11], approaches based on the combination of feature extraction and clustering [12–20], approaches based on region segmentation [21–27], and approaches based on deep learning [28–31].

10.2.1 Threshold Segmentation Methods

Threshold segmentation-based change detection first computes the difference image of bitemporal images and then obtains changed regions using a threshold segmentation technique. Nelson [7] proposed a single-threshold approach used for change detection. Although the approach has a low computational complexity, it is sensitive to noise, and it provides a low accuracy for LIM. To improve the single-threshold approach, Jin [8] proposed a double-threshold segmentation approach and applied it to change detection. Although the double-threshold approach is superior to the single-threshold method for improving the accuracy of change detection, the choice of thresholds is still quite difficult. Moreover, the method has a low robustness for change detection. Based on the previous research, Lv et al. [6] first computes the difference image by employing multi-threshold segmentation and voting strategy. After that, a region-based image segmentation method is used for obtaining a superpixel result of the post-event image. Finally, the changed regions are obtained by computing the ratio of the number of pixels belonging to the foreground and background. The multi-threshold segmentation is able to provide better results for change detection, but it has a high computational complexity and requires more parameters. Clearly, change detection approaches based on threshold segmentation [9, 10] are simple and fast, but they generate changed regions only, depending on the grayscale value of images. They cannot obtain accurate detection results because the statistical features of image and spatial information are missing. Therefore, it is difficult to obtain accurately changed regions using threshold approaches.

10.2.2 Clustering Segmentation Methods

Compared to the first group of approaches used for change detection, the second utilizes the combination of feature transformation and clustering to achieve change detection. Dong et al. [11] proposed a change detection approach based on the combination of discrete wavelet transform (DWT) and fuzzy

c-means (FCM). First, DWT is used to obtain the frequency feature of bitemporal images. The changed regions are generated by computing the minus value and ratio value of images transformed by DWT. After that, the lower-half frequency components of the minus image and the higher-half frequency components of the ratio image are fused to enhance the contrast of the changed regions and the unchanged regions. The inverse discrete wavelet transform (IDWT) is then used to generate the final difference image. Finally, FCM is used to obtain changed regions. However, DWT only extracts image features with a single scale and a single phase. Gabor is more popular than DWT in image transformation since it can extract multi-scale and multi-phase image features. Li et al. [12] employ Gabor to extract the texture features of an image instead of DWT. Then, the difference image is obtained by computing the difference of bitemporal images processed by Gabor. Finally, a clustering method is used to extract the changed regions. Because the local spatial feature of images is exploited, the approach is superior to DWT-based approaches for change detection.

To obtain a better difference image, Zheng et al. [13] applied a logarithm operation to the ratio of bitemporal images. Moreover, the saliency map of a post-event image is used for guiding the detection of changed regions and k-means is used to detect the truly changed regions. Though the approaches based on the combination of feature transform and clustering [14–16] provide better results than threshold segmentation-based approaches for change detection, these approaches are only suitable for images that have a clear difference between high frequency and low frequency. In contrast, they are unsuitable for many remote sensing images because of the complex background, texture, and illumination of images [17]. They usually fail to detect changed regions that have a complex background and blurred edges. Besides, these methods are sensitive to noise because of fuzzy clustering algorithms [18, 19].

10.2.3 Region Segmentation Methods

The third group of approaches achieves change detection using region-based segmentation and the attributes of regions such as the ratio of length and width of objects, the shape of objects, and the slope of objects. The advantage of this kind of method is that it is insensitive to noise and is able to provide better contours. Zhang et al. [20] proposed a method of object-based change detection for high-resolution remote sensing images in urban areas. The method firstly employs region-based image segmentation to obtain the spatial attributes of objects. Then, the relationship between the spatial attributes of objects and characteristics of changed regions is explored to achieve change detection. However, the method proposed by Zhang et al. requires complex region descriptors to constrain changed regions. Ren et al. [21] compared pixel-based methods to region-based methods and demonstrate that the latter is able to provide better results since the region attribute is exploited and falsely changed regions are removed. To overcome the problem, Lv et al. [22] proposed an object-based expectation maximization (OBEM) post-processing method to obtain truly changed regions while removing falsely changed regions. The method uses multi-scale segmentation to obtain region attributes and uses an expectation–maximization (EM) algorithm to compute truly changed regions. Although change detection based on region segmentation has some advantages, the global information of images is missed, and manual feature descriptors are required [23]. More approaches used for change detection are reported in references [24–26].

10.2.4 Segmentation Methods Using Deep Learning

The last group of approaches utilize deep learning methods. Gong et al. [27] proposed a change detection approach using a deep neural network for synthetic aperture radar images. As the proposed network architecture included only a few hidden layers and it adopted full connection without using convolutional operation, the context of images is utilized inefficiently. Moreover, it is difficult to train the network due to the full connection. To obtain better results, Liu et al. [28] proposed a new deep convolutional coupling network that is fully unsupervised without using any labels. To apply convolutional neural networks (CNNs) [29] to landslide recognition, Ding et al. [30] employ a CNN and texture change detection to recognize landslides. Because CNN employs multiple pooling layers and a fully connectional layer to achieve classification tasks, the final result is coarse and has a low recognition accuracy.

10.3 Unsupervised Change Detection for Remote Sensing Images

Change detection approaches based on image segmentation are often used for landslide mapping (LM) from very high-resolution (VHR) remote sensing images. According to the spectral characteristics of landslides, the grayscale value of pixels belonging to landslides is large. Therefore, landslide regions can be obtained using image segmentation. It is well known that threshold segmentation [31] is the simplest and the fastest approach for image segmentation. However, it is also the least effective because it is difficult to obtain a good segmentation result. At present, a large number of intelligent segmentation algorithms, such as the active contour model [32], watershed [33], random walker [34], graph cut [35], spectral clustering [36], FCM [37], Gaussian mixed model [38], and fully convolutional network (FCN) [39], have been widely used in image segmentation. These algorithms can be categorized into four groups: partial differential equation (PDE)–based algorithms, region-based algorithms, graph-based algorithms, and pixel-based algorithms.

Different algorithms are suitable for different types of images. However, these approaches usually have two limitations. One is that they are sensitive to thresholds used for image segmentation and require too many parameters. The other one is that the computational complexity of these approaches depends on the image size, and thus they require a long execution time for VHR remote sensing images. Therefore, it is important to design a accurate yet efficient segmentation algorithm for VHR images.

Based on the above requirement, an unsupervised change detection using fast fuzzy c-means clustering (CDFFCM) for LM is presented in this chapter. This is based on the recently published paper by Lei et al. [40]. The presented CDFFCM has two contributions. The first is that it employ a Gaussian pyramid-based fast FCM clustering algorithm to obtain candidate landslide regions that have a better visual effect due to the utilization of image spatial information. The second is that it use the difference of image structure information instead of grayscale difference to obtain more accurate landslide regions. In experiments, three comparative approaches, edge-based level-set (ELSE), region-based level-set (RLSE), change detection-based Markov random field (CDMRF), and the presented CDFFCM are evaluated in three true landslide cases in the Lantau area of Hong Kong. The experimental results show that the presented CDFFCM is superior to three alternative approaches in terms of higher accuracy, fewer parameters, and shorter execution time.

10.3.1 Image Segmentation Using Image Structuring Information

Although a great number of change detection approaches [6–8, 11–13] have been proposed in the past decades, most of these approaches rely on the difference result obtained by bitemporal images and threshold segmentation. These approaches are thus complex and inefficient. Figure 10.1 shows the bitemporal images and their difference image.

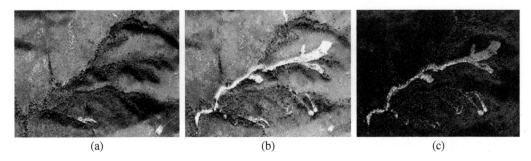

|(a)|(b)|(c)|

Figure 10.1 The bitemporal images and their difference image [40] / MDPI / CC BY 4.0: (a) Pre-event image at t_1. (b) Post-event image at t_2. (c) The difference image.

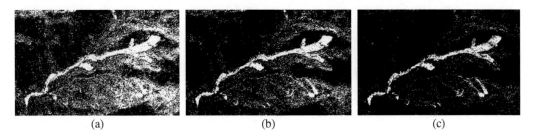

|(a)|(b)|(c)|

Figure 10.2 The difference images using single-threshold [40] / MDPI / CC BY 4.0: (a) The threshold is 0.2. (b) The threshold is 0.3. (c) The threshold is 0.4.

For Figure 10.1c, some popular approaches obtain changed regions using threshold segmentation. The segmentation result is sensitive to the threshold, as shown in Figure 10.2.

To address the issue, Lv et al. [6] employ multi-threshold and voting technologies to achieve change detection. Although the approach is superior to a single threshold, it requires more parameters. Actually, it is a multi-threshold image segmentation approach by introducing local spatial neighboring information. Moreover, this approach is also sensitive to parameter values. Therefore, it is unable to overcome the drawback of a single threshold for change detection. To address this problem, we present a clustering-based image segmentation approach used for change detection. Although the difference image of bitemporal images is complex, as shown in Figure 10.1c, the structuring information of the bitemporal image is similar. We try to compute the difference image using the structuring information of images instead of the grayscale value. We have known that image segmentation is able to address the problem because the segmentation result of an image includes rich spatial structuring information that is useful for change detection.

To show our motivation for change detection, Figure 10.3 shows change detection based on the difference in grayscale value. It is clear that the difference image is sensitive to the change in grayscale value. To overcome the drawback, Figure 10.4 shows the effect of employing image structuring information for change detection.

We can see that Figure 10.3d includes falsely changed regions because the difference image is sensitive to the change of grayscale values. Figure 10.4c shows a better result than Figure 10.3d due to the utilization of the structuring information for the images in Figure 10.4c. Consequently, we can achieve a novel clustering-based image segmentation approach that is able to obtain accurately changed regions by computing the structuring difference between bitemporal images.

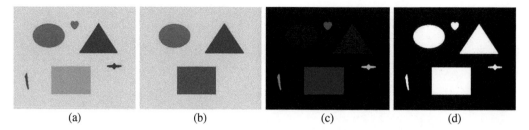

Figure 10.3 Change detection based on the difference of grayscale value [40] / MDPI / CC BY 4.0: (a) Image 1. (b) Image 2. (c) The difference image. (d) Threshold segmentation (the threshold is 0.4).

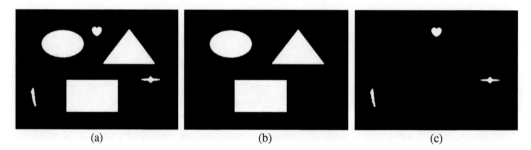

Figure 10.4 Change detection using the difference of image structuring information [40] / MDPI / CC BY 4.0: (a) The structuring information corresponds to Figure 10.3a. (b) The structuring information corresponds to Figure 10.3b. (c) The truly changed regions.

10.3.2 Image Segmentation Using Gaussian Pyramid

In this chapter, FCM is first employed to generate segmentation results of bitemporal images. Second, the segmentation results are used for computing the difference of image structuring to obtain changed regions. Because the computational complexity of FCM is influenced by the image size, it requires a long running time to perform FCM on a VHR image. Although some improved FCM algorithms, such as enhanced fuzzy *c*-means (EnFCM) [41] and fast robust fuzzy *c*-means (FRFCM) [42], are able to reduce the computational complexity of FCM, they are only suitable for grayscale image. Both EnFCM and FRFCM employ a histogram to replace pixels to remove the redundant information of images. However, it is difficult to extend EnFCM and FRFCM to multi-band remote sensing images. The Gaussian pyramid [43] is an excellent approach for addressing the problem because it is able to remove the redundant information while preserving structuring information for multi-band images. We use the Gaussian pyramid to obtain multi-resolution images that have similar structuring information, and then implement FCM on a low-resolution image to speed up the algorithm. After that, we use the obtained clustering centers to compute the final membership on the original VHR image. Figure 10.5 shows the multi-resolution images obtained by the Gaussian pyramid.

Because FCM algorithms aim to obtain the clustering centers and classification labels, we can compute the clustering centers of a low-resolution image instead of the original high-resolution image from the Gaussian pyramid. Since there is a high data redundancy for images, we can obtain similar clustering centers for multi-resolution images obtained by the Gaussian pyramid. Table 10.1 shows the clustering centers obtained by FCM for the image shown in Figure 10.5.

Figure 10.5 Multi-resolution images obtained by Gaussian pyramid [40] / MDPI / CC BY 4.0: (a) The original image of size 923 × 593. (b) The low-resolution image of size 461 × 296. (c) The low-resolution image of size 230 × 148. (d) The low-resolution image of size 115 × 74.

Table 10.1 Clustering centers of multi-resolution images as in Figure 10.5 using FCM (c = 5) [40] / MDPI / CC BY 4.0.

Clustering center	(a)	(b)	(c)	(d)
Center 1	57.4, 43.6, 26.5	64.8, 54.2, 39.3	72.8, 65.3, 52.2	81.3, 77.3, 65.3
Center 2	113.5, 110.2, 87.6	116.4, 113.7, 91.3	118.4, 116.3, 93.6	120.9, 118.8, 95.3
Center 3	149.9, 147.8, 111.4	151.2, 149.2, 112.2	152.4, 150.4, 112.7	154.4, 152.4, 113.9
Center 4	173.7, 179.5, 130.6	172.9, 178.7, 128.9	172.1, 178.0, 127.4	154.4, 152.4, 113.9
Center 5	245.8, 246.3, 240.5	243.0, 242.1, 235.1	239.9, 237.8, 229.5	235.6, 232.3, 222.2

Furthermore, we can obtain the segmentation results of original bitemporal images using clustering centers shown in Table 10.1 Figure 10.6 shows the corresponding segmentation results, where five colors represent five classes.

In Table 10.1, we can see that the clustering centers are similar for four different-resolution images shown in Figure 10.5. Therefore, a fast-clustering algorithm can be achieved by using Gaussian pyramid for VHR images.

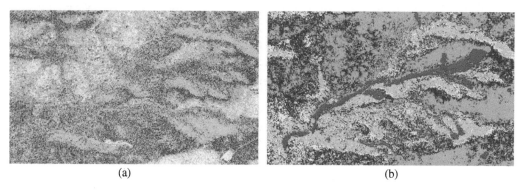

Figure 10.6 Segmentation results of pre- and post-event images shown in Figure 10.5 using FCM (c = 5) [40] / MDPI / CC BY 4.0: (a) Pre-event image. (b) Post-event image.

10.3.3 Fast Fuzzy C-Means for Change Detection

In this chapter, we employ a pixel-based clustering algorithm to achieve image segmentation. Because FCM is one of the most popular clustering algorithms, we applied FCM to bitemporal images to obtain changed regions; the objective function of FCM algorithms is defined as follows:

$$J_m = \sum_{i=1}^{N_f} \sum_{k=1}^{c} u_{ki}^m \left\| x_i - v_k^p \right\|^2, \tag{10.1}$$

where x_i is the gray value of the i-th pixel, v_k^p represents the prototype value of the k-th cluster, and u_{ki} denotes the fuzzy membership value of the i-th pixel with respect to cluster k. N_f is the total number of pixels in an image f, and c is the number of clusters. The parameter m is a weighting exponent on each fuzzy membership that determines the amount of fuzziness of the resulting classification.

According to the definition of J_m and the constraint that $\sum_{k=1}^{c} u_{ki} = 1$ for each pixel x_i, and using the Lagrange multiplier method, the calculation of the membership partition matrix and the clustering centers is given as follows:

$$u_{ki} = \frac{x_i - v_k^{-2/(m-1)}}{\sum_{j=1}^{c} x_i - v_j^{-2/(m-1)}}, \tag{10.2}$$

$$v_k^p = \sum_{i=1}^{N_f} u_{ki}^m x_i \Big/ \sum_{i=1}^{N_f} u_{ki}^m. \tag{10.3}$$

It can be seen from Eqs. (10.1)–(10.3) that FCM has two drawbacks for image segmentation. One is that FCM is sensitive to noise due to the missing spatial information of an image. The other one is that the computational complexity of FCM is influenced by the number of pixels in the image. To address these drawbacks, the local spatial information is often incorporated into the objective function of FCM [28, 44, 45]. Although the introduction of local spatial information is able to reduce the influence of noise for image segmentation, it also leads to a high computational complexity, especially for high-resolution images. The objective function of the improved FCM algorithms is defined as:

$$J_m = \sum_{i=1}^{N_f} \sum_{k=1}^{c} u_{ki}^m \left\| x_i - v_k^p \right\|^2 + \sum_{i=1}^{N_f} \sum_{k=1}^{c} u_{ki}^m G_{ki}, \tag{10.4}$$

where G_{ki} represents a fuzzy factor; it is used to control the influence of neighborhood pixels on the central pixel. Similar to Eqs. (10.2) and (10.3), the membership and clustering centers of improved FCM algorithms are obtained as follows:

$$u_{ki} = \frac{\left(\left\| x_i - v_k \right\|^{-2} + G_{ki} \right)^{-1/(m-1)}}{\sum_{j=1}^{c} \left(\left\| x_i - v_j \right\|^{-2} + G_{ji} \right)^{-1/(m-1)}}, \tag{10.5}$$

$$v_k^p = \sum_{i=1}^{N_f} u_{ki}^m x_i \Big/ \sum_{i=1}^{N_f} u_{ki}^m. \tag{10.6}$$

The improved fuzzy clustering algorithm clearly has a high computational complexity for VHR images. According to Eqs. (10.1) and (10.4), the computational complexity of the FCM algorithm is influenced by the values of N_f and c. The execution time of the FCM algorithm will be longer when the image size is larger or c is larger. Thus, we can reduce the computational complexity of FCM by

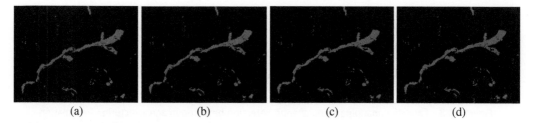

| (a) | (b) | (c) | (d) |

Figure 10.7 Candidate landslides of Figure 10.1b using four groups of clustering centers in Table 10.1: (a) Candidate landslides using clustering centers shown in Table 10.1a [40] / MDPI / CC BY 4.0. (b) Candidate landslides using clustering centers shown in Table 10.1b. (c) Candidate landslides using clustering centers shown in Table 10.1c. (d) Candidate landslides using clustering centers shown in Table 10.1d.

Table 10.2 Execution time (in seconds) of FCM algorithm on multi-resolution images shown in Figure 10.5 (c = 5) [40] / MDPI / CC BY 4.0.

	(a)	(b)	(c)	(d)
Iterations	100	88	66	62
Time	69.84	17.48	3.28	0.98

removing the redundant information of images while preserving the structuring information of images. Inspired by the FRFCM and the Gaussian pyramid, the Gaussian pyramid is first used to generate multi-resolution images, and FCM is then performed on a low-resolution image to achieve fast image segmentation. After that, membership filtering is performed on labels. As a result, the candidate landslide regions are obtained. Figure 10.7 shows candidate landslides for the image shown in Figure 10.1b using the fast FCM algorithm. Table 10.2 shows the comparison of execution time and iteration times for multi-resolution images shown in Figure 10.5.

The results shown in Figure 10.7 are similar because of the similar clustering center values. The regions marked by red color correspond to clustering center 5, i.e. candidate landslides. Table 10.2 shows that a low-resolution image requires shorter execution time to compute clustering centers. Figure 10.7 and Table 10.2 show that the fast FCM algorithm obtains accurate segmentation results while reducing the execution time of the FCM algorithm. In practical applications, the selection of a low-resolution image provided by the Gaussian pyramid is decided by the image resolution of bitemporal images.

10.3.4 Postprocessing for Change Detection

Although changed regions can be obtained by computing the difference results of segmented bitemporal images, there are still a lot of falsely changed regions caused by noise and illumination variation. Figure 10.8 shows candidate landslides of the bitemporal images where the landslide regions have a high brightness in Figure 10.8c. Therefore, it was easy to obtain candidate landslides from the segmentation result according to clustering centers. The candidate landslides are shown in Figure 10.8d. However, many false landslide regions were also detected. It was necessary to remove false landslide regions by using change detection based on image structuring information that is defined as:

$$L^d_{c^c c^r} = L^{post}_{c^c c^r} - L^{pre}_{c^c c^r}, \tag{10.7}$$

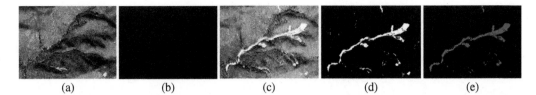

(a) (b) (c) (d) (e)

Figure 10.8 Landslide mapping using change detection based on image structuring information (T_1 = 0.8, the grayscale value of image is normalized from 0 to 1) [40] / MDPI / CC BY 4.0: (a) Pre-event image. (b) Candidate false landslide regions L^{pre}. (c) Post-event image. (d) Candidate landslide regions L^{post}. (e) Detected landslide regions according to L^{pre} and L^{post}, where regions marked by red color represent true landslide regions and regions marked by cyan color represent removed false landslide regions.

$$L_{c^c c^r}^{db} = \begin{cases} 1, & L_{c^c c^r}^{d} > 0 \\ 0, & L_{c^c c^r}^{d} \leq 0 \end{cases}, \tag{10.8}$$

where c^c and c^r are the corresponding column and row, respectively, in the bitemporal images and difference image. L^{pre} and L^{post} represent candidate landslide regions of pre-event and post-event images, respectively. L^d and L^{db} denote difference image and binarized difference image, respectively.

We apply Eqs. (10.7) and (10.8) to two groups of bitemporal images to detect changed regions. Figure 10.8b shows all the false landslide regions that were removed completely because these regions were too small. Therefore, we could remove false landslide regions in Figure 10.8d according to Eqs. (10.7) and (10.8). On the contrary, Figure 10.9 shows false landslide regions could be removed by using Eqs. (10.7)–(10.8).

By comparing Figures 10.8–10.9, we find that there were two cases for the removal of false landslide regions using Eqs. (10.7) and (10.8). In the first case, we could remove false landslide regions

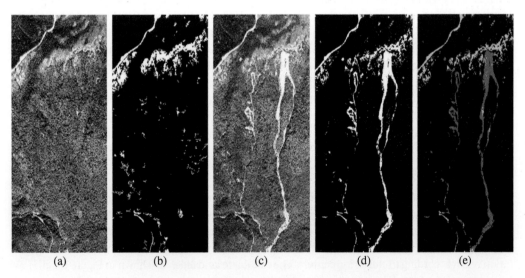

(a) (b) (c) (d) (e)

Figure 10.9 Landslide mapping using change detection based on image structuring information (T_1 = 0.8, the grayscale value of image is normalized from 0 to 1) [40] / MDPI / CC BY 4.0: (a) Pre-event image. (b) Candidate false landslide regions L^{pre}. (c) Post-event image. (d) Candidate landslide regions L^{post}. (e) Detected landslide regions according to L^{pre} and L^{post}, where regions marked by red color represent true landslide regions and regions marked by cyan color represent removed false landslide regions.

when the maximal value of clustering centers of the pre-event images was larger than T1, e.g. Figure 10.9. It was the opposite in the second case, where we could not remove false landslide regions, i.e. the maximal value of clustering centers of pre-event image was smaller than T1, e.g. Figure 10.8. To address the problem of the removal of false landslide regions, we employ a multivariate morphological reconstruction (MMR) [46] to process the pre-event image and obtain false landslide regions using the threshold T1 for the second case. The reconstruction and segmentation results of the pre-event image are shown in Figure 10.10.

Figure 10.10 shows that the presented approach overcame the problem existing in the second case, i.e. false landslide regions were removed from the candidate landslide regions. Based on the analysis above, the true landslide regions were obtained easily by Eqs. (10.7) and (10.8). Figure 10.10e and 10.11c show that true landslide regions were detected and false landslide regions were removed by the presented approach. Therefore, it was easier to achieve LM using the difference of the image structuring information instead of the difference of the grayscale value. However, there were still a lot of false landslide regions. We could employ morphological filters and morphological reconstruction (MR) operations to remove these false landslide regions. Because the detected landslide regions included many holes in large regions and many isolated points, we needed to fill holes while removing isolated points. We designed an effective morphological filter that was able to remove isolated points while filling holes.

Let S denote a set whose elements are eight-connected boundaries, each boundary enclosing a background region (i.e. a hole). Based on the basic morphological dilation and complementation operations, the operation of filling holes can be defined as follows:

$$X_m = \delta(X_{m-1}, E) \cap S^c, \tag{10.9}$$

where $m \in N^+$, X_0. is an initial array (the same size as the array containing S), E is a structuring element that is a disk of size 3×3, and δ denotes the morphological dilation operation. All the holes will be filled until $X_m = X_{m-1}$. The final output denoted by S' is defined as:

$$S' = X_m \cup S. \tag{10.10}$$

If we use the symbol F to denote the operation of filling holes, we propose an effective morphological filtering method that is suitable for smoothing landslide regions,

$$S'' = \varepsilon_m(F(\delta(S, E_c)), E_c), \tag{10.11}$$

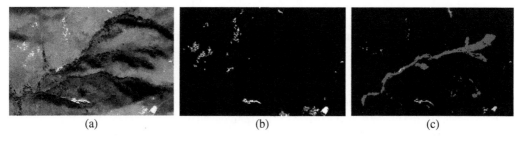

(a)   (b)   (c)

Figure 10.10 Reconstruction and segmentation results of the pre-event image [40] / MDPI / CC BY 4.0: (a) Reconstruction result using MMR (the structuring element is a disk of size 5 × 5). (b) Candidate false landslide regions L^{pre} using the threshold method, $T_1 = 0.8$. (c) Detected landslide regions according to the difference between Figures 10.8d and 10.10b, where regions marked by red color represent true landslide regions and regions marked by cyan color represent removed false landslide regions.

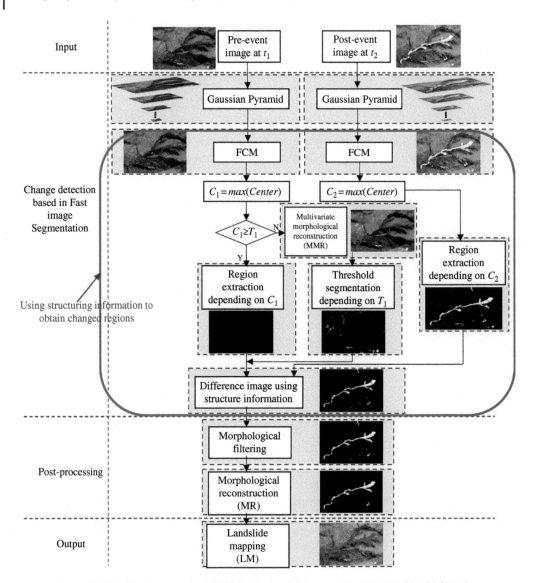

Figure 10.11 The framework of CDFFCM for landslide mapping [40] / MDPI / CC BY 4.0.

where E_c is a structuring element that is used for connecting leaking regions. Generally, the size of E_c increases with the image size. Let r be the radius of E_c, then

$$r = round(\ min\ (W_{bi}, H_{bi})/500) + 1, \qquad (10.12)$$

where W_{bi} and H_{bi} denote the width and height of bitemporal images, respectively. Because the size of E_c is decided by the image size, the presented morphological filter is parameter free. It can be seen from Eq. (10.11) that the presented filtering method is able to fill holes due to $F1$-*score* and the closing operation. However, the method was unable to remove small regions caused by noise or false landslides. Because MR [47] was able to remove small regions while maintaining the large regions in an image, we could employ MR to remove false landslide regions. The reconstructed result of the image was often decided by the size of the structuring element. Generally, larger structuring elements remove smaller regions during morphological dilation reconstruction. The basic morphological erosion and dilation reconstructions are defined as:

$$
\begin{cases}
R^{\varepsilon_m}(f, E_c) = \varepsilon_{m_f}^{(n)}(f_m) \vee f, & f_m = \delta(f, E_c) \\
R^{\delta}(f, E_c) = \delta_f^{(n)}(f_m) \wedge f, & f_m = \varepsilon_m(f, E_c)
\end{cases},
\tag{10.13}
$$

where f_m is a marker image that $f_m \geq f$ for R^{ε_m} and $f_m \leq f$ for R^{δ}. B is the structuring element used in both Eqs. (10.11) and (10.13), where ε_m denotes a morphological erosion operation. R^{ε_m} and R^{δ} denote morphological erosion and dilation reconstruction, respectively. Based on R^{ε_m} and R^{δ}, the combinational MR operations that are more popular in practical applications, i.e. morphological opening and closing reconstructions denoted by R^{O} and R^{C}, are presented as follows:

$$
\begin{cases}
R^{O}(f, E_c) = R^{\delta}(R^{\varepsilon_m}(f, E_c), E_c) \\
R^{C}(f, E_c) = R^{\varepsilon_m}(R^{\delta}(f, E_c), E_c)
\end{cases}.
\tag{10.14}
$$

It can be seen from Eq. (10.13) that morphological dilation reconstruction is able to remove objects that are smaller than E_c while morphological erosion reconstruction has the opposite effect due to duality. Therefore, we employ R^{C}, shown in Eq. (10.14), to optimize the detected landslides S''.

10.3.5 The Proposed Methodology

The framework of the presented CDFFCM used for LM is shown in Figure 10.11. It consists of the following steps:

1) Input bitemporal images and set parameters. T_1 is a threshold used for comparing the values of clustering centers; also it is used for image binarization. c is the number of clusters used for FCM.
2) Fast image segmentation. A VHR image included a large number of pixels, which meant that it was time-consuming to achieve image segmentation for the VHR image. We employed the Gaussian pyramid to obtain low-resolution images that were applied to FCM to reduce the computational burden for the VHR images.
3) Change detection using difference of image structuring information: Because the pre-event and post-event images were segmented by FCM, we could compute the difference image according to the segmented images without using original bitemporal images.
   - The structuring information of post-event image was extracted, i.e. the candidate landslide regions were extracted from the post-event image according to the maximal module value of its clustering centers.
   - The structuring information of pre-event image was extracted, i.e. the false landslide regions were extracted from a pre-event image according to the maximal module value of its clustering centers or the threshold T_1.
   - The difference of structuring information was obtained, i.e. the changed regions were extracted by computing the difference between the candidate landslide regions and the false landslide regions.
4) Post-processing: Because the difference image was coarse and included lots of false landslide regions, morphological filtering and filling technologies were applied to the modification of LM. Furthermore, we employed an MR to remove false landslide regions to improve the accuracy of LIM.
5) Output result.

10.3.6 Experiments

In order to evaluate the presented approach, three groups of bitemporal VHR landslide remote sensing images were used in our experiments. Three popular approaches [9, 10], i.e. ELSE, RLSE [10], and CDMRF [9], were considered as alternative approaches to demonstrate the effectiveness and efficiency of CDFFCM for LM. All approaches were implemented with MATLAB 2017b (http://www.mathworks.com) and performed on a DELL desktop with Intel(R) Core (TM) CPU, i7-6700, 3.4 GHz, and 16 GB RAM (Dell, Shanghai, China).

10.3.6.1 Data Description

Three groups of bitemporal images on area A, area B, and area C were captured by the Zeiss RMK TOP 15 Aerial Survey Camera System (Jena, Germany) at a flying height of approximately 2.4 km in December 2007 and in November 2014. The locations of area A, area B, and area C are shown in Figure 10.12. The geometrical resolution of the bitemporal images is 0.5 m. All images used in the experiments were preprocessed via geometry registration and radiometric correction.

- Area A: Figure 10.13a, b show the pre- and post-event images of area A. The size of bitemporal images was 1252 × 2199. The true boundaries of landslides in images were generated manually by human experts in the field of remote sensing, and they are shown in Figure 10.13c. Figure 10.13d shows the corresponding ground truth segmentation.
- Area B: Figure 10.14a, b show the pre- and post-event images of area B. The size of the images was 923 × 593. The ground truths, i.e. the boundaries and segmentation results of landslides in the post-event image, are shown in Figure 10.14c, d, respectively.

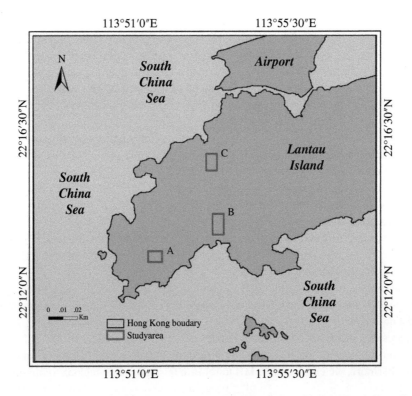

Figure 10.12 Study area locations of area A, area B and area C on Lantau Island, Hong Kong, China [40] / MDPI / CC BY 4.0.

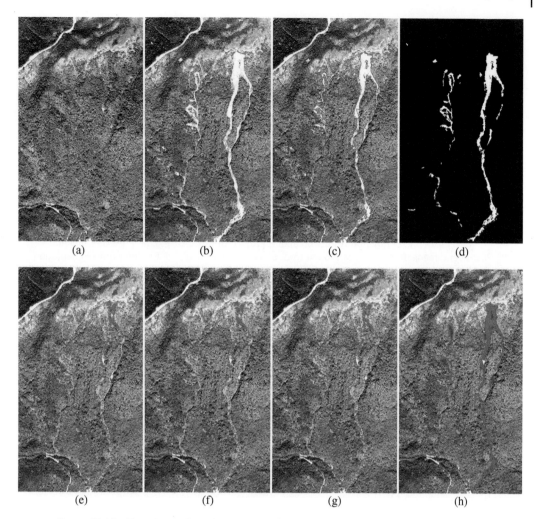

Figure 10.13 Experimental results on area A [40] / MDPI / CC BY 4.0. (a) Pre-event image. (b) Post-event image. (c) Contour of ground truth on post-event image. (d) Ground truth segmentation. (e)–(h) Landslide mapping results using ELSE, RLSE, CDMRF, and CDFFCM overlaid on the post-event image, respectively.

- Area C: Figure 10.15a, b show the pre- and post-event images of area C. The size of the images was 750×950. The ground truths, i.e. the boundaries and segmentation results of landslides in the post-event image, are shown in Figure 10.15c, d, respectively.

10.3.6.2 Experimental Setup

To fairly compare the performance of different approaches, the parameter values used for the experiments follow the original papers. The parameter values $\alpha = 1.5$, $C_0 = 1.0$, and $\sigma_1 = \sigma_2 = 1.0$ were used for ELSE and RLSE, where α is a constant coefficient to reduce the effect of data dispersion on the experimental results by increasing the standard deviation term, and σ is the standard deviation of the Gaussian filter used for preprocessing. The template size of the Gaussian filter was 9×9, and the time step was $\Delta T = 5.0$. The parameter values used for CDMRF were as

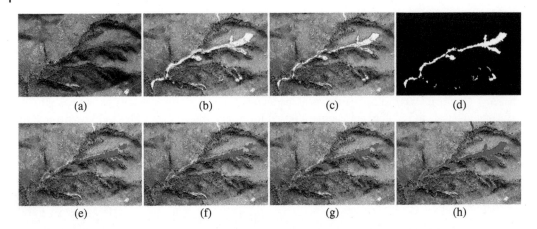

(a)   (b)   (c)   (d)

(e)   (f)   (g)   (h)

Figure 10.14 Experimental results on area B [40] / MDPI / CC BY 4.0. (a) Pre-event image. (b) Post-event image. (c) Contour of ground truth on post-event image. (d) Ground truth segmentation. (e)–(h) Landslide mapping results using ELSE, RLSE, CDMRF, and CDFFCM overlaid on the post-event image, respectively.

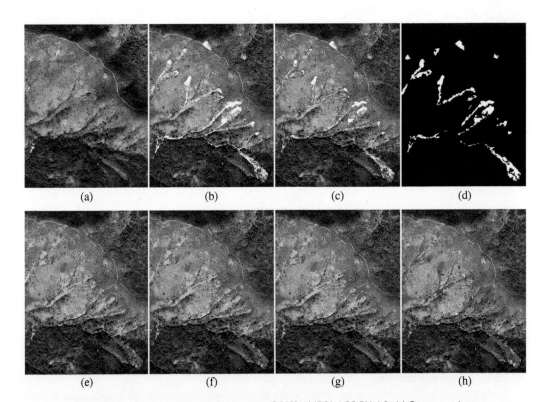

(a)   (b)   (c)   (d)

(e)   (f)   (g)   (h)

Figure 10.15 Experimental results on area C [40] / MDPI / CC BY 4.0. (a) Pre-event image. (b) Post-event image. (c) Contour of ground truth on post-event image. (d) Ground truth segmentation. (e)–(h) Landslide mapping results using ELSE, RLSE, CDMRF, and CDFFCM overlaid on the post-event image, respectively.

follows: $T = 1.0$, $\Delta T = 1.5$, $c_b^w = 50$, and $M = 5$. The value of $(T + \Delta T)$ was consistent with the effect of α, i.e. it aimed to reduce the effect of data dispersion on the experimental results. c_b^w is a weighting coefficient that was used to balance the pairwise potential. M is the number of clusters. To our knowledge, the values of the parameters were the best for the alternative approaches. For the presented CDFFCM, only two parameters, T_1 and c were required, and other parameters could be computed automatically according to the size of the bitemporal images. We set $T_1 = 0.8$ and $c = 5$ in this experiment. Both the structuring elements used for image filtering and MR and the multiple of downsampling used for a Gaussian pyramid were decided by the size of the bitemporal images.

In order to quantitatively compare CDMRF, ELSE, and RLSE with the presented CDFFCM, four quantitative evaluation indices [10] are presented:

$$Precision = P_{lm}/P_l, \tag{10.15}$$

$$Recall = P_{lm}/P_r, \tag{10.16}$$

$$F1\text{-}score = 2 \times \frac{Precision \times Recall}{Precision + Recall}, \tag{10.17}$$

$$Accuracy = P_{lm}/(P_l + P_{rum}), \tag{10.18}$$

where P_{lm} is the total pixel number of the identified landslides that are matched with the corresponding ground truth, P_r is the total pixel number of the ground truth, P_l is the total pixel number of the identified landslides, and P_{rum} is the total pixel number of the corresponding ground truth that is not matched with the identified landslides. P_{over} is the total pixel number of detected false landslides.

10.3.6.3 Experimental Results

1) **Area A**

Figure 10.13a, b shows the pre-event and post-event images for area A, respectively. The landslide regions are clear in Figure 10.13b. Although we could recognize landslide regions visually according to the difference between Figure 10.13a, b, it was difficult to achieve the computer recognition due to a complex background and noise. There was also a clear difference between Figure 10.13a, b for background regions due to the influence of the illumination. Figure 10.13c shows the contour of true landslide regions that was delineated by experts in the field of remote sensing, and Figure 10.13d shows the binary image of landslide regions corresponding to Figure 10.13c. Figure 10.13e–h shows the identification results of landslide regions generated by ELSE, RLSE, CDMRF, and the presented CDFFCM, respectively. Compared to the first three results, the last one provided a better visual effect for LM. Moreover, the last one was closer to the ground truth than the other results. The first three results included a great number of breakage regions due to missed detections. Because the bitemporal images were obtained in December 2007 and in November 2014, respectively, the landslide regions had changed during the seven years due to environmental change. Figure 10.13b shows that some landslide regions were covered by trees, grass, or soil. Besides, two images taken at different times on the same unchanged area were also different due to the influence of the imaging system, environment, and noise. Therefore, the difference result of bitemporal images usually included a large number of falsely changed regions caused by illumination or noise. Meanwhile, the difference image usually missed a lot of truly changed regions as well due to fresh trees or grass on landslide regions.

As traditional change detection approaches, such as ELSE, RLSE, and CDMRF, rely on the difference result of bitemporal images, the final identification results had a clear difference from the ground truth, as shown in Figure 10.13e–g. Compared to ELSE, both RLSE and CDMRF provided a better visual effect because of two reasons. One is that RLSE employs region segmentation instead of the gradient segmentation employed by ELSE. The other reason is that CDMRF had a stronger capability of noise suppression due to the introduction of spatial context information of landslides. Figure 10.13h shows the result obtained by the presented CDFFCM that overcame the shortcomings existing in ELSE, RLSE, and CDMRF. Because the presented CDFFCM employed fuzzy clustering to obtain the spatial structuring information of bitemporal images, the difference result relied on the structuring change other than the change of grayscale values. The presented approach was therefore insensitive to noise. It was not only able to exclude falsely changed regions but was also able to supply the missed landslide regions that were truly changed. In Figure 10.13e–h, the regions marked by red color were true-detected landslide regions, the regions marked by blue were false-detected landslide regions, and the regions marked by cyan were true-missed landslide regions. According to Figure 10.13h, the presented CDFFCM provided the best result for LM because the area of red regions was the largest while the sum of blue and cyan regions was small.

2) **Area B**

Figure 10.14a, b shows the pre-event and post-event image for area B, respectively. The landslide regions shown in Figure 10.14b were clearer than those in A-area. There was a very clear difference between Figure 10.14a and 10.14b due to environmental change. Figure 10.14c shows the contour of true landslide regions, and Figure 10.14d shows the landslide regions corresponding to Figure 10.14c. We can see that there was less bare rock or soil in Figure 10.14a compared to Figure 10.14b. Figure 10.14e–h shows the comparison results generated by ELSE, RLSE, CDMRF, and the presented CDFFCM for LM, where the result generated by the presented CDFFCM included more red regions and less cyan regions in Figure 10.14h. It provided a better visual effect than the first three results for LM. Moreover, it was closer to the ground truth than the three previous results. The four approaches used for LM generated better results for Figure 10.14 than Figure 10.13 since area B included less noise.

Although A-area is different from B-area in image size and texture features, the presented CDFFCM achieved the best identification results in four approaches for both area A and area B. Furthermore, the presented CDFFCM had a stronger robustness for LM.

3) **Area C**

Compared to area A and area B, it was more difficult to obtain accurate landslide regions for area C due to the complexity of landslides. Figure 10.15a, b shows the pre-event and post-event images for area C, respectively. We can see that some landslide regions in Figure 10.15b were covered by the fresh vegetation of the surroundings. Therefore, it was more difficult and took more time to delineate the true landslide regions in area C than for area A and area B. Figure 10.15c shows the contour of true landslide regions of the post-event, and Figure 10.15d shows the landslide regions corresponding to Figure 10.15c. Figure 10.15e–h shows the identification results of landslide regions generated by ELSE, RLSE, CDMRF, and the presented CDFFCM. The presented CDFFCM provided the best result since Figure 10.15h was closer to the ground truth shown in Figure 10.15d than Figure 10.15e–g.

As can be seen from Figure 10.15e–h, there were a lot of cyan regions in Figure 10.15e–g, which indicated a great number of true landslide regions were missed by ELSE, RLSE, and CDMRF. The presented CDFFCM provided better visual effects than three other approaches because Figure 10.15h included more red regions (i.e. true landslide regions) and fewer cyan regions (i.e. missing true landslide regions). However, Figure 10.15h included more blue regions (i.e. detected false landslide regions) as well.

10.3.6.4 Experimental Analysis

1) **Evaluation Index Analysis**

In practical applications, *Precision* and *Recall* are consistent for evaluating algorithm performance, i.e. a high *Precision* corresponds to a high *Recall* when P_r is close to P_l. However, *Precision* and *Recall* are inconsistent in some cases, i.e. a high *Precision* corresponds to a low *Recall* or vice versa when P_r is far smaller or larger than P_l.

- Example 1 *Precision* and *Recall* are consistent for performance evaluation of algorithms. Figure 10.16 and Table 10.3 illustrate Example 1.
- Example 2 *Precision* and *Recall* are dissimilar for performance evaluation of algorithms, i.e. a high *Precision* corresponds to a high *Recall*. However, *F1-score* and *Accuracy* are consistent, larger or smaller.

Generally, P_r is a constant while P_l is usually a variable. *Precision* and *Recall* are inconsistent when P_l decreases as shown in Figure 10.17 and Table 10.4.

Table 10.4 shows that *F1-score* and *Accuracy* become larger or smaller consistently but *Precision* and *Recall* were inconsistent. It is thus *F1-score* and *Accuracy* that were usually used for evaluating the performance of an approach for LM.

2) **Results Analysis**

To further verify the effectiveness of the presented method, the detected landslides were compared with the ground truths by using two popular indices, i.e. *F1-score* and *Accuracy*. The final quantitative evaluation results are presented in Table 10.5 and illustrated by a corresponding bar chart shown in Figure 10.18 for the four approaches, ELSE, RLSE, CDMRF, and the presented CDFFCM.

For area A, area B and area C, ELSE, RLSE, and CDMRF provide higher *Precision* but lower *Recall* than CDFFCM because a larger number of true landslide regions were missed

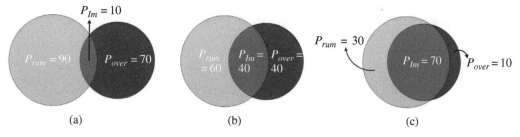

(a)	(b)	(c)

Figure 10.16 Example 1 [40] / MDPI / CC BY 4.0 for *Precision* and *Recall*, where P_r = 100, P_l = 80: (a) Test 1. (b) Test 2. (c) Test 3.

Table 10.3 The variation of *Precision* and *Recall* via increasing P_{lm} [40] / MDPI / CC BY 4.0.

Indices	Test 1	Test 2	Test 3	Trend
Precision	0.125	0.500	0.875	↑
Recall	0.100	0.400	0.700	↑
F1 − score	0.111	0.444	0.778	↑
Accuracy	0.060	0.286	0.636	↑

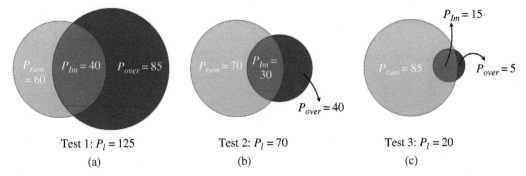

Test 1: $P_l = 125$ Test 2: $P_l = 70$ Test 3: $P_l = 20$

(a)   (b)   (c)

Figure 10.17 Example 2 [48] / IEEE / CC BY 4.0 for *Precision* and *Recall*, where $P_r = 100$ and P_l is variant. (a) Test 1, $P_l = 125$. (b) Test 2, $P_l = 70$. (c) Test 3, $P_l = 20$.

Table 10.4 The variation of *Precision* and *Recall* via decreasing P_l [40] / MDPI / CC BY 4.0.

Indices	Test 1	Test 2	Test 3	Trend
Precision	0.320	0.429	0.750	↑
Recall	0.400	0.300	0.150	↓
F1 − score	0.356	0.353	0.250	↓
Accuracy	0.216	0.214	0.143	↓

Table 10.5 Quantitative evaluation and comparison of different approaches for landslide mapping [40] / MDPI / CC BY 4.0.

Study areas	Methods	Evaluation indices (%)			
		Precision	*Recall*	*F1-score*	*Accurary*
A	ELSE	90.33	51.72	65.78	49.01
	RLSE	89.57	48.78	63.16	46.16
	CDMRF	85.90	53.69	66.08	49.34
	CDFFCM	79.42	93.56	85.91	75.30
B	ELSE	90.69	75.93	82.66	70.44
	RLSE	90.64	72.85	80.78	67.75
	CDMRF	88.44	77.13	82.40	70.07
	CDFFCM	87.86	90.26	89.04	80.25
C	ELSE	89.86	56.75	69.57	53.33
	RLSE	90.17	53.59	67.22	50.63
	CDMRF	86.25	60.10	70.84	54.84
	CDFFCM	76.18	81.66	78.82	65.05

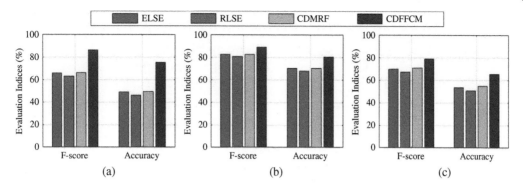

Figure 10.18 Bar-chart corresponding to the values of Table 10.5 [40] / MDPI / CC BY 4.0: (a) area A. (b) area B. (c) area C.

by the first three approaches, i.e. the first three approaches obtained a small value of P_l while CDFFCM obtained a large value of P_l. However, the presented CDFFCM obtained a significantly higher *F1-score* and *Accuracy* than the three other approaches, as shown in Table 10.5, which demonstrates that CDFFCM was superior to the three alternative approaches on three criteria.

The four approaches obtain higher *Precision* and *Recall* for area B than area A and area C because the landslide regions were covered by more grass and fewer trees in area B than in area A and area C. Trees led to the missing of more true landslide regions than grass. Consequently, there were less blue regions in Figure 10.14h than Figures 10.13h and 10.15h, and area B obtained higher values with *F1-score* and *Accuracy* than area A and area C.

To further demonstrate that the presented CDFFCM had a high computational efficiency for LM, Table 10.6 shows the comparison of the execution time of these different approaches on area A, area B and area C. The four approaches were implemented repeatedly more than 30 times, and the final average execution time is shown in Table 10.6. The presented approach required a shorter execution time than the three comparative approaches. The running time of ELSE, RLSE, and CDMRF relied on the image size. For example, ELSE required 80.12 seconds for area A but 29.32 seconds for area B since the size of area A was 1252×2199, but it was 923×593 for area B. The computational complexity of the presented CDFFCM did not depend on the image size because the Gaussian pyramid method is able to obtain difference resolution images. The fuzzy clustering algorithm was only performed on a low-resolution image to obtain clustering centers. Therefore, the presented CDFFCM had the advantage of a faster execution time.

In Table 10.6, area A required a longer execution time than area B and area C for all four approaches because the size of area A was larger than area B and area C. The presented

Table 10.6 Comparison of execution time of four different approaches (in seconds) [40] / MDPI / CC BY 4.0.

	Methods			
Study areas	**ELSE**	**RLSE**	**CDMRF**	**CDFFCM**
A	80.12	57.21	80.10	**16.92**
B	29.32	39.41	29.31	**11.20**
C	51.98	48.03	43.52	**13.57**

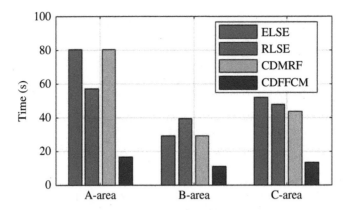

Figure 10.19 Bar-chart corresponding to the values in Table 10.6 [48] / IEEE / CC BY 4.0.

CDFFCM required a shorter execution time than the three comparative approaches because it employed the fast FCM algorithm to achieve image segmentation. Therefore, CDFFCM was clearly superior to the three alternative approaches due to the high computational efficiency, as shown in Figure 10.19.

3) **Parameter Analysis**

In the paper [40], the presented CDFFCM was evaluated for different landslide images under appropriate parameter settings. The CDFFCM includes two required parameters, where T_1 was a threshold used for comparing with the values of clustering centers, and it was used for image binarization as well. c was the number of clusters used for FCM. The structuring element denoted by E_c was used for both image filtering and MR, where r is the radius of structuring element and it could be decided by the image size. The multiple of downsampling denoted by p_m^d is used for Gaussian pyramid. For the presented CDFFCM, T_1 and c are required, but r and p_m^d are not required because the latter two parameters can be computed automatically according to the size of bitemporal images, i.e.

$$p_m^d = round \left(min \left(W_{bi}, H_{bi} \right) / 200 \right). \tag{10.19}$$

The parameter r could be computed according to Eq. (10.12). Therefore, the presented CDFFCM only requires setting the values of T_1 and c.

To test the influence of parameters on the performance of CDFFCM, four quantitative evaluation indices, *Precision*, *Recall*, *F1-score*, and *Accuracy* were analyzed. Taking area B as the research object and setting $c = 5$, Figure 10.20a shows the effect of a change on each index by setting different values of T_1. A small value of T_1 meant too many false landslide regions were detected from the pre-event image, which indicated the detected-true landslide regions were much smaller than the ground truth. Consequently, a high *Precision* and low *Recall*, *F1-score*, and *Accuracy* were obtained in Figure 10.18a. By increasing T_1, the detected result of false landslide regions was closer to the true result, which indicated the detected-true landslide regions were closer to the ground truth. Therefore, the values of *Recall*, *F1-score*, and *Accuracy* got significantly larger, while the value of *Precision* got slightly smaller, as shown in Figure 10.20a. Furthermore, we can see that the four indices were convergent when $T_1 \geq 0.6$. A similar experiment was carried out on area A and area C. We obtained $T_1 \geq 0.65$. Therefore, we considered $T_1 = 0.8$ as an empirical value in our experiments.

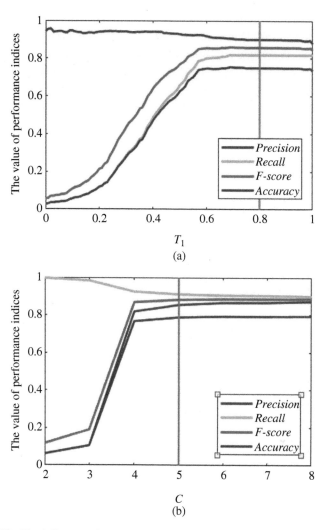

Figure 10.20 The influence of parameters on the performance of the presented CDFFCM
[40] / MDPI / CC BY 4.0. (a) Parameter T_1. (b) Parameter c.

Figure 10.20a shows the influence of T_1 on the performance of CDFFCM. Similarly, we analyzed the influence of c on the performance of CDFFCM, as shown in Figure 10.20b When the
value of c was small, the detected landslide regions were much larger than the ground truth,
resulting in a high *Recall* and low *Precision*, *F1-score*, and *Accuracy*. With the increase of the
value of c, the four indices were convergent, and higher values of *F1-score* and *Accuracy* were
obtained. Because a large c would lead to a high computational complexity for the FCM algorithm, we considered $c = 5$ as an empirical value in our experiments.

In our experiments, we detected the landslides using bitemporal images, which were acquired
at different times. Because the difference image of the bitemporal images mainly included landslides but no other changed information such as land use change, deforestation, and lake water
level change event, the presented CDFFCM provided a high accuracy of landslide identification
in our experiments. However, if the post-event image included other changed information, for

example, a new road, it would be difficult to differentiate a new road and landslides because they have similar image features. Therefore, the performance of the presented CDFFCM relies on the bitemporal images. If we want to identify landslides from changed regions, a classifier is required, and a great number of training samples and label images are necessary as well. However, the collection of a great number of bitemporal images and corresponding label images is difficult to obtain, so we do not provide much discussion of the classification of changed regions.

10.4 End-to-End Change Detection for VHR Remote Sensing Images

Most of the approaches used for LIM rely on traditional feature extraction and unsupervised classification algorithms. However, it is difficult to use these approaches to detect landslide areas because of the complexity and spatial uncertainty of landslides. In this section, we present a novel approach based on a fully convolutional network within pyramid pooling (FCN-PP) for LIM. This is based on the recently published letter by Lei et al. [48]. The presented approach has three advantages. First, this approach is automatic and insensitive to noise because MMR is used for image preprocessing. Second, it is able to take into account features from multiple convolutional layers and explore efficiently the context of images, which leads to a good trade-off between wider receptive field and the use of context. Finally, the selected PP module addresses the drawback of global pooling employed by CNN, FCN, and U-Net and, thus, provides better feature maps for landslide areas. Experimental results show that the presented FCN-PP is effective for LIM, and it outperforms the state-of-the-art approaches in terms of five metrics, *Precision, Recall, Overall Error, F1-score*, and *Accuracy*.

10.4.1 MMR for Image Preprocessing

The difference image of bitemporal images includes lots of noise and falsely changed regions. There are two ways to address the problem. One is to implement image filtering on bitemporal images and then compute the difference image; the other one is to compute the difference image and then implement image filtering. Because the former requires performing image filtering on pre-event and post-event images, it needs a long running time. To speed up the implementation of the presented approach, we adopt the second way, i.e. performing image filtering on the difference image.

Since the frequency domain filtering is unsuitable for removing falsely changed areas and preserving the structuring information of truly changed areas, to achieve preprocessing we use a morphological filter that is an efficient spatial filter. As traditional morphological filters are only effective for grayscale images, we use MMR operators here. MMR has two advantages for multichannel image filtering. First, MMR is able to filter noise while maintaining object details. Second, MMR has a low computational complexity, for fast lexicographic order denoted by $\leq_{PCA}$, is employed.

Let $v(R, G, B)$ represent a color pixel, and $v'(P_f, P_s, P_t)$ denote the transformed color pixel using the principal component analysis (PCA). The operation $\leq_{PCA}$ is defined as follows [49]:

$$v_1' \leq_{PCA} v_2' \Leftrightarrow \begin{cases} P_f^1 < P_f^2, & or \\ P_f^1 = P_f^2, & and\, P_s^1 < P_s^2\, or \\ P_f^1 = P_f^2, & and\, P_s^1 = P_s^2\, and\, P_t^1 < P_t^2 \end{cases}. \tag{10.20}$$

Let $\vec{\varepsilon}_m$ and $\vec{\delta}$ denote multivariate morphological erosion and dilation, respectively. According to (10.20), $\vec{\varepsilon}$ and $\vec{\delta}$ are defined as follows:

$$
\begin{cases}
\vec{\varepsilon}_{mB}(\boldsymbol{f}_{co}) = \wedge_{\substack{PCA \\ E\in\boldsymbol{B}}}(\boldsymbol{f}_{co-E}) \\[2mm]
\vec{\delta}_{B}(\boldsymbol{f}_{co}) = \vee_{\substack{PCA \\ E\in\boldsymbol{B}}}(\boldsymbol{f}_{co-E})
\end{cases},
\tag{10.21}
$$

where $\vee_{PCA}$ and $\wedge_{PCA}$ denote the supremum and infimum based on lexicographical ordering $\leq_{PCA}$, respectively. $\boldsymbol{f}_{co}$ represents a color image, and B is a structuring element. We let $\boldsymbol{f}$ and $\boldsymbol{f}_{co}$ represent a marker and a mask image, respectively, where $\boldsymbol{f} \leq_{PCA} \boldsymbol{f}_{co}$. We propose a morphological closing reconstruction operation, denoted by R^C, and it is defined as follows:

$$
R^C_{\boldsymbol{f}}(\boldsymbol{f}_{co}) = R^{\vec{\varepsilon}_m}_{\boldsymbol{f}}\left(R^{\vec{\delta}}_{\boldsymbol{f}}(\boldsymbol{f}_{co})\right),
\tag{10.22}
$$

where $R^{\vec{\varepsilon}_m}_{\boldsymbol{f}}(\boldsymbol{f}_{co}) = \vec{\varepsilon}^{\langle i\rangle}_{mf}(\boldsymbol{f}_{co})$, $R^{\vec{\delta}}_{\boldsymbol{f}}(\boldsymbol{f}_{co}) = \vec{\varepsilon}^{\langle i\rangle}_{mf}(\boldsymbol{f}_{co})$. As in a conventional MR operation, $\vec{\varepsilon}^{\langle 1\rangle}_{mf}(\boldsymbol{f}_{co}) = \vec{\varepsilon}_m(\boldsymbol{f}_{co})\vee_{PCA}\boldsymbol{f}$, $\vec{\varepsilon}^{\langle i\rangle}_{mf}(\boldsymbol{f}_{co}) = \vec{\varepsilon}_m\left(\vec{\varepsilon}^{\langle i-1\rangle}_{mf}(\boldsymbol{f}_{co})\right)\vee_{PCA}\boldsymbol{f}$, according to the duality of morphological operators, $\vec{\delta}^{\langle 1\rangle}_{f}(\boldsymbol{f}_{co}) = \vec{\delta}(\boldsymbol{f}_{co})\wedge_{PCA}\boldsymbol{f}$, $\vec{\varepsilon}^{\langle i\rangle}_{mf}(\boldsymbol{f}_{co}) = \vec{\varepsilon}_m\left(\vec{\varepsilon}^{\langle i-1\rangle}_{mf}(\boldsymbol{f}_{co})\right)\vee_{PCA}\boldsymbol{f}$, and $2\leq i\leq n$, $i, n\in N^+$, where n satisfies $\vec{\varepsilon}^{\langle n\rangle}_{mf}(\boldsymbol{f}_{co}) = \vec{\varepsilon}^{\langle n-1\rangle}_{mf}(\boldsymbol{f}_{co})$ and $\vec{\delta}^{\langle n\rangle}_{f}(\boldsymbol{f}_{co}) = \vec{\delta}^{\langle n-1\rangle}_{f}(\boldsymbol{f}_{co})$.

We applied MMR to a difference image to remove small image structures while preserving large object structures, which is helpful for subsequent feature learning.

10.4.2 Pyramid Pooling

FCN-PP[48] chooses a three-level PP module that includes three different scales (convolutional kernels: 5×5, 10×10, and 15×15; strides: 5, 10, and 15), where the first scale (5×5) is marked by cyan color, and the second and the third scales (10×10, 15×15) are marked by purple and yellow, respectively. Then, we use a 1×1 convolution to reduce the dimension of the three different size feature maps to achieve upsampling. Here, bilinear interpolation is used for upsampling to obtain feature maps of the same size as the original feature map. The final output of the pyramid module is a fusion result of multilevel feature maps. The average pooling is chosen in the pyramid module as it provides better global information than the max pooling.

To verify the validity of the PP module for landslide feature learning, we extract the outputs from convolutional layer at different scales. Figure 10.21 shows that a large convolutional kernel means a wider received, which that is helpful for global feature representation, while a small convolutional kernel means a narrower received field, which is helpful for local detailed feature representation. We used multiscale features to achieve a stronger feature representation than single-scale feature learning. As landslide areas have a serious spatial uncertainty, it is difficult to learn effective landslide features. The PP module is able to overcome the difficulty, and it is suitable for feature representation of landslide areas.

In Figure 10.21, the input is the same as feature maps of FCN, which are pooled by using different-size convolutional kernels. Then, these results are upsampled and fused with the feature maps from previous convolutional layers, leading to a final output with more accurate localization and better semantic information than the feature maps of FCN. Consequently, the presented FCN-PP

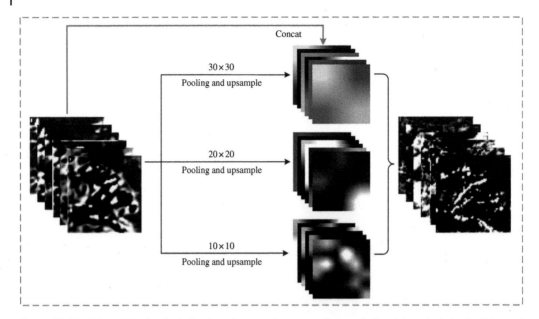

Figure 10.21 PP module that includes three layers, where a small convolutional kernel of size 5 × 5 is used for the first layer, a middle convolutional kernel of size 10 × 10 is used for the second layer, and a larger convolutional kernel of size 15 × 15 is used for the last layer [48] / IEEE / CC BY 4.0.

achieves a trade-off between the use of context and the localization accuracy, making a significant improvement over the CNN approach in LIM. Moreover, the elegant architecture leads to a requirement of a small number of training samples.

10.4.3 The Network Structure of FCN-PP

In practical applications, our purpose is to detect landslide areas in a post-event image. It is clear that we can use image segmentation algorithms to obtain landslide areas from the difference image of bitemporal images. Because the final output only includes changed and unchanged areas, the problem is viewed as a pixel-level binary classification, i.e. a binary segmentation task. The typical use of deep convolutional networks is on classification tasks. However, a pixel-level classification task (semantic segmentation) is more complex due to the requirement of localization.

Although a CNN is able to achieve effective image classifications, it provides a poor result on image segmentation since it employs global pooling and misses the spatial information of images. FCN [39] overcomes the problem by taking into account the combined features from multiple convolutional layers, which results in a better localization and the use of context. Consequently, FCN provides better image segmentation results than CNN. Inspired by the idea of FCN, we present an FCN-PP to obtain better LIM results. The presented FCN-PP is able to capture a wider receptive field, and it eventually overcomes the drawback of global pooling. Figure 10.22 shows the presented network structure.

In Figure 10.22, the FCN-PP similarly yields a U-shape architecture that includes four pooling layers and corresponding four upsampling layers. It also has an elegant structure because pooling layers on the left and upsampling layers on the right are symmetric. Furthermore, the PP module is integrated into FCN-PP to overcome the global pooling problem.

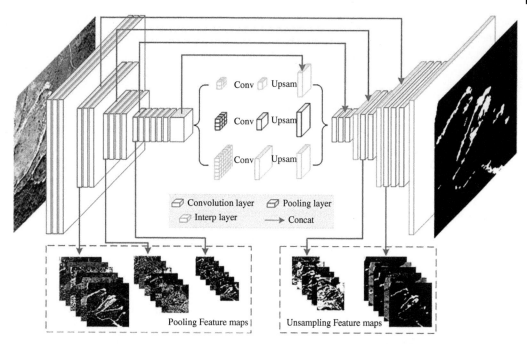

Figure 10.22 Structure of FCN-PP. It is a symmetric fully convolutional network that includes four pooling layers and four deconvolutional layers [48] / IEEE / CC BY 4.0.

10.4.4 Experiments

The presented FCN-PP is compared with two groups of popular approaches, i.e. unsupervised (ELSE [10], RLSE [10], CDMRF [9]), change detection-based fast FCM (CDFFCM) and supervised (CNN [30], FCN [39], and U-Net [50]) change detection for LIM. The first group of these approaches is implemented with MATLAB 2017b, and the second group is implemented with PyTorch on a workstaion with Intel Xeon CPU E5-1620v4, 3.5 GHz, four cores, 64-GB RAM, and double NVIDIA GTX 1080 GPU.

10.4.4.1 Data Description

Five pairs of bitemporal images on areas A-E in Hong Kong were captured by the Zeiss RMK TOP 15 Aerial Survey Camera System at a flying height of approximately 2.4 km on December 2007 and on November 2014, respectively [6]. Due to the geometrical resolution of bitemporal images of 0.5 m, the captured images have a large size. We cropped A–E areas to obtain five interesting areas including different types of landslide areas. The size of A–E areas are 750×950, 1252×2199, 923×593, 1107×743, and 826×725, respectively. Because it is very difficult to build a large data set of bitemporal landslide images, we built a small data set that is considered as a training set. To distinguish the training data and testing data, we first cropped three typical areas from areas A–C, where different kinds of landslide occur. The rest of areas A–C and D-E are then used for training data. The training images are cropped overlappingly. The testing images and training images have the same size of 473×473, and they have no overlapping areas. Finally, we obtain 139 training images that are overlapping. To increase the training data, each image is rotated by $\pm 30°$, horizontally and vertically flipped, sheared by $\pm 30°$, and scaled to 80% and 125% of its original size, but the final image has the same size as the original image. Finally, we obtain 1112 training image pairs of size 473×473 and three testing image pairs.

10.4.4.2 Experimental Setup

In our experiments, the parameter values of the first group of comparable approaches follow the original papers. In the presented FCN-PP, because the convolutional layers before the PP module are used for feature extraction, the corresponding parameters are initialized using the parameters of the first four convolutional layers in the VGG-16. We set a small learning rate that is 1×10^{-7} for pretrained convolutional layers to avoid overfitting. Stochastic gradient descent with a constant learning rate of 1×10^{-4}, weight decay of 0.0005, momentum of 0.99, minibatch size of 4, and epochs of 30 were used to train the presented network. In addition, the structure element employed by MMR is a disk of size 1×1.

To compare quantitatively the existing approaches with the presented FCN-PP, five quantitative evaluation indices are presented: $Precision = P_{lm}/P_l$, $Recall = P_{lm}/P_r$, $Overall\ error = (P_{over} + P_{rum})/P_t$, $F1\text{-}score = \dfrac{2 \times Precision \times Recall}{Precision + Recall}$, and $Accuracy = P_{lm}/(P_l + P_{rum})$, where P_{lm} is the total number of pixels of the detected landslides that are matching with the relevant ground truth, P_r is the total number of pixels of the ground truth, P_l is the total number of pixels of the detected landslides, and P_{rum} is the total number of pixels of the relevant ground truth that is unmatched with the detected landslides. P_{over} is the total number of pixels of detected false landslides. P_t is the total number of pixels of the test image. A large value of *Precision* means a small number of false alarms, and a large value of *Recall* means a small number of missed detections. A small value of *Overall error* (*OE*) means a small sum of false alarms and missed detections. Interestingly, *OE*, *F1-score*, and *Accuracy* reveal the overall detection performance, and a good approach for LIM corresponds to large values of *F1-score*, and *Accuracy* but a small value of *OE*.

10.4.4.3 Experimental Results

Figures 10.23a, b and 10.24a, b show two pairs of bitemporal images. We can see that the landslide areas are simple and continuous in Figure 10.23d, but they are complex and discontinuous in Figure 10.24d. Therefore, it is more difficult to extract the landslide areas in Figure 10.24b than in Figure 10.23b using traditional approaches. Figures 10.23c and 10.24c show the difference images of pre-event and post-event images. It is clear that each difference image includes lots of noise that influence the detection of the true landslides areas. Figures 10.23e–h and 10.24e–h show landslide areas detected by four conventional approaches, ELSE, RLSE, CDMRF, and CDFFCM. Because ELSE, RLSE, and CDMRF employ general image segmentation models to achieve landslide areas detection, they are sensitive to noise.

The detected landslide areas include lots of discontinuous areas that are continuous in ground truths. Although CDFFCM addresses the problem by using image filtering and improved FCM algorithm that incorporates spatial information of images into its objective function, some false landslide areas are detected, as shown in Figures 10.23h and 10.24h. Compared with unsupervised learning approaches, CNN is able to capture the semantic information of landslide areas, but the detected areas are coarse, as shown in Figures 10.23i and 10.24i. FCN provides a better result than CNN and four unsupervised learning approaches. However, the detail of landslide areas is removed in Figs. 10.23i and 10.24i since the global pooling is adopted by FCN. Note that although the experimental results of the area C are

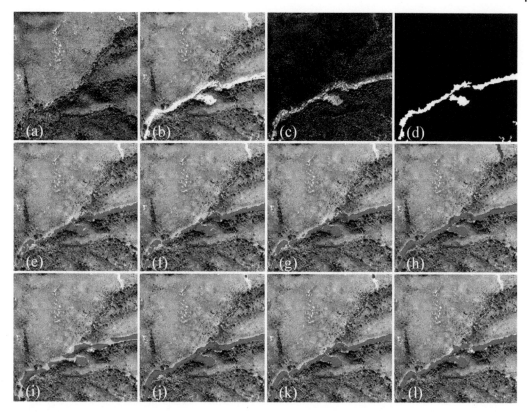

Figure 10.23 Comparison of results on the A-area using different approaches [48] / IEEE / CC BY 4.0, where the red areas are true-detected landslides, the blue areas are false-detected landslides, and the cyan areas are true-missed landslides. (a) Pre-event image. (b) Post-event image. (c) Difference image. (d) Ground truth. (e) ELSE. (f) RLSE. (g) CDMRF. (h) CDFFCM. (i) CNN. (j) FCN. (k) U-Net. (l) FCN-PP.

not shown here, the practical results also show that the presented FCN-PP is superior to the other approaches.

10.4.4.4 Experimental Analysis

For quantitative evaluation of the presented FCN-PP, we compare experimental results with ground truths according to five performance indices. The experiments are shown in Tables 10.7 and 10.8, note that, Larger values are better for *Precision*, *Recall*, *F1-score*, *Accuracy*, while smaller values are better for *OE*. It can be seen that *Precision* and *Recall* are inconsistent for the evaluation of results. ELSE, RLSE, and CDMRF obtain high *Precision* but low Recall values, while the CDFFCM obtains low *Precision* but high *Recall* values. Considering that *OE*, *F1-score*, and *Accuracy* evaluate the overall performance of an approach, CDFFCM is superior to ELSE, RLSE, and CDMRF, according to Tables 10.7 and 10.8. Among the existing supervised approaches, U-Net is superior to CNN and FCN. What is abundantly clear is that the presented FCN-PP obtains the best performance index, as shown in Tables 10.7 and 10.8.

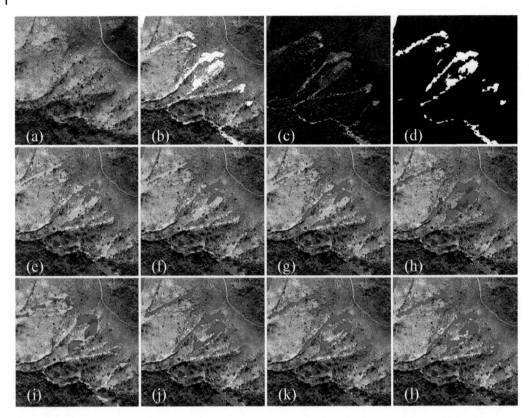

Figure 10.24 Comparison of results on the area B using different approaches [48] / IEEE / CC BY 4.0. (a) Pre-event image. (b) Post-event image. (c) Difference image. (d) Ground truth. (e) ELSE. (f) RLSE. (g) CDMRF. (h) CDFFCM. (i) CNN. (j) FCN. (k) U-Net. (l) FCN-PP.

Table 10.7 Quantitative results of area A [48] / IEEE / CC BY 4.0.[a]

Method	Precision	Recall	OE	F1-score	Accurary
ELSE	86.79	74.02	1.75	79.90	66.53
RLSE	86.61	75.87	1.68	80.89	67.91
CDMRF	84.89	79.15	1.64	81.92	69.38
CDFFCM	79.30	**91.01**	1.54	84.75	73.54
CNN	75.47	62.16	2.73	68.17	51.71
FCN	79.10	88.45	1.64	83.51	71.69
U-net	85.65	83.92	1.42	84.78	73.58
FCN-PP	**89.84**	90.32	**0.93**	**90.08**	**81.95**

[a] The best values are boldfaced.

Table 10.8 Quantitative results of area B [48] / IEEE / CC BY 4.0.[a]

Method	Precision	Recall	OE	F1-score	Accurary
ELSE	91.24	51.29	3.93	65.67	48.89
RLSE	91.14	56.00	3.63	69.37	53.10
CDMRF	87.42	60.01	3.56	71.17	55.24
CDFFCM	73.52	88.42	3.18	80.28	67.06
CNN	71.44	51.44	5.07	59.82	42.67
FCN	83.94	84.39	2.33	84.17	72.66
U-net	87.88	87.93	1.77	87.91	78.42
FCN-PP	**96.03**	**95.62**	**0.61**	**95.82**	**91.99**

[a] The best values are boldfaced.

10.5 Discussion and Summary

Image segmentation methods have been widely used in change detection tasks of remote sensing , including LM. LM is an important research topic in change detection [48, 49]. However, it is difficult to use change detection approaches directly to achieve LM due to the irregular texture, shape, and size of landslide regions.

In this chapter, a change detection approach based on fast FCM clustering (CDFFCM) has been presented and applied to LM. The presented method has two advantages: First, we use the difference of image structure information instead of grayscale difference to obtain more accurate landslide regions. It demonstrates that the CDFFCM method is insensitive to illumination, noise, and environmental change. Second, by introducing a Gaussian pyramid, the presented method has a low computational complexity since the Gaussian pyramid can remove the data redundancy of images for FCM.

Beyond that, a symmetric FCN-PP also was presented for change detection. It is insensitive to noise because MMR has been used to filter noise and non-landslide areas in difference images, which is helpful for improving training and testing images. FCN-PP can automatically obtain change detection results. Owing to an elegant architecture of powerful deep convolutional network, it is able to trade off the use of context and the localization accuracy. The experiments have shown that the presented CDFFCM achieves better change detection results, and the designed FCN-PP is better in learning image features to improve change detection results. They have all been successfully applied to LM of change detection task.

References

1 Chander, G., Hewison, T.J., Fox, N. et al. (2013). Overview of intercalibration of satellite instruments. *IEEE Trans. Geosci. Remote Sens.* **51** (3): 1056–1080.
2 Komarov, A.S. and Buehner, M. (2017). Automated detection of ice and open water from dual-polarization RADARSAT-2 images for data assimilation. *IEEE Trans. Geosci. Remote Sens.* **55** (10): 5755–5769.
3 Pacifici, F., Longbotham, N., and Emery, W.J. (2014). The importance of physical quantities for the analysis of multitemporal and multiangular optical very high spatial resolution images. *IEEE Trans. Geosci. Remote Sens.* **52** (10): 6241–6256.

4 Cartus, O., Siqueira, P., and Kellndorfer, J. (2017). An error model for mapping forest cover and forest cover change using L-band SAR. *IEEE Geosci. Remote Sens. Lett.* **15** (1): 107–111.

5 Zhan, X., Zhang, R., Wang, P. et al. (2017). Monitoring surface type changes with S-NPP/JPSS VIIRS observations. In: *2017 IEEE International Geoscience and Remote Sensing Symposium (IGARSS)*, 1288–1291. IEEE.

6 Lv, Z.Y., Shi, W., Zhang, X., and Benediktsson, J.A. (2018). Landslide inventory mapping from bitemporal high-resolution remote sensing images using change detection and multiscale segmentation. *IEEE J. Selected Topics Appl. Earth Observ. Remote Sens.* **11** (5): 1520–1532.

7 Nelson, R.F. (1983). Detecting forest canopy change due to insect activity using Landsat MSS. *Photogrammetric Eng. Remote Sens.* **49** (9): 1303–1314.

8 Jin, Y.Q. (2005). Change detection of enhanced, no-changed and reduced scattering in multi-temporal ERS-2 SARimages using the two-thresholds EM and MRF algorithms. In: *Proceedings. 2005 IEEE International Geoscience and Remote Sensing Symposium* IGARSS'05, vol. **6**, 3994–3997. IEEE.

9 Li, Z., Shi, W., Lu, P. et al. (2016). Landslide mapping from aerial photographs using change detection-based Markov random field. *Remote Sens. Environ.* **187**: 76–90.

10 Li, Z., Shi, W., Myint, S.W. et al. (2016). Semi-automated landslide inventory mapping from bitemporal aerial photographs using change detection and level set method. *Remote Sens. Environ.* **175**: 215–230.

11 Dong, R. and Wang, H. (2017). A novel VHR image change detection algorithm based on image fusion and fuzzy C-means clustering. arXiv preprint *arXiv*:1706.07157.

12 Li, Z., Shi, W., Zhang, H., and Hao, M. (2017). Change detection based on Gabor wavelet features for very high resolution remote sensing images. *IEEE Geosci. Remote Sens. Lett.* **14** (5): 783–787.

13 Zheng, Y., Jiao, L., Liu, H. et al. (2017). Unsupervised saliency-guided SAR image change detection. *Pattern Recogn.* **61**: 309–326.

14 Shi, W., Shao, P., Hao, M. et al. (2016). Fuzzy topology-based method for unsupervised change detection. *Remote Sens. Lett.* **7** (1): 81–90.

15 Shao, P., Shi, W., He, P. et al. (2016). Novel approach to unsupervised change detection based on a robust semi-supervised FCM clustering algorithm. *Remote Sens. (Basel)* **8** (3): 264.

16 Sharma, A. and Gulati, T. (2017). Change detection in remotely sensed images based on image fusion and fuzzy clustering. *Int. J. Electronics Eng. Res.* **9** (1): 141–150.

17 Gong, M., Zhou, Z., and Ma, J. (2011). Change detection in synthetic aperture radar images based on image fusion and fuzzy clustering. *IEEE Trans. Image Process.* **21** (4): 2141–2151.

18 Ghosh, A., Mishra, N.S., and Ghosh, S. (2011). Fuzzy clustering algorithms for unsupervised change detection in remote sensing images. *Inform. Sci.* **181** (4): 699–715.

19 Gong, M., Su, L., Jia, M., and Chen, W. (2013). Fuzzy clustering with a modified MRF energy function for change detection in synthetic aperture radar images. *IEEE Trans. Fuzzy Syst.* **22** (1): 98–109.

20 Zhang, X., Xiao, P., Feng, X., and Yuan, M. (2017). Separate segmentation of multi-temporal high-resolution remote sensing images for object-based change detection in urban area. *Remote Sens. Environ.* **201**: 243–255.

21 Keyport, R.N., Oommen, T., Martha, T.R. et al. (2018). A comparative analysis of pixel-and object-based detection of landslides from very high-resolution images. *Int. J. Appl. Earth Observ. Geoinform.* **64**: 1–11.

22 Lv, Z., Liu, T., Wan, Y. et al. (2018). Post-processing approach for refining raw land cover change detection of very high-resolution remote sensing images. *Remote Sens. (Basel)* **10** (3): 472.

23 Leichtle, T., Geiß, C., Wurm, M. et al. (2017). Unsupervised change detection in VHR remote sensing imagery–an object-based clustering approach in a dynamic urban environment. *Int. J. Appl. Earth Observ. Geoinform.* **54**: 15–27.

24 Wang, Q., Yuan, Z., Du, Q., and Li, X. (2018). GETNET: a general end-to-end 2-D CNN framework for hyperspectral image change detection. *IEEE Trans. Geosci. Remote Sens.* **57** (1): 3–13.

25 Yu, H., Yang, W., Hua, G. et al. (2017). Change detection using high resolution remote sensing images based on active learning and Markov random fields. *Remote Sens. (Basel)* **9** (12): 1233.

26 Lv, Z. and Zhang, W. (2018). Contextual analysis based approach for detecting change from high resolution satellite imagery. *J. Indian Soc. Remote Sens.* **46** (1): 43–50.

27 Niu, X., Gong, M., Zhan, T., and Yang, Y. (2018). A conditional adversarial network for change detection in heterogeneous images. *IEEE Geosci. Remote Sens. Lett.* **16** (1): 45–49.

28 Lei, T., Jia, X., Zhang, Y. et al. (2018). Significantly fast and robust fuzzy c-means clustering algorithm based on morphological reconstruction and membership filtering. *IEEE Trans. Fuzzy Syst.* **26** (5): 3027–3041.

29 Gong, M., Zhao, J., Liu, J. et al. (2015). Change detection in synthetic aperture radar images based on deep neural networks. *IEEE Trans. Neural Netw. Learn. Syst.* **27** (1): 125–138.

30 Liu, J., Gong, M., Qin, K., and Zhang, P. (2016). A deep convolutional coupling network for change detection based on heterogeneous optical and radar images. *IEEE Trans. Neural Netw. Learn. Syst.* **29** (3): 545–559.

31 van Aarle, W., Batenburg, K.J., and Sijbers, J. (2011). Optimal threshold selection for segmentation of dense homogeneous objects in tomographic reconstructions. *IEEE Trans. Med. Imaging* **30** (4): 980–989.

32 Huo, G., Yang, S.X., Li, Q., and Zhou, Y. (2016). A robust and fast method for sidescan sonar image segmentation using nonlocal despeckling and active contour model. *IEEE Trans. Cybern.* **47** (4): 855–872.

33 Bai, M. and Urtasun, R. (2017). Deep watershed transform for instance segmentation. In: *Proceedings of the IEEE Conference on Computer Vision and Pattern Recognition*, 5221–5229.

34 Grady, L. (2006). Random walks for image segmentation. *IEEE Trans. Pattern Anal. Mach. Intell.* **28** (11): 1768–1783.

35 Ju, W., Xiang, D., Zhang, B. et al. (2015). Random walk and graph cut for co-segmentation of lung tumor on PET-CT images. *IEEE Trans. Image Process.* **24** (12): 5854–5867.

36 Chen, J., Li, Z., and Huang, B. (2017). Linear spectral clustering superpixel. *IEEE Trans. Image Process.* **26** (7): 3317–3330.

37 Ji, Z., Liu, J., Cao, G. et al. (2014). Robust spatially constrained fuzzy c-means algorithm for brain MR image segmentation. *Pattern Recogn.* **47** (7): 2454–2466.

38 Ji, Z., Huang, Y., Xia, Y., and Zheng, Y. (2017). A robust modified Gaussian mixture model with rough set for image segmentation. *Neurocomputing* **266**: 550–565.

39 Long, J., Shelhamer, E., and Darrell, T. (2015). Fully convolutional networks for semantic segmentation. In: *Proceedings of the IEEE conference on computer vision and pattern recognition*, 3431–3440.

40 Lei, T., Xue, D., Lv, Z. et al. (2018). Unsupervised change detection using fast fuzzy clustering for landslide mapping from very high-resolution images. *Remote Sens. (Basel)* **10** (9): 1381.

41 Zhang, H., Wang, Q., Shi, W., and Hao, M. (2017). A novel adaptive fuzzy local information c-means clustering algorithm for remotely sensed imagery classification. *IEEE Trans. Geosci. Remote Sens.* **55** (9): 5057–5068.

42 Memon, K.H. and Lee, D.H. (2018). Generalised kernel weighted fuzzy C-means clustering algorithm with local information. *Fuzzy Set. Syst.* **340**: 91–108.

43 Guo, F.F., Wang, X.X., and Shen, J. (2016). Adaptive fuzzy c-means algorithm based on local noise detecting for image segmentation. *IET Image Process.* **10** (4): 272–279.

44 Szilagyi, L., Benyo, Z., Szilágyi, S.M., and Adam, H.S. (2003). MR brain image segmentation using an enhanced fuzzy c-means algorithm. In: *Proceedings of the 25th annual international conference of the IEEE engineering in medicine and biology society* (IEEE Cat. No. 03CH37439), vol. **1**, 724–726. IEEE.

45 Lan, Z., Lin, M., Li, X. et al. (2015). Beyond gaussian pyramid: Multi-skip feature stacking for action recognition. In: *Proceedings of the IEEE Conference on Computer Vision and Pattern Recognition*, 204–212.

46 Lei, T., Zhang, Y., Wang, Y. et al. (2017). A conditionally invariant mathematical morphological framework for color images. *Inform. Sci.* **387**: 34–52.

47 Chen, J.J., Su, C.R., Grimson, W.E.L. et al. (2011). Object segmentation of database images by dual multiscale morphological reconstructions and retrieval applications. *IEEE Trans. Image Process.* **21** (2): 828–843.

48 Lei, T., Zhang, Y., Lv, Z. et al. (2019). Landslide inventory mapping from bitemporal images using deep convolutional neural networks. *IEEE Geosci. Remote Sens. Lett.* **16** (6): 982–986.

49 Ding, A., Zhang, Q., Zhou, X., and Dai, B. (2016). Automatic recognition of landslide based on CNN and texture change detection. In: *2016 31st Youth Academic Annual Conference of Chinese Association of Automation (YAC)*, 444–448. IEEE.

50 Ronneberger, O., Fischer, P., and Brox, T. (2015). U-net: Convolutional networks for biomedical image segmentation. In: *International Conference on Medical image computing and computer-assisted intervention*, 234–241. Cham: Springer.

11

Image Segmentation for Material Analysis

11.1 Introduction

Scanning electron microscope (SEM) [1] is a surface analysis technique with a wide range of applications, mainly using secondary electron signals to scan the surface of the sample to obtain the corresponding high magnification image, and the scale of analysis can be up to the nanometer level, which can effectively observe the microstructure and state of the sample [2–4], so it has been widely used in materials, chemical, and biological fields.

Porous metal materials [5] can be divided into two categories according to their pore morphology: independent pore type and continuous pore type metal materials. The independent pore type metal material has the characteristics of small specific gravity, good rigidity, specific strength, good vibration absorption, and sound absorption performance. In addition to the above characteristics, the continuous pore type metal material also has the characteristics of permeability and good ventilation. By changing the microscopic pore parameters of metal materials [6, 7], a variety of materials with special functions can be obtained. In the field of high-tech applications, porous materials have shown great development prospects. In order to evaluate better the objective properties of porous metal materials, researchers mainly use interactive methods to obtain pore information of porous metal materials. However, the interactive measurement of all pore information is very time-consuming and labor-intensive. Researchers usually only measure a few pores information and use their averaged results to estimate the overall physicochemical properties of the material. Due to human subjectivity, the simplicity of measurement methods, and the incompleteness of measurement data, the data measured by different experimental analysts are quite different. Thus, it is challenging to evaluate the real and objective properties of porous metal materials. Therefore, research on image segmentation algorithms for porous metal materials is crucial for accurately analyzing their physical properties [8].

The foam material is a material or a solid with a certain number of holes formed by a network structure of interconnected or closed holes, and the boundary or surface of the holes is composed of pillars or flat plates. The pore structure is two-dimensional and is formed by the aggregation of a large number of polygonal pores on a plane, which is called a "honeycomb" material due to its shape resembling the hexagonal structure of a honeycomb. More common is the three-dimensional structure formed by a large number of polyhedral-shaped pores in space, which is usually called "foam" material [9] and is regarded as an ideal material for many new optoelectronics because of its excellent performance. Foam material also has the characteristics of low water absorption, good sound insulation, and heat insulation, making it an ideal material for fields requiring high strength and low density. It is widely used in wind power generation, rail transit, ships, aerospace,

Image Segmentation: Principles, Techniques, and Applications, First Edition. Tao Lei and Asoke K. Nandi.
© 2023 John Wiley & Sons Ltd. Published 2023 by John Wiley & Sons Ltd.

building energy conservation, and other fields. A large number of directional or random holes are dispersed in the foam material, and the diameter of these holes is about 2 μm to 3 mm. Due to the different ways of using materials, the design requirements of the holes are different. The calculation and feature extraction of the pore size of the foam material play an extremely important role in the identification of the properties of the foam and the identification of the tissue structure [10, 11]. Due to the particularity of foam materials, it is usually necessary to image them under an SEM to analyze their physical properties, and it is of great significance to study image segmentation algorithms for foam materials to analyze their physical properties effectively.

The SEM image of ceramic materials consists of the material area (i.e. crystal grains) and the gaps between the crystal grains (i.e. grain boundaries). Since ceramic is an insulating material, it does not have the property of conducting electricity and is easily broken down by high-voltage electricity during imaging [12]. In order to avoid this kind of situation, it is necessary to control the grain size as small as possible, that is, there are more grain boundaries in the same size area so that high-voltage electricity can be drawn from the grain boundaries, and the ceramic sample is protected from breakdown. However, the grain size directly determines the performance of the ceramic material [13], so it is necessary to count the distribution of the grain size in the SEM image and then indirectly establish the corresponding relationship between the experimental conditions and the performance of the ceramic material. The current grain analysis in SEM images mainly relies on artificial means, and the measurement results have obvious limitations. First, a ceramic SEM image contains a large number of grains, which is time-consuming and labor-intensive for manual statistics, and the measurement is difficult and inefficient. Second, the size and shape of the crystal grains are irregular, and manual measurement is easy to be affected by subjective factors and causes large errors. Therefore, research on a clustering algorithm that can automatically measure the grain size with high accuracy and fast calculation speed is of great significance to the analysis of the physical properties of ceramic materials.

11.2 Related Work

With the continuous development of materials science, more and more materials are discovered by researchers. Most materials have the characteristics of low density, high strength, sound insulation, heat insulation, and high permeability, which are widely used in many ideal materials such as new optoelectronic components [2]. The number, size, area, and porosity of its pores play an important role in material preparation, production, and evaluation. With the increase in research needs and the development of the instrumentation industry, the emergence of scanning electron microscopy has gradually played an important role in the study of the morphology and properties of materials.

Using image segmentation technology, the surface morphology and properties of materials can be studied, and the physical properties of objects can be reflected at the microscopic level, thereby providing a reliable basis for exploring the properties of objects. Therefore, it is necessary to carry out research such as image feature extraction and image segmentation for SEM images of these materials [1].

11.2.1 Metal Materials

The resolution of SEM images of porous metal materials is usually large, and the target holes are bonded and overlapping. At the same time, the boundaries are blurred, leading to the hole parameters not being extracted efficiently. To solve the problem, Vincent et al. [14] proposed a watershed segmentation algorithm based on the distance transform. Although the algorithm can accurately identify the boundaries of large holes, it cannot effectively capture the information of small holes. As a result, the final segmentation result is not ideal. In order to solve this problem, researchers [15] proposed a threshold-based target segmentation algorithm, but this kind of algorithm cannot effectively extract the close-distributed hole pixels to meet the high-precision segmentation requirements. Subsequent researchers proposed a hole extraction scheme based on spectral clustering [16]. Although this type of algorithm improves the ability to separate holes, the spectral clustering algorithm needs to construct a similarity matrix, leading to a long execution time of the algorithm. The fuzzy c-means (FCM) clustering algorithm proposed by Dunn and improved by Bezdek [17] has good applicability to porous metal material images. However, when the FCM algorithm is directly applied to image data, each pixel is regarded as an independent sample point. Although the FCM algorithm can reasonably divide the pixels, it ignores the spatial structure information of the image, which makes the FCM more sensitive to the uneven distribution of pixels and noise information. In order to improve the effect of FCM on image segmentation, researchers have proposed a variety of improvement strategies. However, they still face the problems of low accuracy and long execution time of hole segmentation, which needs further research and exploration.

11.2.2 Foam Materials

The main features of the foam SEM data set are as follows: The foam images are mainly obtained by scanning electron microscopy [1], and the image details are rich, but the foam pores are intricate and irregular in shape, and the image contrast is low and segmentation is difficult [18]. Based on the above characteristics, to realize the analysis of the properties of foam materials based on image segmentation technology, initially, people mostly use the Image series software [19] to measure manually and analyze the pores. But the depth of field of the SEM is not suitable for the shallower mesopores (aperture diameter of 2–50 nm), while large pores (diameter > 50 nm) will form an illusion on grayscale, making them difficult to distinguish. As they manually identify the pixel value and enter the software for identification, the human factor is large, and misjudgment is prone to occur in subsequent aperture identification. To solve this problem, researchers carried out 3D modeling of the foam material [20] and used a 3D algorithm to calculate its main foam material skeleton to realize the pore size identification of the foam material. However, the 3D modeling process is complicated, and the error is large. This has brought great difficulties to follow up in research. With the development of image segmentation technology, researchers found that using the watershed algorithm [9] to segment foam materials can greatly reduce the workload. Due to the interference of noise information in the foam SEM images, the watershed algorithm is prone to over-segmentation. With the rapid development of clustering algorithms [21] in the industrial field, researchers have applied clustering algorithms to the segmentation of foam materials, thus realizing the segmentation of pore sizes.

11.2.3 Ceramics Materials

Because of the problem of ceramic material grain segmentation, since the SEM image of ceramic materials mainly presents the outline information of the crystal grains and lacks texture detail information, it usually requires a lot of preprocessing and interactive processing. Therefore, it is difficult to apply directly mainstream image segmentation algorithms to grain segmentation. In this regard, Jiang et al. [22] proposed an automatic grain segmentation method for sandstone analysis. This study uses a combination of superpixels and fuzzy clustering to segment sandstone images and obtains better sandstone grains segmentation results. However, this method relies on the SLIC (simple linear iterative clustering) superpixel segmentation results and the subsequent region merging algorithm. It is difficult for SLIC to obtain good pre-segmentation results on the SEM images of ceramic materials, so this method is not suitable for the grain segmentation of SEM images of ceramic materials. On this basis, Banerjee et al. [23] proposed an automatic grain segmentation and grain size method. The method first detects the edge of the image and performs binarization processing and then uses morphological closing operations and dilation operations to obtain closed contours. It uses small area removal and contour refinement operations to obtain a single-line closed contour. Compared with the previous method, this method is simpler to calculate, and it is easy to obtain good segmentation results for SEM grain images with more uniform gray values. Because of the high degree of detailed information, it is prone to miss-segmentation of complex grain images. The above methods all use unsupervised image segmentation technology to achieve grain segmentation. When image segmentation technology based on supervised learning is applied to SEM image grain segmentation, it usually faces high manual annotation costs, limited training samples that can be obtained, and deep learning. The image segmentation results have problems such as low edge accuracy. Many limitations make it difficult for the current image segmentation technology to effectively solve the problem of grain segmentation in SEM images of ceramic materials.

11.3 Image Segmentation for Metal Material Analysis

Due to the problems of high noise and blurred hole contours in the SEM images corresponding to porous metal materials, it is difficult for mainstream area-based image segmentation methods to achieve better hole segmentation. The pixel-based image segmentation method can identify the category of each pixel, which can effectively solve the problem of metal porous material segmentation. However, the mainstream FCM algorithm [24] is sensitive to noise and lacks spatial information of the image, so the segmentation accuracy of metal porous materials is low. To solve the problem, the improvement strategies mainly include a fuzzy clustering algorithm based on spatial neighborhood information, a fuzzy clustering algorithm based on regularization constraints, and a fuzzy clustering algorithm based on histogram information. The fuzzy clustering algorithm based on spatial neighborhood information mainly uses neighborhood pixels to modify the category of central pixels to achieve the purpose of correct classification. Recently, Li et al. [25] proposed an L_P clustering algorithm based on fuzzy local information. The algorithm uses the L_P norm as the similarity measurement criterion and embeds spatial fuzzy factors into its objective function. Although the segmentation effect of the algorithm is improved to a certain extent, it has poor robustness to noise interference.

In response to this problem, Zhang et al. [26] proposed a spatially constrained fuzzy clustering algorithm based on deviation sparse (DSFCM_N), which completes data correction during iterations and avoids the smoothing of the original data by the filtering strategy. The DSFCM_N uses regularization constraints to correct the original data and integrates the spatial information of the neighborhood to avoid noise interference, which effectively improves the image segmentation effects.

However, the above algorithms need to calculate the spatial neighborhood information of the image in each iteration, leading to the high time complexity of the algorithm. To address this issue, researchers [27] presented a fast fuzzy clustering algorithm based on histogram information. This kind of algorithm firstly suppresses noise interference by filtering and secondly uses histogram information to achieve fast image segmentation, which reduces its computational complexity.

Based on the above analysis, this chapter adopts the fast and robust fuzzy c-means clustering [27] (FRFCM) as the pre-segmentation algorithm, which not only effectively improves the image segmentation result but also reduces the computational complexity. Although the FRFCM presented in Chapter 5 can better achieve the segmentation of porous metal materials, it is still unable to extract different levels of hole information.

11.3.1 Segmentation of Porous Metal Materials

Considering the problem that the traditional improved algorithm is time-consuming for porous metal material images, this chapter adopts the FRFCM algorithm presented in Chapter 5 to achieve porous material segmentation. The FRFCM first uses morphological filtering to remove the interference noise of the test image, then completes the initial clustering on the histogram, then further corrects the numerical distribution of its membership matrix through median filtering, and finally outputs the segmentation result. Figure 11.1 shows the segmentation results of the FRFCM algorithm for porous metal materials.

As can be seen in Figure 11.1, the metal porous material not only has a high resolution but also has blurred hole boundaries. We utilize the FRFCM algorithm to obtain the ideal segmentation result of metal porous material, which can provide technical support to classification holes.

Compared with the traditional FCM algorithm, the adopted FRFCM algorithm has two advantages. On the one hand, FRFCM adopts morphological reconstruction, which can better protect the image details while filtering out the noise and has better performance against different types of noise. On the other hand, FRFCM directly corrects the category of the pixel through membership filtering, which avoids the repeated calculation of the neighborhood information in the iteration and further reduces the computational complexity of the algorithm.

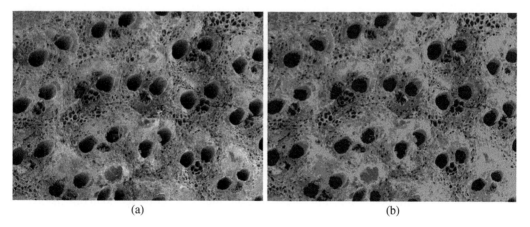

(a)   (b)

Figure 11.1 Segmentation result of porous metal material, $c = 4$, c represents the number of cluster centers. (a) Test image. (b) Segmentation result of FRFCM.

11.3.2 Classification of Holes

Although the FRFCM can segment metal porous materials, it is still difficult to count the hole information directly using the segmentation results. In order to classify holes according to different areas, a simple binary morphological reconstruction method is used to realize the rapid classification of holes. The method is summarized as follows: (1) Firstly, the data distribution corresponding to the minimum clustering center is extracted as the hole output to obtain the binary hole image. (2) Secondly, the binary hole image is morphologically filled, and then the holes are automatically classified according to the area by using multi-scale morphological closed reconstruction. (3) Finally, we computed the number and area of holes with different sizes for porous metal material images.

Based on the pre-segmentation of FRFCM, the subsequent attribute evaluation is completed through the hole classification strategy presented in this chapter. The specific steps of hole classification are as follows:

1) Initialization: the radius of the disc-shaped structural element is B_1, the step size is $\hat{s}$, the hole level is $\hat{p}$; then $1 \leq i \leq \hat{p}$, the radius of the largest structural element is $B_1 + \hat{s} \times (\hat{p} - 1)$, the scale parameter is $Scale = D_g/D_a$, D_g represents the distance on the map, and D_a represents the actual distance;
2) Extract the minimum cluster center obtained by FRFCM as the hole output, and fill it with holes to obtain the image f^I;
3) Perform morphological closed reconstruction of f^I by different structural elements $B_i = B_1 + \hat{s} \times (i-1)$ to obtain an image $f^{(\delta\varepsilon)}$, and $1 \leq i \leq \hat{p}$;
4) Calculate the difference between f^I and $f^{(\delta\varepsilon)}$, which can screen out images of holes of different sizes;
5) Count hole attributes such as the number, area ratio, and actual average area of holes of different sizes from hole output images $\left(f^{R_1}, f^{R_2}, \cdots, f^{R_{\hat{p}}}\right)$.

Then a connected domain label is utilized to obtain the number N_i of the i-typed hole, and then the average area S_i and area ratio R_i of the corresponding hole are calculated. The calculation formulas are as follows:

$$S_i = \frac{1}{N_i} \sum_{m=1}^{M} \sum_{N=1}^{N} f^{R_i}, \tag{11.1}$$

$$R_i = \frac{1}{M \times N} \sum_{m=1}^{M} \sum_{N=1}^{N} f^{R_i}. \tag{11.2}$$

According to the scale parameter $Scale$, the actual average area $\hat{S}_i = S_i/Scale$ of the i-typed of hole is obtained, and finally, the relevant information of different levels of holes is obtained, namely, the histogram of the number of multi-level holes $N = (N_1, N_2, \cdots, N_{\hat{p}})$, the histogram of the proportion of multi-level holes $R = (R_1, R_2, \cdots, R_{\hat{p}})$, and the histogram of the average area of multi-level holes $\hat{S}_i = (\hat{S}_1, \hat{S}_2, \cdots, \hat{S}_{\hat{p}})$.

Figure 11.2 shows the practicability of the hole classification strategy presented in this chapter on the FRFCM segmentation result. It is not difficult to find from Figure 11.2 that the hole classification strategy can effectively screen the distribution of holes with different sizes. By analyzing the objective properties of holes with different sizes, it lays a solid foundation for the performance evaluation of porous metal materials.

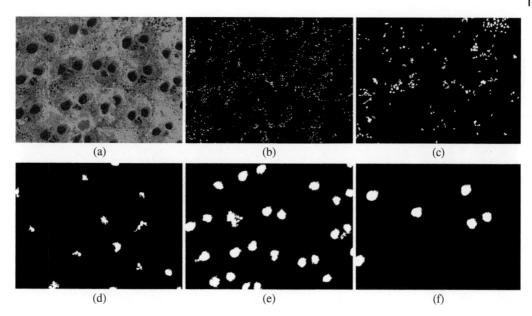

Figure 11.2 The classification results of holes, (a) The segmentation result of FRFCM. (b) The distribution of holes in the first category. (c) The distribution of holes in the second category. (d) The distribution of holes in the third category. (e) The distribution of holes in the fourth category. (f) The distribution of holes in the fifth category.

11.3.3 Experiment Analysis

In order to verify the applicability of the different algorithms to SEM images, a porous metal material image of size 1024×1536 is selected as the test data. The initialization parameters are set as follows: B_1, $\hat{s}$, and $\hat{p}$ are all set to 5, $r_1 = 2$, and the scale of the experiment is $Scale = D_g/D_a = 375/500 = 0.75\ pixel/\mu m$; although the fuzzy c-means clustering algorithms can detect the hole target, they all ignore the hole classification strategy. Figure 11.3 shows the hole classification result using the multi-scale morphological closed reconstruction scheme.

As can be seen from Figure 11.3, the hole classification strategy can accurately screen holes of different sizes, which can effectively avoid the subjectivity of artificial marking, and the accurate hole classification lays the foundation for the subsequent attribute evaluation. Figure 11.4 shows the basic properties of different sizes of holes. Using the evaluation result can help researchers to explore the physical properties of the test materials, which will pave the way for the development of light industry technology.

Figure 11.5 shows the segmentation effect of different algorithms on metal material images. It can be seen from Figure 11.5 that DSFCM_N obtains more hole connectivity results than FLIL and FRFCM, resulting in a higher over-detection rate. However, the FRFCM is more reasonable for the division of hole boundary and hole size, which provides a guarantee for the subsequent accurate measurement of hole information.

In order to quantify the practicability of the three algorithms for metal materials, F-score and comparison scores (CS) are used as measurement criteria:

$$\text{F-score} = 2(Acc \times R)/(Acc + R), \tag{11.3}$$

$$CS = \sum_{k=1}^{\hat{p}} A_k \cap G_k/A_k \cup G_k, \tag{11.4}$$

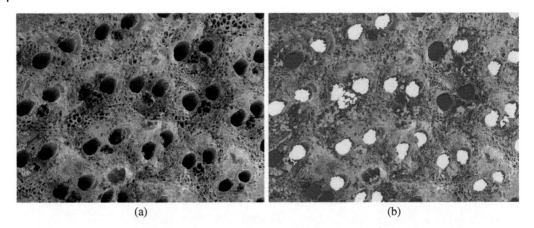

(a) (b)

Figure 11.3 The hole classification results from the pre-segmentation of the FRFCM algorithm. (a) The original test image. (b) The hole classification result.

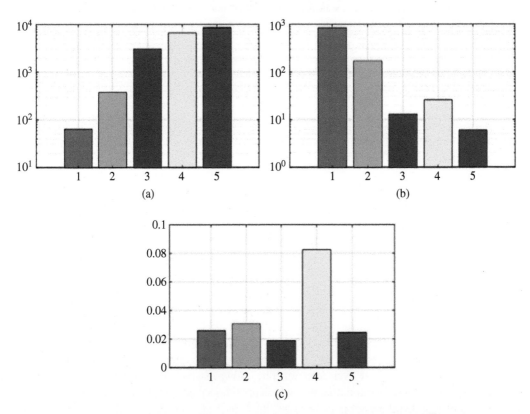

Figure 11.4 Property evaluation of different pore sizes of porous metal materials. (a) The average area of different sizes of pores. (b) Diagram of the total number of different sizes of pores. (c) The area ratio of different sizes of pores.

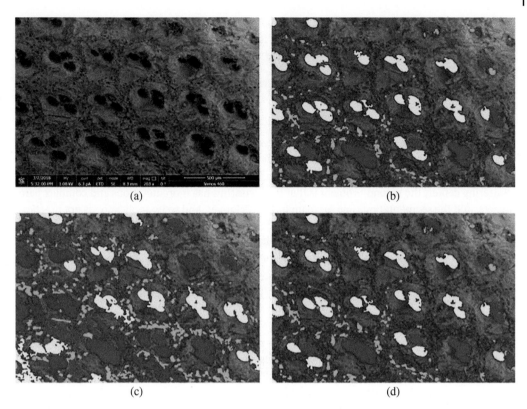

Figure 11.5 Classification results of holes in porous metal materials by different algorithms. (a) Original image, (b) Classification result of holes based on FLIL pre-segmentation (c) Classification result of holes based on DSFCM_N pre-segmentation. (d) Classification result of holes based on FRFCM pre-segmentation.

where Acc is the accuracy rate and R is the recall rate, and the corresponding calculation formulas are $Acc = \sum_{k=1}^{\hat{p}} A_k \cap G_k / \sum_{j=1}^{\hat{p}} A_j$, $R = \sum_{k=1}^{\hat{p}} A_k \cap G_k / \sum_{j=1}^{\hat{p}} G_j$, A_k represents the type k of hole obtained by the test algorithm, G_k represents the standard segmentation result of the type k of hole, $\cap$ represents intersection operation, $\cup$ means union operation.

Since the holes of the first and the second types are difficult to label, the experiment only marked the third to fifth types of holes and tested the performance of different algorithms. Table 11.1 shows the performance metrics and running time of the three comparison algorithms for porous metal materials in Figure 11.5. It can be seen from Table 11.1 that DSFCM_N has the lowest F-score and CS due to the serious over-detection rate. FLIL and FRFCM obtain relatively close test indicators. FRFCM considers the gray level information of the image, so it has the shortest running time. The

Table 11.1 The evaluation index of different algorithms for metal material image.

Index	FLIL	DSFCM_N	FRFCM
F-score	0.94	0.75	0.95
CS	0.88	0.60	0.90
Running time (s)	67.43	187.67	5.56

test results show that the post-processing method can objectively present the hole positions of the porous metal material, which lays a foundation for the property evaluation of the test material.

11.4   Image Segmentation for Foam Material Analysis

11.4.1   Eigenvalue Gradient Clustering

Although the foam image can be segmented effectively using the FCM algorithm, there are several problems: one of them is that the clustering center of its FCM algorithm needs to be obtained empirically by humans, and the subsequent segmentation results are influenced by the selection of the clustering center; the second problem is that the algorithm has more iterations and complex parameters, and the segmentation time for foam images is longer, and so its application is not practical. To solve the above two problems, we present a fast and automatic image segmentation algorithm employing superpixel-based graph clustering (FAS-SGC). This is based on the recently published article by Lei et al. [28]. FAS-SGC uses a morphological reconstruction of the regional minimum image to remove the initial structural element parameters and provide fast image segmentation and employs eigenvalue clustering to find potential clustering centers, thus greatly reducing the computational complexity.

Although many improved spectral clustering algorithms have been proposed [29], few of them focus on automatic spectral clustering. Some researchers employ the maximum intervals of eigenvalues [30] to estimate the potential clustering centers. However, this method often suffers from failures, as shown in Figure 11.6. Here, we present the eigenvalue gradient clustering (EGC) to improve the prediction accuracy of the potential number of clusters. Firstly, we analyze the eigenvalues of spectral clustering. As we employ a watershed transform based on AMR and regional minimum removal (AMR-RMR-WT) [28] to generate superpixel results, the corresponding data set is defined as $V = \{v_1, v_2, ...v_n\}$, and

$$v_j = \frac{1}{\sum_{x_p \in \partial_j} \varphi_j(x)} \sum_{p=1}^{\sum_{x_p \in \partial_j} \varphi_j(x)} x_p, \tag{11.5}$$

where ∂_j denotes the j-th region in a superpixel image and v_j is the average gray-scale value of pixels in ∂_j.

$$\varphi_j(x_p) = \begin{cases} 1 & x_p \in \partial_j \\ 0 & otherwise \end{cases}. \tag{11.6}$$

According to V, we can get the affinity matrix $A_m \in R^{n \times 3}$, and

$$A_{mji} = \begin{cases} exp\left(-\dfrac{\|v_j - v_i\|^2}{2\sigma^2}\right) & j \neq i \\ 0 & j = i \end{cases}, \tag{11.7}$$

where σ^2 is the scaling parameter of A_m. Furthermore, we can compute the degree matrix denoted by D_m, which is a diagonal matrix, and the Laplacian matrix is defined as

$$L_m = D_m^{-1/2} A_m D_m^{1/2}. \tag{11.8}$$

The eigenvalue set of A_m is $\lambda = \{\lambda_1, \lambda_2, ...\lambda_n\}$, $\lambda_1 = 1$, and $\lambda_1 \geq \lambda_2 \geq \cdots \geq \lambda_n$. Generally, the first c eigenvalues and their corresponding eigenvectors are used for k-means clustering to obtain the final clustering result. However, it is difficult to set the value of c. By analyzing the eigenvalue

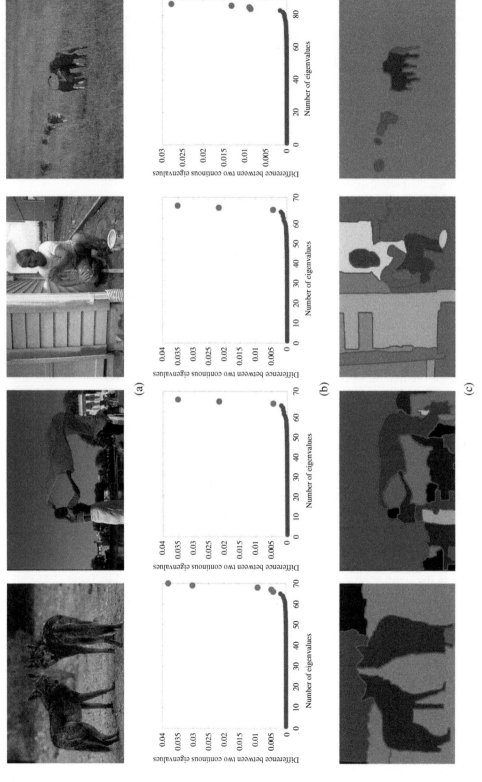

Figure 11.6 Automatic clustering results using AMR-RMR-WT and EGC [28] / IEEE / CC BY 4.0. (a) Original image. (b) Decision graph is used for estimating the number of clusters. (c) Segmentation result.

distribution in Figure 11.6 we can see that most of the eigenvalues are small and a few of them are large, which indicates that there is a large number of redundancies in an image. How to remove redundant eigenvalues and preserve useful ones is a problem. Here, we use the idea of clustering to replace the maximal eigenvalue interval. Assume that the eigenvalues of an affinity matrix could be grouped into three groups, where the first group is redundant and useless due to very small eigenvalues, and the second group may be important and useful for classification since it has larger values than the first group clearly, and the last group is similar to the second group but it has higher values than the second. However, it will take a long execution time to perform clustering on eigenvalue sets due to the many iterations required. To decrease the number of iterations and improve the clustering accuracy, we perform clustering on eigenvalue gradient sets. As eigenvalue gradients can reduce the number of different values in λ, it is easier to implement clustering on eigenvalue gradient sets than on eigenvalue sets.

We present the eigenvalue gradient set λ^g,

$$\lambda^g = \{\lambda_1^g, \lambda_2^g, ... \lambda_{n-2}^g\}, \tag{11.9}$$

where $\lambda_i^g = \lambda_{i+1} - \lambda_{i+2}$, $1 \leq i \leq n-2$. As $\lambda_1 = 1$ we usually remove λ_1 from λ.

We perform FCM on λ^g; the objective function is

$$J = \sum_{j=1}^{n-1} \sum_{k=1}^{c} u_{kj}^{m'} \left\| \lambda_j^g - y_k \right\|^2, \tag{11.10}$$

where y_k represents the prototype value of the k-th cluster, u_{kj} and denotes the membership value of the j-th sample with respect to a cluster k. $U = [u_{kj}]^{c \times (n-2)}$ is the membership partition matrix. The parameter c is the number of clusters. The parameter m' is a weighting exponent on each fuzzy membership that determines the amount of fuzziness of the classification results.

Figure 11.7 shows the comparison of eigenvalue clustering and EGC. Table 11.2 shows the comparison of eigenvalue clustering and EGC. Figure 11.7b shows a better clustering results than Figure 11.7a, which demonstrates that EGC is superior to eigenvalue clustering for finding potential clustering centers. Table 11.2, Y_1 and Y_2 represent the variance of inter-class in the first class and the second class, respectively. Eigenvalue clustering requires more iterations than EGC, and the former obtains a larger variance of inter-class than the latter, which means that the latter provides more accurate classification results and requires fewer iterations.

11.4.2 The Algorithm

According to the above analysis, the specific algorithm of EGC is proposed in Table 11.3.

In algorithm 1, y_k denotes the k-th clustering center, where $k \leq 3$. $C(y_1)$ and $C(y_2)$ denote the number of elements classified into y_1 and y_2, respectively.

Based on the above analysis, we present the detailed steps of FAS-SGC below.

Step 1: Compute two superpixel results using algorithm 1 AMR-RMR-WT;

Step 2: Compute the data set to be classified $V = \{v_1, v_2, ... v_n\}$ and two affinity matrixes A_{mji}^1 and A_{mji}^2 according to (11.7);

Step 3: Implement algorithm EGC on A_{mji}^1 to obtain the number of clusters c';

Step 4: Compute $L_m = D_m^{-1/2} A_{mji}^2 D_m^{1/2}$;

Step 5: Perform eigenvalue decomposition on L_m to obtain eigenvectors;

Step 6: Perform k-means on top c' eigenvectors;

Step 7: Reshape label image and output segmentation result.

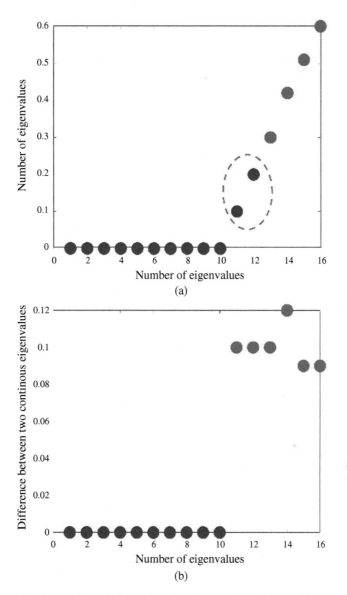

Figure 11.7   Comparison of eigenvalue clustering and EGC [28] / IEEE / CC BY 4.0.
(a) Eigenvalue clustering. (b) EGC.

Table 11.2   Comparison of eigenvalue clustering and EGC [28] / IEEE / CC
BY 4.0.

	Iterations	J	Y_1	Y_2
Eigenvalue clustering	9	0.07370	0.0039	0.0164
EGC	5	0.00059	0	0.0001

Table 11.3 EGC algorithm [28] / IEEE / CC BY 4.0.

	Algorithm 1 (EGC algorithm)
	Input: λ (Eigenvalue set, where $\lambda_1 \geq \lambda_2 \geq \cdots \geq \lambda_n$)
	Output: c' (The number of clusters used for spectral clustering)
1	Initialization: set $m' = 2$, $T = 50$, $\eta = 10^{-5}$ and $c = 3$
2	Compute λ^g and initialize randomly the membership partition matrix $U^{(0)}$
3	For $t = 1$ to T
4	Update the clustering centers $v_k = \sum_{i=1}^{n-1} u_{ki}^{m'} x_i / \sum_{i=1}^{n-1} u_{ki}^{m'}$
5	Update the membership partition matrix $u_{ki} = \sum_{j=1}^{c} \left(\frac{\|x_i - y_k\|^2}{\|x_i - y_j\|^2} \right)^{-1/(m'-1)}$
6	If $max\{U^{(t)} - U^{(t-1)}\} < \eta$, then
7	Break
8	Else
9	$t = t + 1$ and go to Step 4
10	End if
11	End for
12	Sort y_k in descending, $y_1 \geq y_2 \geq y_3$
13	Count the number of samples that belongs to the first two classes y_1 and y_2
14	Output $c' = C(v_1) + C(v_2)$

11.4.3 Experiment Analysis

To show the presented FAS-SGC algorithm is useful for some special images, we now apply FAS-SGC to an SEM image with a very high resolution of size 1278×892. SEM is an imaging device that generates a topological image of samples using a beam of electrons to achieve a much higher spatial resolution than optical microscope. The device can capture the surface morphology of samples, and thus it is widely used in scientific research fields such as medicine, biology, materials science, chemistry, and physics. Generally, SEM can provide a range of magnification varying from about 15 to 50 000. Here, an SEM image of porous material is considered as the test image, as shown in Figure 11.8, where dark areas denote holes and brighter areas denote connections. Researchers want to know the size and distribution of holes to analyze the physical and mechanical properties of porous materials. Traditionally, they first select one or two holes and then compute the size of holes manually. However, obtained results are often inaccurate and lack statistical significance. We try to use image segmentation technology to obtain accurate results of hole distribution. We applied a set of comparable algorithms and the presented FAS-SGC algorithm on the SEM image. Segmentation results are shown in Figure 11.8.

In this experiment, we first performed FAS-SGC to obtain the value of the parameter c'. Figure 11.8 shows that all these algorithms, except FNCut, are able to detect holes. These detected results, provided by HMRF-FCM [31], FLICM [32], KWFLECM [33], FRFCM [27], and DSFCM_N [26], include too many small areas as these algorithms are sensitive to noise. Liu's method [34], SFFCM [35], AFCF [36], and FAS-SGC [28] generate detection results by applying superpixel algorithms.

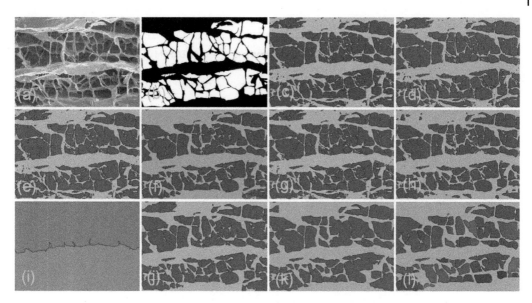

Figure 11.8 Segmentation results on a high-resolution SEM image [28] / IEEE / CC BY 4.0. (a) Image. (b) Ground truth. (c) HMRF-FCM. (d) FLICM. (e) KWFLICM. (f) Liu's algorithm. (g) FRFCM. (h) DSFCM_N. (i) FNCut. (j) SFFCM. (k) AFCF. (l) FAS-SGC.

Table 11.4 shows the performance comparison of different algorithms. Both AFCF and FAS-SGC are automatic image segmentation algorithms, and they show better performance than the other algorithms compared. Also, we can see that FAS-SGC obtains the best performance indices, demonstrating that FAS-SGC is effective for SEM image segmentation.

In practical applications, researchers can obtain the data of hole distribution according to detection results. The data are significant for the analysis of material properties. Table 11.5 shows the average area of holes in Figure 11.8 obtained by different algorithms. Note that Table 11.5 does

Table 11.4 Performance comparison of different algorithms on the high-resolution SEM image [28] / IEEE / CC BY 4.0.

Algorithms	PRI↑	CV↑	VI↓	GCE↓
HMRF-FCM [31]	0.7676	0.7636	1.1354	0.2316
FLICM [32]	0.7619	0.7580	1.1494	0.2354
KWFLICM [33]	0.7675	0.7636	1.1342	0.2313
Liu's algorithm [34]	0.8198	0.8179	0.9251	0.1783
FRFCM [27]	0.7666	0.7625	1.1392	0.2327
DSFCM_N [26]	0.8076	0.8052	0.9603	0.1889
FNCut [37]	0.5104	0.4007	1.9678	0.4883
SFFCM [35]	0.8017	0.7992	1.0074	0.1978
AFCF [36]	0.8253	0.8235	0.8920	0.1716
FAS-SGC [28]	**0.8267**[a]	**0.8252**	**0.8514**	**0.1665**

[a] The best values are highlighted.

Table 11.5 Average area of holes in Figure 11.8 obtained by different algorithms [28] / IEEE / CC BY 4.0.

	GT	HMRF-FCM	FLICM	KWFLICM	Liu's method	FRFCM	DSFCM_N	SFFCM	AFCF	FAS-SGC
area	17 374	17 830	20 169	19 831	18 707	17 749	16 144	19 925	20 454	**17 3**[a]**69**

[a] The best values are highlighted.

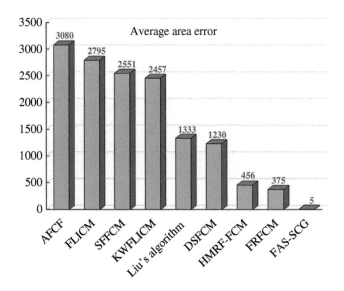

Figure 11.9 Average area error among different algorithms [28] / IEEE / CC BY 4.0.

not contain the data obtained by FNCut since it fails to detect holes in the SEM image. Furthermore, Figure 11.9 shows the error comparison of the average area among different algorithms. We consider the difference value between the average area of holes in the segmentation result and the average area of holes in ground truth as the error of the average area. FAS-SGC obtains the minimal error, which demonstrates that FAS-SGC is able to obtain more accurate data on hole distribution on the SEM image than comparable algorithms.

11.5 Image Segmentation for Ceramics Material Analysis

In order to effectively solve the problem of grain segmentation in ceramic material SEM images, this chapter adopts a joint driving method based on data and model and presents a ceramic material grain segmentation algorithm based on the robust watershed transform (RWT) combined with a lightweight rich convolution feature network. This algorithm first solves the problem of uneven grayscale caused by reflections on the surface of the material through image preprocessing. Next, the robust watershed transformation is used to realize the pre-segmentation of the grains in images and solve the problems of over-segmentation caused by traditional watershed algorithms and the difficulty of balancing the number of segmented regions and the accuracy of boundaries. Finally, the grain contour from the lightweight rich convolution feature network is employed to optimize the pre-segmentation result from RWT. Compared with mainstream image segmentation algorithms, on the one hand, the presented algorithm [38] uses RWT to achieve more accurate grain region positioning; on the other hand, the fusion of low- and high-level image features is performed to obtain more accurate grain boundaries. The experimental results show that the presented algorithm not only realizes the accurate calculation of the grain size of ceramic materials but also shows high calculation efficiency, which provides objective and accurate data for analyzing the physical properties of ceramic materials.

The algorithm presented in this chapter is mainly composed of three parts: firstly, the image is preprocessed to solve the problem of the uneven gray value of images. Secondly, RWT is used to achieve image pre-segmentation. Finally, the morphological contour optimization is implemented, and the pre-segmentation results are improved by using the contours from the convolutional neural network.

11.5.1 Preprocessing

The principle of SEM imaging is to strike an electron beam on the surface of the sample and interact with the surface of the sample. Due to the insulating properties of the ceramic material, the resulting image is poor with uneven gray value caused by the grain reflections. In order to solve this problem in the industry, a metal coating method is used to plate metal with a small resistivity (such as gold) on the surface of the sample, and then the quality of SEM images is effectively improved. However, the metal coating method is expensive. Therefore, we use the multi-scale retinex (MSR), algorithm proposed by Jobson et al. [39], to solve the above problem. MSR is expressed as:

$$R_{MSR} = \sum_{n=1}^{N} w_n R_n, \tag{11.11}$$

$$F_n(x,y) = \mu_N \exp\left(-\frac{x^2 + y^2}{c_n^2}\right), \tag{11.12}$$

where R_{MSR} is the reflection image output by the MSR algorithm, and usually $N = 3$, which means three scales of low, medium, and high. w_n is the weight coefficient, $w_1 = w_2 = w_3 = 1/3$. $F_n(x,y)$ is the Gaussian wrap function, and μ_N is the normalization factor, whose value satisfies $\iint F_n(x,y)dxdy = 1$. c_n represents the Gaussian surrounding space constantly; we usually set $c_1 = 15$, $c_2 = 80$, and $c_3 = 200$.

The MSR algorithm is used for image preprocessing, and the preprocessing result is shown in Figure 11.10. It can be seen that the overall gray value of images after preprocessing is relatively uniform, which is beneficial to the subsequent algorithm to achieve accurate grain segmentation.

11.5.2 Robust Watershed Transform

The preprocessed images need to have coarse segmentation implemented. The watershed transform is simple, and it is thus easy to be implemented. It mainly relies on the gradient information of an image to achieve image segmentation. Its disadvantage is that the texture information of images cannot be used well, and the SEM image of ceramic materials just lacks texture information. It is suitable to use the watershed transform to achieve coarse segmentation. The traditional watershed transformation easily leads to image over-segmentation. To solve this problem, scholars have proposed the morphological gradient reconstruction–based watershed transform (MGR-WT) [40]. First, the gradient is calculated on a preprocessed image, then the gradient is reconstructed to remove the useless local minima, and finally, the watershed transformation is performed to obtain segmentation results. Generally, morphological gradient reconstruction (MGR) can effectively smooth the local minimum area of an image, reduce the number of local minimums in the gradient

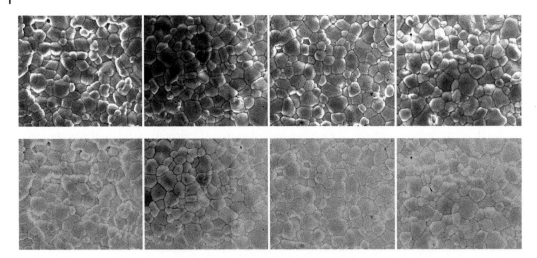

Figure 11.10 Comparison results between original images and the pre-processing results [38] / with permission of Editorial Office of Acta Automatica Sinica. The preprocessing algorithm can effectively solve the problem of uneven grayscale of ceramic SEM images. The first row shows the original images, the second row shows the preprocessed images.

image, and then suppress the image over-segmentation caused by the watershed transformation. MGR operators usually involve the selection of structural elements. The size and shape of structural elements will affect the final reconstruction result and segmentation effect. When the size of the structure elements is suitable, segmentation accuracy and over-segmentation can be balanced, but it is difficult to determine the best size for structural elements. In Figure 11.11 (using morphological closed reconstruction to act on a gradient image, the gradient image is generated by the SE algorithm [41]. r_{cs} represents the radius of the circular structure element selected by the morphological closed reconstruction), when the value of the structure element parameter r_{cs} used for gradient reconstruction is small ($r_{cs} = 1$), it is easy to cause over-segmentation; the segmentation results contain a large number of small regions. When the value of r_{cs} is large ($r_{cs} = 10$), it is easy to cause under-segmentation, multiple target segmentation results are merged, and the contour accuracy is low. When the value of r_{cs} is moderate ($r_{cs} = 5$), the segmentation results can take into account the number of segmented regions and the contour accuracy. However, the value of r_{cs} is

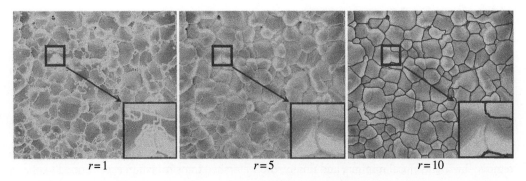

$r = 1$　　　　　　$r = 5$　　　　　　$r = 10$

Figure 11.11 Comparison of image segmentation results for MGR-WT using different parameters, r_{cs} is the radius of the circular structure element used for gradient reconstruction [38] / with permission of Editorial Office of Acta Automatica Sinica.

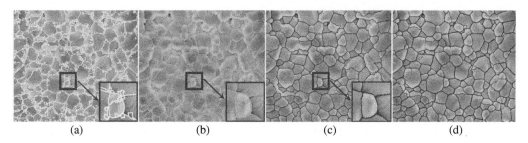

Figure 11.12 Comparison of segmentation results of ceramic material grains between RWT and MGR-WT [38] / with permission of Editorial Office of Acta Automatica Sinica. (a) Segmentation results of a morphological watershed algorithm without removing small regions. (b) MGR-WT segmentation results. (c) RWT segmentation results. (d) Comparison of overall segmentation results (the blue line represents the RWT result, the green line represents the MGR-WT result, and the pink line represents the ground truth result).

empirical, and reducing the number of divided regions is at the expense of contour accuracy. To solve the above problems, RWT is presented in [12].

Let f^m and g denote the label image and the mask image of the constraint transformation, respectively. The parameter-free adaptive MGR of g reconstructed from f^m is shown in formula (11.13), which m represents the scale of the largest structure element, usually, $m > 10$, and the multi-scale structure elements satisfy the relationship $b_1 \subseteq ...b_i \subseteq b_{i+1}... \subseteq b_m$.

$$\psi(g, m) = V_{1 \leq i \leq m}\left\{R_g^\phi(f^m)_{b_i}\right\}. \tag{11.13}$$

Using the above formula to reconstruct the image g and perform the watershed transformation on it, the segmentation result still contains more small areas, as shown in Figure 11.12a. The main reason is that the value of i starts from 1. Although setting a larger value to i can reduce the number of small areas, it also reduces the contour accuracy. In order to get better reconstruction results, these small areas need to be removed. Suppose H is the gradient image, I is the regional minimum image of H, and W is the segmentation result obtained by the watershed transformation. $I = (I_1, I_2, ..., I_n)$, I_j represents the j-th connected component in the image I, $1 \leq j \leq n$. Similarly, $W = W_1 \cup W_2 \cup ... \cup W_n$, W_j represents the j-th segmented area in W, and thus:

$$\sum_{p \in W_j} \theta(x_p) \geq \sum_{q \in I_j} \theta(x_q), \tag{11.14}$$

where $W_{j_1} \cup W_{j_2} \neq \varphi, 1 \leq j_1, j_2 \leq n, j_1 \neq j_2$, x_p is the p-th pixel in W, x_q is the q-th pixel in I.

$$\theta(x_i) = \begin{cases} 1 & x_i \in W_j \text{ or } I_j \\ 0 & \text{otherwise} \end{cases}. \tag{11.15}$$

Formulas (11.14) and (11.15) show that by removing the smaller connected components in the image I, the purpose of merging smaller segmented regions is achieved, and the smaller connected components are removed by using formula (11.16) to realize the region merging according to the segmentation results.

$$I^r = R_I^\delta(\varepsilon(I)b_k), \tag{11.16}$$

where k is a parameter used for structure elements; if the value of k is larger, then more small areas will be merged in the image W.

RWT firstly constructs the parameter-free adaptive MGR formula, then calculates the local minimum, and uses binary morphological reconstruction to optimize the local minimum. It can be seen from (11.13) that RWT uses multi-scale structure elements to achieve gradient reconstruction, and the reconstruction result is convergent as m increases. Therefore, m is a constant, which solves the problem that the single-scale gradient reconstruction results are easily affected by structure element parameters. In addition, it can be seen from (11.16) that the change of parameter k will lead to the change in the number of area minimums. However, this change will only affect the number of final segmented areas and does not affect the final contour accuracy. It can be seen from Figure 11.12b that MGR-WT may divide a complete grain into two parts due to a mistake. In Figure 11.12c, RWT overcomes the mistake and achieves the correct grain segmentation. Therefore, RWT presented in this chapter can achieve better grain segmentation than MGR-WT.

11.5.3 Contour Optimization

Although Figure 11.12 shows RWT provides good and robust segmentation results and is not easily affected by environment and parameters, RWT still has the following problems: On the one hand, the gap between different grains is too large, which leads to the problem of double-line contour. On the other hand, because the segmentation results are excessively dependent on the quality of gradient images, RWT lacks the use of image semantic information, which leads to the problem of inaccurate contour positioning.

In order to solve the above-mentioned double-line contour problem, a double-line elimination strategy based on morphological contour optimization is presented. Firstly, we check the area covered by each tag to ensure that each tag in the image can only cover one area. Secondly, we eliminate the areas that are smaller than a given structure element. Based on this processing idea, the contour optimization method based on morphology is selected here to process the image. The specific steps are given as follows:

1) Ensuring that each label in the image can only cover one area.
2) Performing morphological opening operations on each labeled area. Subtracting the result of the opening operation from the original image, as shown in formula (11.17)

$$f_m = U_{l=1}^{L} b_l - \delta(\varepsilon(b_l)). \tag{11.17}$$

3) Reallocating f_m to adjacent areas, ensuring each area has a different label and relabeling all areas.

In order to further improve the accuracy of segmentation contours and avoid the segmentation results relying too much on the gradient image obtained by SE [41], an image contour prediction model based on a CNN is introduced. The traditional neural network edge detection model cannot extract high-level semantic information from images. Most edge detection models based on CNN networks use the last layer of the convolutional network to achieve edge detection, owing to the shallow detail information, which is missed and easily causes a vanishing gradient. RCF (richer convolutional features network) is an excellent network based on VGG16, which employs the combination of all the hierarchical features from the convolutional layers to detect edge information. In RCF, all parameters can automatically learn multi-scale and multi-level features from images. Before predicting the image contour, we first change the size of the original image, build a set of image pyramids, and then input these images to the RCF for forward transmission. Then, RCF uses bilinear interpolation to restore the feature maps to the original image resolution. Finally, the average value of these feature maps is outputted as the final edge detection result.

The RCF network can provide more accurate grain contour, but it suffers from problems such as a mountain of parameters and slow training speed. Therefore, RCF consumes too many computing

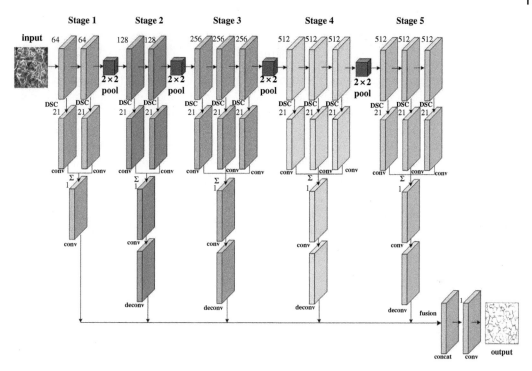

Figure 11.13 LRCF network architecture [38] / with permission of Editorial Office of Acta Automatica Sinica.

and memory resources. In order to solve this problem, the standard convolution in RCF is replaced by depth-wise separable convolution (DSC) to obtain a lightweight and richer convolutional feature network (LRCF), as shown in Figure 11.13. LRCF is applied to the prediction of ceramic grain contour, combined with RWT to achieve the ceramic grain segmentation based on the joint drive of data and model. Figure 11.14 shows a comparison of the segmentation results. It can be seen that the SE-RWT can provide good segmentation for most grains but obtains inaccurate contours for a few grains. In contrast, LRCF-RWT provides better grain segmentation results than SE-RWT, as shown in Figure 11.14b, c.

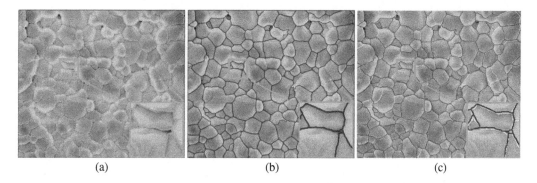

(a)   (b)   (c)

Figure 11.14 Comparison results between SE-RWT and LRCF-RWT [38] / with permission of Editorial Office of Acta Automatica Sinica. (a) The segmentation result using SE-RWT. (b) The segmentation result using LRCF-RWT. (c) The comparison results. (The pink line is the result from LRCF-RWT, the green line is the result from SE-RWT, and the yellow line is ground truth).

11.5.4 Experiment Analysis

In order to verify the performance of LRCF-RWT segmentation algorithm proposed in this chapter, we have compared a series of popular segmentation algorithms, including clustering-based segmentation algorithms, the algorithm proposed by Liu et al. [34], random walker (RW) [42], SLIC [43], LSC [44], Banerjee, et al. [23] algorithm, SE gradient combined with watershed transform method (SE-WT) [41], SE gradient combined with adaptive morphological reconstruction and watershed transform method (SE-AMR-WT) [45], RCF combined with watershed transform (RCF-WT) method [46] and so on. We use four algorithm indicators to test the segmentation results, which are the covering (CV), the variation of information (VI), global consistency error (GCE), and the boundary displacement error (BDE). The larger the value of CV, the better the segmentation result; the smaller the value of VI, GCE, and BDE, the better the segmentation result. Figure 11.15 shows the comparison of the segmentation results of different segmentation algorithms, and Table 11.6 shows the performance indicators of different segmentation algorithms.

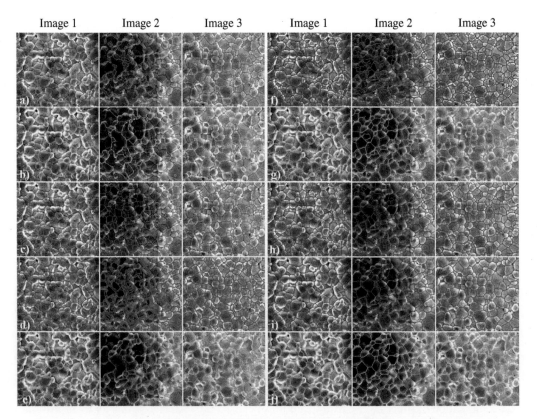

Image 1   Image 2   Image 3   Image 1   Image 2   Image 3

Figure 11.15 Comparison of segmentation results of different segmentation algorithms [38] / with permission of Editorial Office of Acta Automatica Sinica. (a) Liu's-MGR. (b) RW. (c) SLIC. (d) LSC. (e) Banerjee's. (f) SE-MGR-WT. (g) SE-AMR-WT. (h) RCF-MGR-WT. (i) LRCF-RWT. (j) Ground truth.

Table 11.6 Comparison of performance indexes of different segmentation algorithms for ceramic grain segmentation [38] / with permission of Editorial Office of Acta Automatica Sinica.

Methods	CV↑	VI↓	GCE↓	BDE↓
Liu's-MGR [34]	0.2889	3.4270	0.4742	7.3230
Random Walker [41]	0.3556	2.9003	0.1407	13.2147
SLIC [42]	0.3547	3.0524	0.4396	10.1678
LSC [43]	0.3455	2.8820	0.3563	7.5911
Banerjee's [23]	0.5959	2.1992	0.2031	3.9182
SE-MGR-WT [40]	0.4680	2.3887	0.1364	5.0346
SE-AMR-WT [44]	0.8287	1.1280	0.1122	**1.6261**[a]
RCF-MGR-WT [45]	0.6636	1.4952	0.0955	3.5651
LRCF-RWT [38]	**0.8697**	**0.8701**	**0.0763**	1.6262

[a] The best values are highlighted.

11.6 Discussion and Summary

Considering the problem of long execution time caused by the high resolution of porous metal materials, the chapter presents a fast fuzzy clustering algorithm based on histogram information. The method first introduces morphological reconstruction filtering to obtain local spatial information, then uses histogram technology to reduce redundant calculation, and finally uses a membership filter to correct the partition matrix. The method can not only effectively improve the segmentation effect of the algorithm but also improve its computational efficiency. In order to extract the information parameters of different holes, this chapter further designs a hole extraction scheme. By classifying holes of different sizes through morphological operations, we can extract the corresponding hole parameters. The experimental results show that compared with the traditional segmentation algorithm, the algorithm in this chapter not only effectively identifies the tiny holes, which is difficult for human eyes to distinguish, but also achieves high-precision measurement of hole data of porous metal materials. Simultaneously, the execution time of the algorithm is shorter.

Considering the segmentation problem of scanning electron microscopy images of porous foam materials, an FAS-SGC algorithm is proposed. This algorithm uses a morphological reconstruction of the smallest area image to remove the initial structure element parameters and uses feature value clustering to find potential clustering centers, thereby reducing the computational complexity, effectively solving the effect of human factors on the clustering centers of the porous foam material segmentation, making the algorithm achieve better segmentation results on the hole image with a short execution time, and realize subsequent pore size calculation and analysis. Experiments show that the FAS-SGC algorithm is the most effective for the division of holes and the calculation of the area on the porous foam image. Although FAS-SGC can achieve good segmentation results on foam materials without human-computer interaction, FAS-SGC cannot provide good segmentation results obtained by supervised image segmentation algorithms. In the future, we will explore the combination of supervised learning and unsupervised learning algorithms for foam materials to achieve supervised foam image segmentation.

Considering the problem of low efficiency and large error in manual measurement of ceramic material grain size, this chapter presents a ceramic material grain segmentation algorithm driven by data and model. This algorithm solves the problems of over-segmentation caused by the traditional watershed transform and the difficulty in balancing the number of segmented regions and contour accuracy. It also improves the segmentation accuracy by introducing a lightweight CNN and realizes the correct grain segmentation in ceramic material SEM images, which is convenient to compute the size of grains and count the number of different-size grains. Experimental results show that compared with traditional segmentation algorithms, the presented algorithm can achieve accurate segmentation of crystal grains in SEM images for different types of ceramic materials. However, for the ungold-plated SEM image, because individual areas are severely affected by light, the gray value of this area is still too different from other areas after preprocessing, resulting in segmentation errors. In addition, the grain boundaries in some images are not obvious, and the gray value is similar to the crystal grains, which causes the crystal grains to fail to be segmented correctly.

In conclusion, image segmentation technology has broad application prospects in the analysis of metal materials, foam materials, and ceramic materials. Using image segmentation technology can not only quickly obtain hidden attribute information in SEM images but also provide more accurate and objective indicators. Especially with the rapid development of deep learning technology, image semantic segmentation based on the convolutional neural networks will play an increasingly important role in the field of material analysis, laying a solid foundation for the development of high-tech materials.

References

1 Goldstein, J.I., Newbury, D.E., Michael, J.R. et al. (2017). *Scanning Electron Microscopy and X-Ray Microanalysis*. Springer.

2 Hafner, B. (2007). Scanning electron microscopy primer. In: *Characterization Facility, University of Minnesota-Twin Cities*, 1–29.

3 Cazaux, J. (2005). Recent developments and new strategies in scanning electron microscopy. *J. Microsc.* **217** (1): 16–35.

4 Eberle, A.L., Mikula, S., Schalek, R. et al. (2015). High-resolution, high-throughput imaging with a multibeam scanning electron microscope. *J. Microsc.* **259** (2): 114–120.

5 Yuan, W., Tang, Y., Yang, X., and Wan, Z. (2012). Porous metal materials for polymer electrolyte membrane fuel cells – a review. *Appl. Energy* **94**: 309–329.

6 Ünal, F. A., Timuralp, C., Erduran, V., and Şen, F. (2021). Porous metal materials for polymer electrolyte membrane fuel cells. In *Nanomaterials for Direct Alcohol Fuel Cells* (pp. 187–207). Edited by: Şen F. Elsevier.

7 Yang, X.J., Liu, Y., Li, M., and Tu, M.J. (2007). Preparation and application of the porous metal material. *Mater. Rev.* **21**: 380–383.

8 Qin, J., Chen, Q., Yang, C., and Huang, Y. (2016). Research process on property and application of metal porous materials. *J. Alloys Compd.* **654**: 39–44.

9 Malcolm, A.A., Leong, H.Y., Spowage, A.C., and Shacklock, A.P. (2007). Image segmentation and analysis for porosity measurement. *J. Mater. Process. Technol.* **192**: 391–396.

10 Sarker, D.K., Bertrand, D., Chtioui, Y., and Popineau, Y. (1998). Characterisation of foam properties using image analysis. *J. Texture Stud.* **29** (1): 15–42.

11 Stevens, S.M., Jansson, K., Xiao, C. et al. (2009). An appraisal of high resolution scanning electron microscopy applied to porous materials. *JEOL News* **44** (1): 17–22.

12 Yuan, Q., Yao, F., Wang, Y. et al. (2017). Relaxor ferroelectric 0.9 BaTiO 3–0.1 Bi (Zn 0.5 Zr 0.5) O 3 ceramic capacitors with high energy density and temperature stable energy storage properties. *J. Mater. Chem. C* **5** (37): 9552–9558.

13 Yang, Z., Gao, F., Du, H. et al. (2019). Grain size engineered lead-free ceramics with both large energy storage density and ultrahigh mechanical properties. *Nano Energy* **58**: 768–777.

14 Vincent, L. and Soille, P. (1991). Watersheds in digital spaces: an efficient algorithm based on immersion simulations. *IEEE Trans. Pattern Anal. Mach. Intell.* **13** (06): 583–598.

15 Yuan, X., Martínez, J.F., Eckert, M., and López-Santidrián, L. (2016). An improved Otsu threshold segmentation method for underwater simultaneous localization and mapping-based navigation. *Sensors* **16** (7): 1148.

16 Chang, H. and Yeung, D.Y. (2005). Robust path-based spectral clustering with application to image segmentation. In: *Tenth IEEE International Conference on Computer Vision (ICCV'05) Volume 1*, vol. **1**, 278–285. IEEE.

17 Pal, N.R. and Bezdek, J.C. (1995). On cluster validity for the fuzzy c-means model. *IEEE Trans. Fuzzy Syst.* **3** (3): 370–379.

18 Washburn, E.W. (1921). Note on a method of determining the distribution of pore sizes in a porous material. *Proc. Natl. Acad. Sci. U. S. A.* **7** (4): 115.

19 Roux, S., Hild, F., Viot, P., and Bernard, D. (2008). Three-dimensional image correlation from X-ray computed tomography of solid foam. *Compos. A: Appl. Sci. Manuf.* **39** (8): 1253–1265.

20 Wagner, B., Dinges, A., Müller, P., and Haase, G. (2009). Parallel volume image segmentation with watershed transformation. In: *Proceedings of the Scandinavian Conference on Image Analysis*, 420–429.

21 Montminy, M.D. (2001). *Complete Structural Characterization of Foams Using Three-Dimensional Images*. University of Minnesota.

22 Jiang, F., Gu, Q., Hao, H. et al. (2018). A method for automatic grain segmentation of multi-angle cross-polarized microscopic images of sandstone. *Comput. Geosci.* **115**: 143–153.

23 Banerjee, S., Chakraborti, P.C., and Saha, S.K. (2019). An automated methodology for grain segmentation and grain size measurement from optical micrographs. *Measurement* **140**: 142–150.

24 Bezdek, J.C., Ehrlich, R., and Full, W. (1984). FCM: the fuzzy c-means clustering algorithm. *Comput. Geosci.* **10** (2–3, 203): 191.

25 Li, F. and Qin, J. (2017). Robust fuzzy local information and L_p L p-norm distance-based image segmentation method. *IET Image Process.* **11** (4): 217–226.

26 Zhang, Y., Bai, X., Fan, R., and Wang, Z. (2018). Deviation-sparse fuzzy c-means with neighbor information constraint. *IEEE Trans. Fuzzy Syst.* **27** (1): 185–199.

27 Lei, T., Jia, X., Zhang, Y. et al. (2018). Significantly fast and robust fuzzy c-means clustering algorithm based on morphological reconstruction and membership filtering. *IEEE Trans. Fuzzy Syst.* **26** (5): 3027–3041.

28 Jia, X., Lei, T., Liu, P. et al. (2020). Fast and automatic image segmentation using Superpixel-based graph clustering. *IEEE Access* **8**: 211526–211539.

29 Arias-Castro, E., Lerman, G., and Zhang, T. (2017). Spectral clustering based on local PCA. *J. Mach. Learn. Res.* **18** (1): 253–309.

30 Nie, F., Wang, X., Jordan, M., and Huang, H. (2016). The constrained laplacian rank algorithm for graph-based clustering. In: *Proceedings of the AAAI conference on artificial intelligence*, vol. **30**. No. 1.

31 Chatzis, S.P. and Varvarigou, T.A. (2008). A fuzzy clustering approach toward hidden Markov random field models for enhanced spatially constrained image segmentation. *IEEE Trans. Fuzzy Syst.* **16** (5): 1351–1361.

32 Krinidis, S. and Chatzis, V. (2010). A robust fuzzy local information C-means clustering algorithm. *IEEE Trans. Image Process.* **19** (5): 1328–1337.

33 Gong, M., Liang, Y., Shi, J. et al. (2012). Fuzzy c-means clustering with local information and kernel metric for image segmentation. *IEEE Trans. Image Process.* **22** (2): 573–584.

34 Liu, G., Zhang, Y., and Wang, A. (2015). Incorporating adaptive local information into fuzzy clustering for image segmentation. *IEEE Trans. Image Process.* **24** (11): 3990–4000.

35 Lei, T., Jia, X., Zhang, Y. et al. (2018). Superpixel-based fast fuzzy C-means clustering for color image segmentation. *IEEE Trans. Fuzzy Syst.* **27** (9): 1753–1766.

36 Lei, T., Liu, P., Jia, X. et al. (2019). Automatic fuzzy clustering framework for image segmentation. *IEEE Trans. Fuzzy Syst.* **28** (9): 2078–2092.

37 Kim, T.H., Lee, K.M., and Lee, S.U. (2012). Learning full pairwise affinities for spectral segmentation. *IEEE Trans. Pattern Anal. Mach. Intell.* **35** (7): 1690–1703.

38 Lei, T., Li, Y., Zhou, W. et al. (2020). Grain segmentation of ceramic materials using data-driven jointing model-driven. *Acta Autom. Sin.* **46**: 1–16.

39 Jobson, D.J., Rahman, Z.U., and Woodell, G.A. (1997). A multiscale retinex for bridging the gap between color images and the human observation of scenes. *IEEE Trans. Image Process.* **6** (7): 965–976.

40 Jackway, P.T. (1996). Gradient watersheds in morphological scale-space. *IEEE Trans. Image Process.* **5** (6): 913–921.

41 Dollár, P. and Zitnick, C.L. (2014). Fast edge detection using structured forests. *IEEE Trans. Pattern Anal. Mach. Intell.* **37** (8): 1558–1570.

42 Grady, L. (2006). Random walks for image segmentation. *IEEE Trans. Pattern Anal. Mach. Intell.* **28** (11): 1768–1783.

43 Achanta, R., Shaji, A., Smith, K. et al. (2012). SLIC superpixels compared to state-of-the-art superpixel methods. *IEEE Trans. Pattern Anal. Mach. Intell.* **34** (11): 2274–2282.

44 Li, Z. and Chen, J. (2015). Superpixel segmentation using linear spectral clustering. In: *Proceedings of the IEEE Conference on Computer Vision and Pattern Recognition*, 1356–1363.

45 Lei, T., Jia, X., Liu, T. et al. (2019). Adaptive morphological reconstruction for seeded image segmentation. *IEEE Trans. Image Process.* **28** (11): 5510–5523.

46 Liu, Y., Cheng, M.M., Hu, X. et al. (2017). Richer convolutional features for edge detection. In: *Proceedings of the IEEE Conference on Computer Vision and Pattern Recognition*, 3000–3009.

Index